Manual of Water Supply Practices

M32

Computer Modeling of Water Distribution Systems

Fourth Edition

American Water Works Association

Manual of Water Supply Practices—M32, Fourth Edition

Computer Modeling of Water Distribution Systems

Disclaimer
The authors, contributors, editors, and publisher do not assume responsibility for the validity of the content or any consequences of its use. In no event will AWWA be liable for direct, indirect, special, incidental, or consequential damages arising out of the use of information presented in this book. In particular, AWWA will not be responsible for any costs, including, but not limited to, those incurred as a result of lost revenue. In no event shall AWWA's liability exceed the amount paid for the purchase of this book.

If you find errors in this manual, please email books@awwa.org. Possible errata will be posted at www.awwa.org/resources-tools/resource.development.groups/manuals-program.aspx.

Senior Managing Editor/Project Manager: Melissa Valentine
Cover art: Melanie Yamamoto
Production: Janice Benight
Manual Specialist: Sue Bach

Library of Congress Cataloging-in-Publication Data

Names: Cooper, James P., author. | Robinson, Laredo. Computer modeling of water distribution systems. | American Water Works Association, issuing body.
Title: M32 computer modeling of water distribution systems / by James P. Cooper.
Other titles: Computer modeling of water distribution systems
Description: Fourth edition. | Denver, CO : American Water Works Association, [2018] | Revised edition of: Computer modeling of water distribution systems / by Laredo Robinson, Jerry A. Edwards, Lindle D. Willnow. | Includes bibliographical references and index.
Identifiers: LCCN 2017049002 | ISBN 9781625762528 (alk. paper)
Subjects: LCSH: Water--Distribution--Data processing. | Network analysis (Planning)--Data processing.
Classification: LCC TD481 .C66 2018 | DDC 628.1/440285--dc23 LC record available at https://lccn.loc.gov/2017049002

Printed in the United States of America

ISBN: 978-1-62576-252-8
eISBN: 978-1-61300-439-5

American Water Works Association

American Water Works Association
6666 West Quincy Avenue
Denver, CO 80235-3098
awwa.org

Contents

This page intentionally blank.

Figures

Tables

This page intentionally blank.

Acknowledgments

The AWWA Engineering and Construction Division and the Technical and Educational Council acknowledge the following individuals for their persistence and dedication in developing the fourth edition of this manual:

Chair—James P. Cooper, P.E., *Arcadis, Akron, Ohio*

Authors

Chapter 1 Introduction to Distribution System Modeling

Sharavan V. Govindan, Bentley Systems Inc., Exton, Penn.
Adam Simonsen, IDModeling Inc., Waterbury, Conn.

Chapter 2 Building and Preparing the Model

Christie Patel, P.E., Brown and Caldwell, Philadelphia, Penn.
Mike Rosh, Bentley Systems Inc., Sayre, Penn.

Chapter 3 Hydraulic Tests and Measurements

James P. Cooper, P.E., Arcadis, Akron, Ohio
Douglas Lane, P.E., City of Bellevue, Bellevue, Wash.

Chapter 4 Hydraulic Calibration

Douglas Harrold, P.E., City of Santa Clara Water and Sewer Utilities, Santa Clara, Calif.
Saša Tomic´, P.E., Ph.D., HDR Engineering Inc., New York, N.Y.

Chapter 5 Steady-State Simulation

Paul Hsiung, P.E., Innovyze, Shawnee Mission, Kans.
Amanda Schwerman, P.E., Black & Veatch, Tampa, Fla.

Chapter 6 Extended-Period Simulation

Patrick B. Moore, P.E., Innovyze, Edgewood, N.M.
Lindle D. Willnow, P.E., AECOM, Wakefield, Mass.

Chapter 7 Water Quality Modeling

James P. Cooper, P.E., Arcadis, Akron, Ohio
Megan G. Roberts, P.E., Hazen and Sawyer, P.C., Greensboro, N.C.

Chapter 8 Storage Tank Mixing and Water Age

Bob Hawboldt. P.E., Ph.D., Associated Engineering, Saskatoon, Sask.
Ferdous Mahmood, P.E., Arcadis, Dallas, Texas

Chapter 9 Model Maintenance

Jeff Cowburn, City of Abbotsford, Abbotsford, B.C.
Mike Rosh, Bentley Systems Inc., Sayre, Penn.
Saša Tomic´, P.E., Ph.D., HDR Engineering Inc., New York, N.Y.

Chapter 10 Transient Analysis

Ferdous Mahmood, P.E., Arcadis, Dallas, Texas
Lindle D. Willnow, P.E., AECOM, Wakefield, Mass.

Chapter 11 Advanced Modeling Applications

Rajan Ray, Innovyze, Wakefield, R.I.
Adam Simonsen, IDModeling Inc., Waterbury, Conn.
Thomas M. Walski, Ph.D., P.E., Bentley Systems Inc., Nanticoke, Penn.

The authors would like to acknowledge the following individuals who provided editorial and technical comments and/or contributed in other ways:

Paul F. Boulos, Ph.D., Innovyze, Broomfield, Colo.
Jeff Cowburn, City of Abbotsford, Abbotsford, B.C.
Nass Diallo, P.E., Las Vegas Valley Water District, Las Vegas, Nev.
Jerry A. Edwards, P.E., Bohannan Huston, Englewood, Colo.
Jeff Frey, P.E., Optimatics, Chicago, Ill.
Walter M. Grayman, P.E., Ph.D., W.M. Grayman Consulting Engineer, Cincinnati, Ohio
Douglas Harrold, P.E., City of Santa Clara Water and Sewer Utilities, Santa Clara, Calif.
Paul M. Hauffen, IDModeling Inc., Arcadia, Calif.
J. Erick Heath, P.E., Innovyze, Arcadia, Calif.
Jonathan Keck, California Water Service, San Jose, Calif.
Douglas Lane, P.E., City of Bellevue, Bellevue, Wash.
Patrick B. Moore, P.E., Innovyze, Edgewood, NM.
Myron Nealey, P.E., Denver Water, Denver, Colo.
Christie Patel, P.E., Brown and Caldwell, Philadelphia, Penn.
Theresa O'Grady, P.E., Crawford, Murphy & Tilly, Aurora, IL.
Rajan Ray, Innovyze, Wakefield, R.I.
Jeffrey Rosenlund, DOWL, Sheridan, Wyo.
Mike Rosh, Bentley Systems Inc., Sayre, Penn.
Amanda Schwerman, P.E., Black & Veatch, Tampa, Fla.
Jennifer Suttles, P.E., Hazen & Sawyer, Atlanta, Ga.
Jim Uber, CitiLogics, Covington, Ky.
Bryon Wood, P.E., HDR Engineering, Inc., Ann Arbor, Mich
Andy Yang, San Jose Water Company, San Jose, Calif.
Samual Ziemann, C3 Water, Breslau, Ont.
Jill Zimmerman, Albemarle County Service Authority, Charlottesville, Va.

The following individuals provided peer review of the entire manual. Their knowledge and efforts are gratefully appreciated:

James P. Cooper, P.E., Arcadis, Akron, Ohio
Theresa O'Grady, P.E., Crawford, Murphy & Tilly, Aurora, Ill.
Patrick F. Parault, P.E., Arcadis, Long Island City, N.Y.
Jennifer Santini, American Water Works Association, Denver, Colo.
Jim Siriano, American Water Works Association, Denver, Colo.
Saša Tomić, P.E., Ph.D., HDR Engineering Inc., New York, N.Y.
Thomas M. Walski, Ph.D., P.E., Bentley Systems Inc., Nanticoke, Penn.
Lindle D. Willnow, P.E., AECOM, Wakefield, Mass.

This page intentionally blank.

Chapter 1

Introduction to Distribution System Modeling

1.1. INTRODUCTION

Water utilities seek to provide customers with a safe, reliable, continuous supply of high-quality drinking water while managing costs. This water is often delivered through a complex network of pipelines, numerous pumps, regulating valves, and storage reservoirs. The performance of these water systems is often difficult to measure and understand because of their physical size, complexity, underground location (out of sight, out of mind), and the large amount of system data needed to fully grasp how they function. Sometimes, key pieces of information needed to understand a system are missing or incomplete. One tool that has evolved over time to help water system designers, operators, planners, and managers meet their goals of delivering a safe, reliable, cost-effective water supply is a water distribution system model (herein referred to as a *model*).

Distribution system modeling (commonly referred to as *hydraulic modeling*) involves the use of a computerized mathematical model to predict the performance of a water system to solve a wide variety of issues including, but not limited to, design, operations, system planning, water quality, water loss, energy management, and emergency response. Models can help water utilities by enabling more proactive and responsive decision making, delivering better customer service, and ensuring employee satisfaction. Water utilities can use models to understand the infrastructure and operations needed to support a network of growing residential and commercial consumers. For example, a model can predict pressures and flows within a water system in order to evaluate a design and compare the system performance against design standards. Models are also used in operational studies to answer questions related to storage capacity, control schemes, water delivery, water quality, and more. Water quality models are used to predict water age, track disinfectant residuals, and reduce disinfection by-products in a distribution system.

Computerized system modeling began with the advent of analog computers and has evolved as software and hardware have advanced to become more powerful and easier to use. Models that contain hundreds or thousands of miles of pipeline are now common and can be used on a wide variety of computer platforms. Models that once took hours to run are now run in seconds or fractions of a second. Originally, models were used only to evaluate system hydraulic grade lines, system pressures, and flows.

Historically, model building was an expensive and labor-intensive process due to lack of available data to build from and to support relevant applications. Now, models can effectively share data with geographic information systems (GIS), computer-aided design and drafting (CADD) systems, supervisory control and data acquisition (SCADA) systems, customer information systems (CIS), computerized maintenance management systems (CMMS), and asset management system (AMS) software, thus reducing the effort needed to create, update, calibrate, and maintain a model. Information obtained from a model study can be filtered, organized, and presented using a variety of graphical and tabular methods so that results can be more easily understood and communicated. These advances in technology have broadened the uses of distribution system modeling from just an infrastructure planning tool to an integrated design-and-operations analysis system.

1.2. PURPOSE OF THIS MANUAL

The Engineering Modeling Applications Committee of the American Water Works Association developed this manual. The purpose of this manual is to share the committee's collective experience with distribution system modeling to help educate and communicate the benefits of a model to water utilities and water customers everywhere.

The manual is intended to be a basic-level or primer reference guide to provide new to intermediate modelers with a foundation for, and practical benefits of, water distribution system modeling. The manual takes users through the modeling process from development to system analysis as shown in Figure 1-1. The manual delivers in-depth discussion on

- Model construction and development,
- Field data collection and testing,
- Model calibration,
- Steady-state analysis,
- Extended-period simulation,
- Water quality analysis,
- Storage tank mixing and water age analysis,
- Model maintenance,
- Transient analysis, and
- Advanced modeling applications.

M32 is designed to help modelers use water models as effective tools to plan, design, and operate a water distribution system; maintain acceptable water quality; and reduce both operating costs and water loss within their water distribution systems.

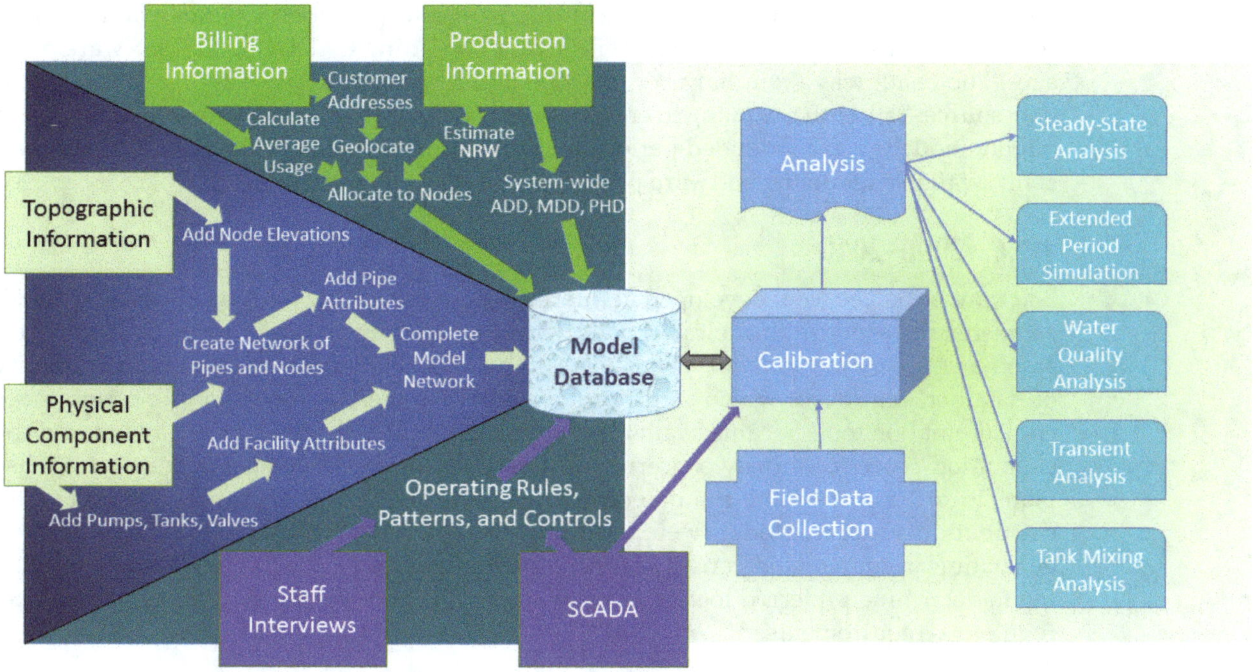

Figure 1-1 Overview of an example modeling process

Abbreviations: ADD (average day demand), MDD (maximum daily demand), PHD (peak hourly demand), SCADA (supervisory control and data acquisition), NRW (non-revenue water).

1.3. HISTORICAL DEVELOPMENT OF DISTRIBUTION SYSTEM MODELING

1.3.1. Pre-1970

Manual engineering calculations for small-pipe systems were used through the 1960s. The Hardy Cross method was sufficient for single-loop systems. However, without the aid of a computer, this method was impractical for more complex systems with multiple loops. In 1950, McIlroy simulated the behavior of water distribution systems using electronic circuitry. However, these physical models were large, expensive, and difficult to use. In the 1960s, digital computer models appeared, and the FORTRAN programming language was primarily used to develop various models that became available to practicing engineers.

1.3.2. 1970–1990

Modeling software was developed with increasing capabilities and more advanced simulations, including extended period simulation and water quality analysis. Graphical user interface capabilities were incorporated to draw the water system and display output. Software packages that contained several compatible modular components were marketed. Some packages used other complementary software for data entry, display, and reporting of results.

1.3.3. 1990s

The 1990s experienced exponential growth of distribution system modeling capabilities. EPANET, a water modeling simulation engine and software developed by the US Environmental Protection Agency (USEPA) to support ongoing research, was made available to the public. Some commercial software vendors have taken the EPANET simulation engine and added an improved user interface and enhanced capabilities. Commercial modeling software was developed to interface with industry-leading CADD and GIS software packages. The result was a familiar user interface and the ability to use existing software and data sources rather than having to create a new water model manually from scratch. Water quality modeling and extended period simulations became standard features in commercially available modeling software packages.

1.3.4. 2000–2010

Software packages were developed to interface more effectively within GIS software environments in response to the rapid adoption of GIS as the standard database format for assets and asset management. The use of GIS had become more common, and the quality of data had improved, which significantly reduced the effort required to develop models. Automation tools became available for optimizing model development and aiding the calibration process. Security concerns resulted in studies to develop emergency response plans in order to evaluate the potential impacts various disasters might have on water distribution systems. Models were used increasingly for water quality analyses, such as evaluating water age and constituent concentrations. USEPA also allowed use of modeling to determine preferred locations for water quality monitoring sites necessary to meet regulatory requirements.

1.3.5. The Present

The availability of tremendous amounts of information from various data sources, real-time instrumentation, faster computers, high computer network bandwidth, and well-funded commercial and academic research and development has expanded the uses and applications of models to newer areas. Drought and water scarcity have raised awareness of water loss in distribution systems, and models are being recognized as a tool to identify and manage water leakage detection to reduce nonrevenue water. Climate change, measuring a utility's carbon footprint, and mandatory conservation in some arid regions have increased the importance of operational efficiency and pump energy management, placing a spotlight on the water distribution system model as a means to analyze and optimize a utility's pumping operations. Additionally, transient analysis studies have resulted in the need to design for transient conditions to reduce potential pipe breaks and water contamination. As concern for water quality within water storage reservoirs increases, modeling of these facilities has become more important to ensure proper mixing and to meet more stringent water quality regulations. Unidirectional flushing programs are being developed to enhance system water quality. The use of machine learning, such as genetic algorithms for system optimization and calibration, and other advanced modeling tools are now in frequent use. Water quality and system security planning continue to be critical applications because of increased regulations and national threat levels. Software packages now have the sophistication to perform network calibrations as well as help determine where closed valves exist within the system. Current water models are helping utilities sustain infrastructure in challenging economic times by allowing system owners to quantify and reduce operating hours, water loss, pipe breaks, energy usage, and a host of other related costs.

1.4. DISTRIBUTION SYSTEM MODELING APPLICATIONS

1.4.1. Benefits of Water Distribution System Modeling

When properly implemented, models become an integral part of the decision-making process for planning, designing, and operating a water distribution system. Water system engineers and operators are ultimately responsible for decisions based on the results that computer models provide.

Distribution system modeling software generally falls into four categories of application: planning, engineering design, system operations, and water quality improvement.

1.4.2. Planning

A primary planning application of distribution system analysis software is used for assisting in the development of near- and long-term capital improvement projects, which include scheduling, staging, sizing, and establishing preliminary routing and location of future facilities. Other applications include planning for water main rehabilitation and system improvement. Rehabilitation plans identify and prioritize mains that need to be cleaned and/or lined. Distribution system improvement plans identify where installation of new mains, storage facilities, and pump stations are necessary to keep pace with growth and/or new utility standards and regulations. The following are examples of several specific system analysis planning applications.

1.4.2.1. Capital Improvement Program. Water utilities usually develop a master plan that identifies future capital improvements necessary to support projected community growth and to replace aging infrastructure. These plans typically consider planning horizons of 5 to 20 years or more. A model is used to help identify and schedule near- and long-term capital improvements.

1.4.2.2. Conservation Impact Studies. Water conservation is desirable for most communities to extend limited water supplies or to reduce water use so that some capital improvements can be delayed or eliminated. A model is useful to predict effects of various conservation measures on future planning periods to evaluate the potential for success.

1.4.2.3. Water Main Rehabilitation Program. A model is used to identify specific water mains that tend to bottleneck the system, due either to increased demands or tuberculated pipes. The model is used to determine the potential hydraulic effects of replacing, upsizing, or rehabilitating aging mains to evaluate the effectiveness of various alternatives.

1.4.2.4. Tank Siting. A water tank should be sited in a location that optimizes water replenishment, effectively meets peak demands, and can refill efficiently during off-peak demand periods.

1.4.3. Engineering Design

Engineering design applications include the sizing of various types of facilities including pipelines, pump stations, control valves, tanks, and reservoirs. These facilities are sized using pressures and flows that result from distribution system modeling. Practical considerations such as accessibility, obstructions, special structures (e.g., hospitals, schools), proximity to alternate water sources, and other circumstances can be factored into system design. In addition, system performance can be analyzed under fire flow conditions and adjustments made to meet fire demand. The following are examples of engineering design problems that can be solved using computer models.

1.4.3.1. Fire Flow Studies. The model is used to simulate fire flow demands at hydrant locations throughout a locality to determine how much water can be delivered to specific fire hydrants within the prescribed fire flow requirements. Where deficiencies are discovered, distribution system improvements such as main reinforcements or looping can also be evaluated using the model. These studies are also used to demonstrate compliance with fire protection standards and guidelines.

1.4.3.2. Control Valve Sizing. A distribution system often has pressure-regulating or pressure-sustaining valves to direct flow to different hydraulic pressure zones. Distribution systems may also have throttle valves to direct flow within a zone to different reservoirs or storage tank locations. The model is used to determine how much flow is expected through these valves so that the valves are sized appropriately.

1.4.3.3. Tank Sizing. Tanks are often sized by estimating the total diurnal flow, fire flow, and emergency storage requirements within a particular zone. However, tank capacity should also consider the rate of water delivery to the tank location and the size of the distribution area. A model is useful to evaluate inflows and outflows to a tank to determine an optimal size for a particular location and/or to specify other improvements so that the preferred tank site is adequately served by transmission mains and pumping stations.

1.4.3.4. Pump Station/Pump Sizing. Models are used to calculate system curves of distribution systems so that appropriate pumps are selected to provide the necessary flow and head. Once sized, the proposed pumps are then used in the model under a wide range of system settings to determine how well they perform under various operating conditions.

1.4.3.5. Estimation of Pressure and Flow at Particular Locations. A water distribution system must provide adequate amounts of water at appropriate pressures within a range typically specified by standards used in the water industry. Models are used to predict pressures under specific demand conditions and under a wide variety of scenarios to identify low pressures and to select infrastructure that will improve flow or lessen pressure deficiencies. Additionally, hydraulic models are increasingly being used to evaluate transient scenarios and develop capital or operational alternatives to mitigate them.

1.4.3.6. Zone Boundary Selection. Most water distribution systems deliver water to customers located at a range of elevations. Such distribution systems are separated into pressure zones that follow consistent elevation contours to keep pressures within reasonable ranges. Models are useful to evaluate potential zone boundaries and to determine the adequacy of infrastructure delivering water to each zone.

1.4.4. System Operations

Model applications for operations include assisting in the development of operating parameters and strategies, operator training programs, and system troubleshooting guidelines. Emergency conditions, energy management, water availability, and other factors may drive operating strategies. For example, contingency plans are developed in the event of a key facility component failure such as a pump station failure. Distribution system modeling is also used to develop operational strategies for energy management and water quality guidelines. Strategies for shifting supply between treatment plants are developed to determine the most efficient use of available water. Optimizing these strategies results in efficient use of pipeline capacities, tank levels, and required treatment plant production, among other things.

1.4.4.1. Personnel Training. Models are used to train personnel who operate the distribution system. System operators can experiment with the model and determine how the system responds to changes in operating parameters and conditions.

1.4.4.2. Troubleshooting. Models are used to troubleshoot potential causes of various problems such as low pressure, water circulation issues, and events that would otherwise be inexplicable.

1.4.4.3. Water Loss Calculations. In the event of a major main break, the model is used to estimate the amount of water lost through the break, which may be required for damage assessments.

1.4.4.4. Emergency Operations Scenarios. Water distribution systems often have critical components; if these components fail, water delivery is interrupted or the level of service is reduced. A model is useful to evaluate the potential impact of a failure and to devise ways to reduce the damage or impact of a critical component failure.

1.4.4.5. Source and Storage Management. Water treatment plants and tanks are sometimes taken out of service for repairs or because the water supply is unavailable for a time. In addition, the quality of water at one source may be better at certain times of the year, so the use of the high-quality source can be maximized. The model is useful for devising operating scenarios that best use multiple water sources to achieve desired system objectives.

1.4.4.6. Model Calibration. Model calibration is a critical step in developing a useful model. The model results are compared with field measurements, and adjustments are made to the model to improve agreement between the two. However, the calibration process is also useful to operations staff in discovering anomalies in the distribution system, such as closed valves, tuberculated pipes, leaks, or false or incomplete infrastructure data. This information, once discovered through the calibration process, can explain operational difficulties and identify distribution system problems that require the development of solutions to resolve and improve system operation.

1.4.4.7. Main Flushing Programs. A model is a useful tool for developing a main flushing program. The model helps identify flow paths in the distribution system so that appropriate flushing locations and sequences can be established. Models are especially useful for developing unidirectional flushing programs.

1.4.4.8. Outage Areas. Water utilities frequently need to isolate a specific area of the distribution system for maintenance or other work by taking a pipeline or facility out of service. Often, it is helpful to identify those customers whose service will be interrupted by the outage event. In addition, those planning the event need to know which valves to close in order to minimize impacts of the outage. Models are a valuable tool for accomplishing this task.

1.4.4.9. Energy Cost Management. Energy costs associated with pumping typically account for 75 to 80 percent of the power consumed by water utilities. With energy costs being such a high percentage of total operating costs, utilities are trying to find ways to reduce their overall energy consumption. Many current models have features to help quantify energy consumption and costs and to optimize pump operation to reduce electrical usage and take advantage of off-peak energy rates.

1.4.5. Water Quality Improvement

Regulations, standards, and expectations for water quality continue to evolve and expand, increasing the demand for water quality analysis. Following are examples of how distribution system modeling is used to improve water quality.

1.4.5.1. Contaminant Tracing. If a contaminant enters the distribution system through a treatment plant, well, reservoir, or other location, it can quickly spread throughout the distribution system, affecting the quality of water consumers receive. The contaminated water may also mix with water from other sources. A model can be used to determine the point of entry of the contaminant and to predict contaminant levels and zones of influence within the distribution system. This allows a utility to identify which customers might be affected by the contaminant and the portions of the distribution system that might need to be flushed.

1.4.5.2. Water Source Tracing. A model can be used to trace the source of water to any point in the model when multiple sources are present. A source trace analysis provides the percentage of water from each source present. When multiple sources serve an area, distribution system modeling also helps devise operating strategies to reduce water age where possible.

1.4.5.3. Water Age. Water age, or hydraulic residence time, is another important water quality parameter in a distribution system, often used as a surrogate for other water quality parameters. Chlorine levels decay over time, increasing the tendency for disinfection by-product (DBP) levels to increase as chlorine reacts with organic compounds in the water and along the pipe wall. To maintain adequate water quality, water utilities strive to minimize water age.

1.4.5.4. Chlorine Levels. A model is used to predict chlorine decay in a distribution system. This helps to determine a target level for chlorine entering the distribution system and to select sites where it may be necessary to boost chlorine levels. Other constituent levels can also be predicted with a model. Many models can analyze more than one constituent at a time.

1.4.5.5. Water Quality Monitoring Locations. Some water quality regulations under the Safe Drinking Water Act (e.g., DBP rules) require utilities to select sites within a distribution system where they will collect samples to demonstrate compliance. A model can help identify the most appropriate locations for these water quality monitoring locations.

1.5. HYDRAULIC MODELS

A model is composed of two parts: a database for the specific water system and a computer program to solve hydraulic and water quality equations. The database contains information that describes the system's infrastructure, demands, and operational characteristics. The computer program solves a set of energy, continuity, transport, or optimization equations to predict pressure flows, tank levels, valve position, pump status, water age, or water chemical concentrations under a variety of scenarios. The computer program also helps create and maintain the database and presents model results in graphical and tabular forms.

1.5.1. Model Data

A model consists of two types of elements: links and nodes. Depending on the model, the primary components include pipes, junctions, pumps, tanks, hydrants, and valves that are represented by either a link or a node. Once built, the model database becomes a valuable asset to the user and is the result of substantial effort in data collection, entry, and quality control. A key to maintaining the model's value is making sure it is updated to reflect changes in infrastructure components, system demands, or operating parameters. Creators of models should consider how the model will be maintained over time to ensure continued confidence in the model's results.

1.5.2. Modeling Software

1.5.2.1. Modeling Equations. At the heart of water distribution system modeling software is a simulation engine that can solve a series of numerical calculations (e.g., hydraulic and water quality equations). Depending on the problem, several kinds of equations are solved.

1.5.2.2. Water Model Analysis

1.5.2.2.1. *Steady-State Analyses.* A steady-state analysis provides a snapshot of pipe system conditions at any instant in time. Steady-state analyses are typically used to evaluate maximum day, average day, peak hour, and fire flow conditions.

1.5.2.2.2. *Extended-Period Simulation.* An extended-period simulation is a series of steady-state simulations executed at specified intervals and performed over a specified time period. This capability may be used, for instance, to model the operation of a water system over a 24-hour period. Such a simulation is useful in modeling variations in demand, reservoir operations, water quality, emergency responses, energy management, and water transfers through transmission pipelines. Extended-period simulation requires that the modeling software model flow and pressure variations, incorporate diurnal water demand patterns, simulate operational controls for pumps and control valves, and allow for varying tank configurations.

1.5.2.3. Specialized Model Analysis

1.5.2.3.1. *Automated Fire Flow Calculation.* Some distribution system modeling packages automatically calculate the available fire flow at each node. These calculations are useful in identifying areas with weak firefighting capability.

1.5.2.3.2. *Water Quality.* Utilities are increasingly interested in modeling the water quality within a distribution system, particularly the decay of chlorine residual and water age. The ability to perform water quality analysis should be a standard part of any modeling package.

1.5.2.3.3. *Energy Analysis.* Energy analysis, available in most modeling software, can help identify inefficient pumps and determine better operational strategies. Energy costs are a significant portion of the total expense for most utilities. Some electricity providers have variable rates, and electricity costs can vary depending on when pumps are operated. Energy consumption and energy costs can be quantified through this type of analysis.

1.5.2.3.4. *Transient Analysis.* Transient pressures (water hammer or surge) can cause pipe breaks, contamination, joints to shift and leak, collapse of pipes, and other serious damage to water distribution networks. A transient event is caused by a sudden change in flow velocity that can be created by a valve that is closed too quickly, pump failure or a pump simply shutting down, a mishandled fire hydrant, and similar activities. After identifying a transient condition (valve closure), water-transient analysis software can identify where a transient event is likely to happen and evaluate multiple transient protection devices that can help mitigate or prevent damage.

1.5.2.4. Modeling Software Package Functionality

1.5.2.4.1. *Scenario Generation.* Distribution systems with any level of complexity are modeled more easily by applying various combinations of demands, facilities, and operating parameters, such as regulating-valve and pumping-unit settings. System modeling packages may allow variation and combination of these three types of data in a simulation for specific or combined model access.

1.5.2.4.2. *Selective Reporting of Results.* The user can specify the results to be reported for each simulation. In some cases, a user may only be interested in results from a specific portion of the system. By limiting results to specific portions of the system, the user saves hard disk space and paper while speeding the user's review time.

1.5.2.4.3. *Data Management.* Modelers can export and import data and model results to and from other applications such as spreadsheets, databases, and GIS systems. These capabilities are widely available and are an important part of any modeling package.

1.5.3. Related Software Systems

Information management within utilities is moving toward greater information sharing, enabling staff to have appropriate information to make decisions. This is often done by using common databases and files that are shared by a multitude of software applications. Distribution system modeling uses different types of information about physical assets, customers, billing, geographical information, and operational information. Furthermore, modeling results can benefit many groups within the utility, strengthening the ability to communicate and share information. Brief descriptions of software systems, information systems, and/or industry standard databases that are in some way related to distribution system modeling are provided here.

1.5.3.1. Geographic Information System. A GIS stores and displays geographically referenced information, that is, information that is easily understood through map display. Spatial relationships between entities are significant for most information stored in GIS. GIS has the potential to store vast amounts of information that is useful for system analysis, including pipe assets, customer meter locations, zoning and land parcel data, aerial photography and other land bases, street locations, digital terrain models (DTMs), digital elevation models (DEMs), and jurisdictional boundaries. Spatial analysis capabilities of a GIS also benefit models for demand allocation, topography extraction, and results presentation, as examples.

GIS has further evolved in the areas of data capacity, structure, integrity, and results presentation. Information in GIS is saved in file formats that most modeling software packages can access. Alternatively, information in a GIS database can be translated into a format that can be imported into the model database. Some GIS land-based information and other data layers can be displayed directly within some modeling packages. The usefulness of GIS pipe data is often dependent on the way the information is collected and stored in the GIS database. GIS data is used to build and/or update models to get a detailed representation of the water distribution network.

1.5.3.2. Computer-Aided Design and Drafting. CADD systems are sometimes used to manage maps of water distribution systems, making them a source of pipe information that can be transferred to the model. In addition, CADD systems are useful as a means to display model information and results.

1.5.3.3. Supervisory Control and Data Acquisition. SCADA systems are used to remotely monitor and control the operation of tanks, pump stations, valves, and other system infrastructure. They are also useful for collecting data such as pressures, flows, water tank levels, valve positions, pump status and speed, chlorine levels, and other information useful in monitoring the distribution system. This information is collected at regular intervals and stored for extended periods of time. SCADA is a good source of operational information as well as data to calibrate a model. SCADA data are also used to define the starting point for operational analyses; this is done using the data to define boundary conditions entered into the model.

1.5.3.4. Customer Information System. CIS is useful for developing water demands based on actual customer water consumption. Typically, average annual water usage and customer rate classes are extracted from CIS and then linked to the model via GIS, modeling, or customized tools. The specifics of how CIS data can be linked to the model are highly dependent on the data and software used by the utility.

1.5.3.5. Laboratory Information Management System. Laboratory information management system (LIMS) instruments, protocols, standards, and software are vital for monitoring water quality and ensuring safety and regulatory compliance. Utilities continuously sample water for certain parameters and are expected to accurately report water quality issues immediately. Water models can assist with developing flushing routines, chlorination, tracking contaminants, determining water age, and independently locating the likely source of water quality issues.

1.5.3.6. Computerized Maintenance Management System. A computerized maintenance management system (CMMS) can act as an access point to post and retrieve system information and costs for operations, maintenance, and rehabilitation. Typically, a CMMS is used to store asset inventory, condition assessments, parts inventory, preventative maintenance activities, service requests, and work orders, all of which can be linked to the object by location and imported into models for a variety of applications. A pipe's maintenance history can be useful for modeling as it can be related to the pipe's roughness.

1.5.3.7. Asset Management System. Similar to CMMS, asset management software is becoming more prevalent as a strategic planning tool for water utilities. This management tool combines long-range planning, life-cycle costing, proactive operations and maintenance, and capital replacement plans based on cost–benefit analyses. These applications are used to meet a wide range of management challenges that water utilities face today. Modeling is an integral part of the long-range decision-making process and can be incorporated into a comprehensive asset management program for achieving system sustainability. AMS seeks to provide informed, timely, and cost-effective decision making for both day-to-day operations and long-term planning.

Many of these systems also enable users to integrate different applications and data sets already in use, including GIS, CMMS, and other work orders as well as modeling/analysis software.

1.6. TRENDS

Several significant trends in distribution system modeling have been identified through network modeling surveys, presentations, discussions at conferences, and *Journal AWWA* articles. Some trends are now well established while others are still in their infancy.

1.6.1. Common Databases

In some utilities, computer systems are organized around a type of architecture where common databases are shared by many applications. In such a framework, a distribution system's modeling package extracts data from a large enterprise database that is shared by many work groups. A database system that supports multiple work groups gets input from key stakeholders and integrates interdepartmental work flows. Asset management, billing, customer service, work management, facilities, engineering, GIS, CADD, modeling, document management, permitting, water quality testing, and other work groups have information that can be linked geospatially. An enterprise database reduces duplication of efforts, inaccuracies, and outdated information.

1.6.2. Energy Analyses

Water distribution system models are increasingly used to improve energy efficiency through better pump scheduling and operations. As energy rates increase, there is greater emphasis on reducing energy consumption and resultant energy costs. Models are used to help identify pumps that need maintenance or replacement, evaluate new pumping control strategies, and operate the water system more cost effectively. In addition, they are used to evaluate pump combinations that work well and to choose between variable-speed and constant-speed pumps.

1.6.3. Advanced Metering Infrastructure

Advanced metering infrastructure (AMI) is used to automatically collect data, including demand and status information, from water meters. This data is then transferred to a database for billing and analysis. Utility providers collect this information in real time (wired or wireless) without having to travel to each physical location to read a meter. AMI helps reduce errors in the demand allocation process.

1.6.4. Transient Analysis

A water transient is an intense pressure wave or surge that occurs when water in motion is abruptly stopped or changes direction. An example of this is water hammer. Transients can be caused by a pump startup, pump failure, valve closure, or other reasons. Awareness that pipe breaks are not necessarily caused by aging infrastructure and that installing a water surge-protection device is less expensive than replacing a pipe network has caused an increase in transient analysis. There is a greater focus in educating field crews, contractors, and fire departments that closing a valve or hydrant quickly can result in a water transient at a nearby location. Today, being able to quantify the magnitude of transients has helped size and identify the right type of surge-protection device.

1.6.5. Water Quality Analyses

Many water modeling software packages model water quality parameters in pipe networks, water tanks, and reservoirs. Utilities find this valuable in response to water quality regulations and heightened public awareness regarding water quality. In addition to being able to model water age, trace contamination sources, and measure constituent concentrations (e.g., chlorine, chloramine) in a water distribution network, some models also analyze disinfection by-product formation and perform other complex chemical analyses. Models can also simulate settling and resuspension of sediments within pipes.

1.6.6. Tank Mixing

Driven by the need to improve water quality and reduce storage residence times, tank performance and mixing is gaining popularity. Tanks can be mixed properly through better design of inflow and outflow ports and control devices, and mechanical mixers, and by improved operational strategies. Computational fluid dynamics (CFD) modeling is a key tool for assessing tank mixing.

1.6.7. Water Security

Water system design and operational decisions are increasingly based on risk and vulnerability assessments. The ability to track contaminants and to isolate and flush potentially contaminated areas is important to securing safe water systems. Water models are used to simulate and evaluate multiple contaminant intrusion scenarios at vulnerable and publicly accessible locations in distribution networks. This aids in the placement of water quality sensors and helps in planning for system security breaches.

1.6.8. Emergency Planning

Emergency planning ranges from planning for water main breaks to contamination to pump failure to power outages. Planning ahead can help mitigate potential disasters. Criticality and segmentation modeling uses the model to systematically analyze segment failures and determine critical pipes and valves in the system whose failure might be especially disruptive or costly. Planning protocols, including the US government's National Incident Management System and several state emergency management systems, are available to assist with structured emergency planning. System modeling plays an integral role in long-term emergency planning efforts.

1.6.9. Real-Time Modeling

In the past, distribution system modeling packages were typically too slow and unwieldy for system operators to generate operating strategies and test "what-if" scenarios. Real-time modeling is an emerging advancement. High-speed processing and data input available from SCADA allow utilities to deliver models to their water system operators with a greater level of confidence in the model's input and results. Careful consideration must be given to the user interface in this regard, and simplified models may be required.

1.7. SUMMARY

Water utilities seek to deliver a safe, reliable, continuous supply of high-quality water to customers every day through a complex distribution network while managing costs. Once developed and calibrated, a water distribution model can predict the behavior of a water distribution system, providing an effective tool to help utility service providers meet their goals.

This chapter provided a fundamental overview of modeling, including a timeline of distribution modeling development, water modeling applications, essential and desirable features in modeling software, and emerging trends. The following chapters provide more detail and guidance in implementing model development and using water models in various applications to analyze, design, and improve the performance of water distribution systems.

1.8. ADDITIONAL RESOURCES

Anderson, J.L., Lowry, M.V., & Thomte, J.C., 2001. Hydraulic and Water Quality Modeling of Distribution Systems: What Are the Trends in the U.S. and Canada? In *Proceedings of the AWWA Annual Conference*. Denver, CO: AWWA.

AWWA Water Distribution Model Survey Subcommittee of the Engineering Modeling Applications Committee of the Engineering and Construction Division Under the Technical & Educational Council, 2014. Committee Report: Trends in Water Distribution System Modeling. *Journal of the American Water Works Association*, 106:10:51–59.

Benedict, R.P., 1910. *Fundamentals of Pipe Flow*. New York: John Wiley & Sons.

Bhave, P.R., 1991. *Analysis of Flow in Water Distribution Systems*. Lancaster, PA: Technomics Publications.

Canning, M.E., 2002. Field Report—Field-Testing of Infrastructure for Asset Evaluation. *Journal of the American Water Works Association*, 94:12:53–57.

Casey, R., 2001. Integrating Modeling, SCADA, GIS, and Customer Systems to Improve Network Management. In *Proceedings of the AWWA Information Management and Technology Conference*. Denver, CO: AWWA.

Cruickshank, J.R., 2010. Hydraulic Models Shed Light on Water Age. *Opflow*, 36:6:18–21.

Draper, T.M., 2008. Automation Works—Leverage Utility Applications With SCADA Data. *Opflow*, 34:5:10–11.

Edwards, J., Koval, E., Lendt, B., & Ginther, P., 2009. GIS and Hydraulic Model Integration: Implementing Cost-Effective Sustainable Modeling Solutions. *Journal of the American Water Works Association*, 101:11:34–42.

Eggener, C.L., & Polkowski, L., 1976. Network Modeling and the Impact of Modeling Assumptions. *Journal of the American Water Works Association*, 68:4:189–196.

Gessler, J., 1980. Analysis of Pipe Networks. In *Closed Conduit Flow*. Chaudry, M.H., & V. Yevjevich, editors. Littleton, CO: Water Resources Publications.

Govindan, S., Walski, T.M., & Cook, J., 2009. Hydraulic Models—Helping You Make Better Decisions. *NRWA Magazine*, First Quarter, 34–40.

Gupta, R., & Bhave, P., 1996. Comparison of Methods for Predicting Deficient Network Performance. *Journal of Water Resources Planning and Management*, 122:3:124.

Hauser, B.A., 1993. *Hydraulics for Operators*. Ann Arbor, MI: Lewis Publishers.

Jeppson, T.W., 1976. *Analysis of Flow in Pipe Networks*. Ann Arbor, MI: Ann Arbor Science Publishers.

Lindley, T.R., & Buchberger, S.G., 2002. Assessing Intrusion Susceptibility in Distribution Systems. *Journal of the American Water Works Association*, 94:6:66–79.

Martin, A., Klamroth, K., Lang, J., Leugering. G., et al., 2012. Mathematical Optimization of Water Networks, *International Series of Numerical Mathematics* 162. Basel, Switzerland: Springer.

Mays, L.W., 1999. *Water Distribution Handbook*. New York: McGraw-Hill.

McIlroy, Malcolm S., 1950. Direct-Reading Electronic Analyzer for Pipeline Networks. *Journal of the American Water Works Association*, 42:4:347–366.

Murphy, Brian, & Kirmeyer, Gregg, 2005. Improving Distribution System Security. *Opflow*, 31:10:18–23.

Nayar, M.L., 1992. *Piping Handbook*. New York: McGraw-Hill.

Seidler, M., 1982. Obtaining an Analytical Grasp of Water Distribution Systems. *Journal of the American Water Works Association*, 74:12:628–630.

Stephenson, M., 1976. *Pipeline Design for Water Engineers*. Amsterdam and New York: Elsevier Scientific.

US Department of Energy. 2006. *Energy Demands on Water Resources—Report to Congress on the Interdependencies of Energy and Water.* Washington, DC: US Department of Energy.

Walski, T.M., 1984. *Analysis of Water Distribution Systems*. New York: Van Nostrand Reinhold.

Walski, T.M., 2006. A History of Water Distribution. *Journal of the American Water Works Association*, 98:3:110–121.

Walski, T.M., Bezts, W., Posluszny, E.T., Weir, M., & Whitman, B.E., 2006. Modeling Leakage Reduction Through Pressure Control. *Journal of the American Water Works Association*, 98:4:147–155.

Walski, T.M., Chase, D.V., Savic, D.A., Grayman, W.M., Beckwith, S., & Koelle, E., 2007. *Advanced Water Distribution Modeling and Management*. Exton, PA: Bentley Systems.

Walski, T.M., Gessler, J., & Sjostrom, J.W., 1990. *Water Distribution—Simulation and Sizing*. Ann Arbor, MI: Lewis Publishers.

Water Research Centre plc., 1989. *Network Analysis—A Code of Practice*. Swindon, UK: Water Research Centre.

Wilkes, D.R., 2010. From Plant to Tap: Optimize Quality. *Opflow*, 36:8:20–23.

Williams, G.S., & Hazen, A., 1920. *Hydraulic Tables*. New York: John Wiley & Sons.

Wood, D.J., & Charles, C.O.A., 1972. Hydraulic Analysis Using Linear Theory. *Journal of the Hydraulics Division*, American Society of Civil Engineers, 98:7:1157–1170.

Wood, D.J., & Rayes, A.G., 1981. Reliability of Algorithms for Pipe Network Analysis. *Journal of the Hydraulics Division*, American Society of Civil Engineers, 107:10:1247–1248.

Wu, Z.Y., & Song, Y., 2014. Optimizing Selection of Fire Hydrants for Flow Tests in Water Distribution Systems. *Procedia Engineering*, Elsevier, 70:1745–1752.

Water Utility Council, 2012. *Buried No Longer: Confronting America's Water Infrastructure Challenge.* Denver, CO: AWWA.

Chapter 2

Building and Preparing the Model

2.1. INTRODUCTION

A primary objective of computer modeling for water systems is to develop and maintain a model that sufficiently emulates the performance of the distribution system. This gives the utility confidence in the model's ability to represent the system under different conditions. The principal tasks associated with creating and managing a model include model construction, calibration, system analysis, and model maintenance. The primary goal of model calibration is to achieve an acceptable level of confidence in the model's ability to reasonably represent the performance of the water system. Following its creation and calibration, the model requires periodic maintenance to remain representative of system operations.

Models should be created and maintained so that the standards of accuracy, completeness, and level of detail are appropriate for the intended model application. Model development begins with careful planning to define requirements and to investigate the sources of information available. A modeling project requires planning by a modeler and oversight, at a minimum, by an engineer and an operator experienced with hydraulics, utility infrastructure, and operations.

Many types of data are used to build hydraulic models. Documentation during model construction and calibration tasks are essential to managing these data and providing for effective, long-term model maintenance. This is especially important when transferring the model to other modelers.

2.1.1. Hardware and Software

A wide array of water distribution modeling software packages are available, ranging from open-source software to commercially available software with different features and costs. A software selection process may be needed prior to selecting a software package.

Selection criteria should be based on a combination of the following: ease of use, modeling features, vendor history, data integration options, maintenance policies, licensing options, costs, technical support, computational engine, vendor location, training options, utility preferences, and current hardware and software requirements and others.

Most software packages used in hydraulic network analysis typically require the same basic system data to set up, refine, and calibrate. Taking the time to gather high-quality data for the model yields a number of benefits to the utility. Better data and a better model can result in savings elsewhere in the organization as a result of improved system knowledge that, in turn, leads to better decisions. This investment in data should be protected by ensuring that the data can be easily migrated from one modeling package to another modeling package if the utility's modeling software requirements change. Most software vendors allow export from their modeling packages to a common, interchangeable file format; EPANET file format is an example of this.

Hardware requirements differ only slightly for the various modeling software packages. As computer and server technology improves, the modeling software should as well. In general, most modeling packages have a minimum hardware specification and an optimal or recommended specification for processor speed, RAM (random access memory), graphics card, and hard-drive space. Some real-time models require additional server hardware or leverage cloud-based storage and shared processing platforms.

2.1.2. Geographic Information Systems

The role of GIS in the water and wastewater industries continues to grow and influence the evolution of modeling. Advances in GIS database design, software, and hardware allow more robust interoperability with modeling databases, improved performance with large data sets, and advanced post-processing tools.

The model shares a mutually beneficial existence with GIS. Just as GIS enhances the usefulness of the model, the model can also extend the functionality of GIS. For utilities that have established a GIS, the added functionality of interfacing with a model can prove valuable from enterprise, customer, and operational perspectives. This interfacing can add further value to both the model and GIS databases. For example, the process of building a model from GIS and the subsequent calibration may reveal errors within the GIS database, such as improperly connected or unconnected pipes, missing attributes, and diameter discrepancies. These adjustments could be identified in the model and should be incorporated into the utility's GIS.

Integration of the model with the GIS requires the highest-quality GIS data. While errors can be identified during the model construction and calibration process, data cleanup efforts should be undertaken within the GIS to make the GIS "model ready." This process includes the creation of a geometric network with predefined topology rules and criteria before model construction as well as a common identifier field between the model database and the GIS database. Advancements in model processing and greater complexity of infrastructure representations within GIS have also led to an increase in the number of models being developed with a one-to-one or near one-to-one relationship with the GIS.

2.1.3. Enterprise Applications

A model can be enhanced with database links to other enterprise-wide applications such as a customer information system (CIS), computerized maintenance management system (CMMS), automatic meter reads (AMR), advanced metering infrastructure (AMI), laboratory information management systems (LIMS), and/or supervisory control and data acquisition (SCADA) systems. Links to these data sources, if available, should be prepared during the building process.

The CIS contains customer data critical to establishing base demands within the model. To be useful, the CIS needs to have x–y coordinates for each customer in the same coordinate system as the model. The CIS also provides the ability to identify customers who may be out of service during a pipe outage by giving them identifiers (IDs) or parcel numbers. CMMSs are used to generate field work order requests and can be a valuable source of attribute information for pipes, hydrants, valves, and other components of distribution systems. This information may include customer complaints, maintenance records, and field observations as well as basic attributes (e.g., size, material, manufacturer, date installed) in case the information is not available in GIS. The model should be set up with unique asset numbers that are consistent with the utility's CMMS to facilitate import of CMMS data during and after model building. Model integration with SCADA offers a variety of mutually beneficial model applications, such as establishing operating ranges for given facilities, identifying boundary conditions necessary for model calibration, historical event system troubleshooting, and assistance with pump scheduling and optimization. However, the utility's security protocols and data firewall limitations should be considered when setting up links to SCADA; this could preclude live two-way communication. Alternatively, connection to a SCADA historian may resolve security and access concerns. In any case, preprocessing of SCADA data is necessary either via software algorithms or manual efforts to import it into modeling software.

Communicating with and relating to various available data sources and departments throughout the utility can avoid inaccuracies or rework in model development, calibration, and maintenance.

GIS and other enterprise applications make building and maintaining models more efficient, and they support a wide array of analysis and reporting functions. These applications enhance the value, efficiency, and accuracy of modeling. Increased adoption of enterprise applications allows for a more sustainable model versus a model that needs to be repeatedly rebuilt and recalibrated.

2.2. PLANNING THE HYDRAULIC MODEL CONSTRUCTION AND DEVELOPMENT PROCESS

While most water distribution systems have many of the same types of assets, each water system's size, configuration, and operations are unique. The modeler must consider the system's distinctive structure and specific requirements when developing the plan for the model.

2.2.1. Establish Purpose and Need

Before developing or updating a model, its current and potential future applications should be identified (see section 1.4 for example model uses). The desired applications dictate the model's level of detail, its required level of accuracy, and the data that will need to be collected for model building and calibration.

After the utility's vision of model application has been established, decisions on such topics as model data loading and accuracy of calibration naturally follow. By planning the modeling objectives first, a utility can determine any necessary integration plans and what is needed in terms of resources and software.

The degree of detail in the model is greatly influenced by the intended purpose of the model and the available data. A model used for long-range planning, transmission main sizing, and reservoir site evaluation may only require the inclusion of major transmission mains and the use of theoretical peaking factors for estimating water demands. Where sufficient detail is lacking, it is critical that the modeler be aware of the model's limitations. These limitations should be clearly stated in any report documenting analysis results.

On the other hand, a model used for fire flow capacity evaluations at the neighborhood level, leakage control studies, or water quality modeling requires considerably more detail, up to the possible inclusion of every transmission and distribution pipe – excluding service lines – in the distribution system. These types of models require the modeler to allocate demands in a more detailed manner.

The amount of time and financial resources that are available also influences the level of detail incorporated into the model. The modeler may be required to conduct a cost–benefit analysis to obtain managerial support for extending extra effort in developing a more detailed model.

2.2.2. Model Structure

It may not be necessary or practical to create the model with the level of detail found in the GIS. Models can be categorized based on their level of detail and by the extent to which the model elements (pipes, pumps, valves, tanks, and nodes) match the corresponding features in GIS. To understand the various model types, it is necessary to define the following two model development terms:

- *Skeletonization:* deleting or excluding unimportant (usually small-diameter) pipes in the model.
- *Reduction:* removing intermediate (hydraulically insignificant) nodes along pipe reaches of similar diameter, material, and construction date and subsequent combining of the adjoining pipe segments into a single pipe segment.

Descriptions for various model types and sizes, along with their advantages and disadvantages, are presented in the following sections.

2.2.2.1. All-Pipes Model. An all-pipes model preserves the same level of detail as the GIS and maintains a one-to-one relationship between individual GIS and model elements. There is no skeletonization or reduction of GIS data to construct an all-pipes model. This one-to-one relationship reduces the effort required to maintain the all-pipes model and establishes data-sharing opportunities between the two databases. However, because of the larger size of all-pipes models, they require longer processing time, large computer storage space, and often a more expensive modeling software license. All-pipes models for larger utilities can exceed 100,000 pipe links.

2.2.2.2. All-Pipes Reduced Model. An all-pipes reduced model is similar to an all-pipes model with the exception that intermediate, hydraulically insignificant nodes along pipe reaches of similar diameter, material, and construction date are dissolved (reduced). The entire length of pipe in the GIS is represented in an all-pipes reduced model, but the model pipe segments are comprised of multiple GIS pipe segments (one-to-many relationship). An all-pipes reduced model requires more model building and maintenance effort and more coordination than an all-pipes model due to this one-to-many relationship. However, because of the reduced number of pipes, an improvement in model loading times and/or file sizes may be achieved. Reduction can decrease the number of pipe segments in an all-pipes model by 50 percent or more. Reduction can also simplify the use of the model because fewer pipes need to be manipulated for what-if analyses, for example. Any pipe reduction should maintain a relationship with GIS for future model updates.

2.2.2.3. Skeletonized Model. A skeletonized model includes all pipe segments equal to or above a specified diameter, typically 12 in. or 16 in. (30 cm or 40 cm) and all hydraulically significant, smaller-diameter pipes. Because skeletonized models represent only larger pipes, they are sometimes referred to as *transmission* or *backbone models*. A one-to-one relationship can be maintained between pipes in the skeletonized model and GIS. A skeletonized model may include only 10 to 20 percent of pipes found in an all-pipes model of the same system. While system performance is improved in comparison to an all-pipes model, detailed evaluations and spatial allocation at a neighborhood level are sacrificed.

2.2.2.4. Skeletonized Reduced Model. A skeletonized reduced model is the same as a skeletonized model except intermediate (hydraulically insignificant) nodes are dissolved (reduced). Skeletonized reduced models represented the standard modeling practice in the 1980s and 1990s when models were manually digitized from water system paper maps. Many utilities have replaced their skeletonized reduced models with more detailed models of their distribution systems.

Figure 2-1 shows examples of the four basic types of model structures.

Variants of the four basic model structures are common. Two of the more common variants are referred to as a *detailed model* and a *hybrid model*. A detailed model is either a skeletonized or a skeletonized reduced model with some smaller-diameter pipes included (typically some 6-in. and 8-in. [15.3-cm and 20.3-cm] pipes). The Stage 2 Disinfectants and Disinfection Byproducts Rule, which was promulgated by the US Environmental Protection Agency (USEPA) in 2006, allowed disinfection by-product sample sites to be selected based on modeling of distribution system water age. If models were used to select sample sites, the rule required that they must include at least 50 percent of the total pipe length and 75 percent of the total pipe volume in the distribution system. To satisfy these requirements, some of the 6-in. and 8-in. (15.3 -cm and 20.3-cm) pipes in the model would have to be included. A detailed model can effectively evaluate water quality concerns in the distribution system. However, the manual selection of small-diameter pipes to include in the detailed model, combined with reduction in pipe data, can result in significant increased model maintenance requirements.

A hybrid model combines large transmission mains (typically with diameters of 16 in. [40.6 cm] and larger) throughout the system with all pipes in select pressure zones or regions of the system. In this way, the model can be used to simulate detailed hydraulics and water quality in those portions of the system that are of concern. Hybrid models are typically only used in large distribution systems. This type of model improves performance and file size efficiency compared to all-pipes or all-pipes reduced models. However,

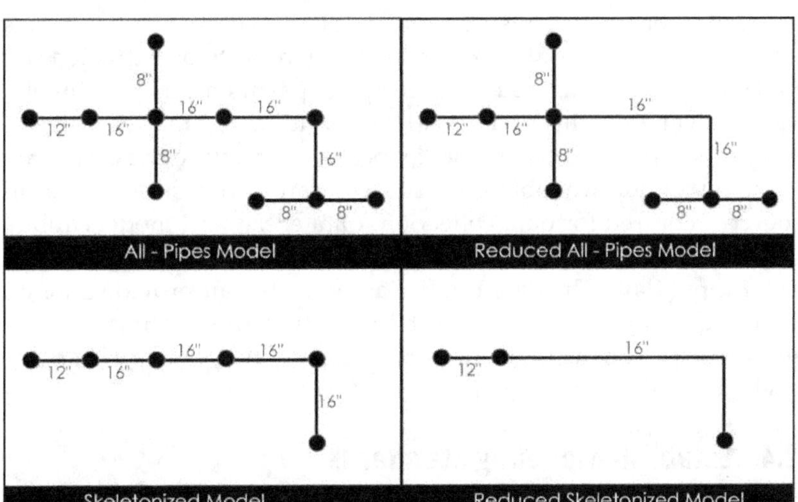

Figure 2-1 Basic model structures

hybrid models require additional time to configure the mix of system-wide transmission facilities with detailed, localized distribution system facilities.

All types of model structures can be constructed using a utility's GIS. However, model structures dictate model maintenance procedures. For all types of reduced models, a one-to-many relationship must be established between the model and GIS to conduct automated update procedures. This requires a unique understanding of the related features between the model and GIS to maintain features that are updated, added, or removed. Various methods for working with one-to-many–related features to support model maintenance can be used, though they typically require additional development in the GIS (i.e., dynamic segmentation) or programming in order to be effective. All GIS features are maintained in an all-pipes model that facilitates an easier automated transaction of data between the GIS and model.

In addition to initial development and ongoing maintenance, other factors must be considered when selecting a model structure. An all-pipes model best facilitates automated maintenance procedures. However, all-pipes models generally require a larger investment in hardware and modeling software and may cost more in initial development. The large amount of data in all-pipes models can significantly increase processing and solution time, especially for extended-period simulations (EPSs). Additionally, storage requirements for all-pipes models can significantly exceed the requirements of skeletonized or reduced models. However, processing and storage considerations become lesser concerns due to the ever-increasing advancements in computer processing, storage devices, and software capabilities.

2.2.3. Data Availability and Quality

Building and maintaining a model usually requires collecting data from, and collaborating with, multiple departments or groups. Water system asset data for facilities and distribution system piping are frequently provided by GIS or related asset management tools and can be supported by engineering, operations, production, customer billing, planning, and topography data. Control-valve settings (including pressure-reducing valves [PRVs], pressure-sustaining valves [PSVs], flow-control valves [FCVs], throttle-control valves [TCVs], and others) and pump station information, including pump curves, pump tests, and set points, may come from the operations department. Demand data may be established from meter records held by the billing department. The utility's field crews or local fire departments may need to be involved if flow testing is required for model calibration. In addition, operations staff may be able to comment on typical system operation, for example, valve closures. Initial settings may need to be extracted from the SCADA historian database. Planning and engineering departments may be involved in validating pipe roughness values in the model. With larger utilities, it may be a significant task to navigate through the various departments to locate and retrieve the best available data.

A review of available data sources can also assist in determining the effort and resources required for data collection, data entry, and model calibration. As the data are being located and gathered, it is important to objectively rate the quality of each category of input data. The data-quality assessment can provide a long-term plan on how to improve the quality of each data set. In addition, these ratings can be used to better understand the basis for the model and provide a roadmap to guide further refinement of the model during the calibration phase.

2.2.4. Establish Modeling Standards

A crucial aspect of the model planning process is the establishment of modeling standards – how facilities will be graphically represented in the model, what level of detail will be included in the model, and what naming convention will be used for everything from pipes to queries to scenarios. Procedures should be established for how these standards will be documented, communicated, and reinforced for both current and future users of the model. Consideration should be given to model documentation that includes a log to record changes to the model after initial development and calibration.

2.2.5. GIS Data Import and Conversion

The current GIS stores facilities within a shapefile or geodatabase, which categorizes elements as feature classes and subtypes within the feature class. Feature classes represent element geometry types such as point, line, or polygon; subtypes of the feature class include more specific facilities or appurtenances of the water system.

Tanks, fire hydrants, valves, and customer meters are examples of elements that may be stored as different subtypes within a node (or point) feature class. Distribution system mains, overflow or drain lines, hydrant laterals, and service laterals are examples of elements that may be stored as different subtypes within a line feature class, although typically distribution mains are the only line features used in a model. It is important for service laterals to be a class that is separate from the distribution mains for the purpose of building models. Land uses, zoning, demographics, and soil types are examples of features that would be stored as different subtypes within a polygon feature class. Typically, point and line feature classes are those features considered for import to the model, with polygon features used as supporting data for a range of model considerations and applications.

When importing elements into the model, it is important to recognize the different native storage structures of the two databases (GIS and model) and to plan accordingly. Unique element identifiers (IDs) are required to maintain a unique connection between both databases to avoid the duplication of any one or more elements. This represents just one of a number of necessary considerations. If a utility wishes to reference the unique element identifier from its GIS or CMMS within the model, it is important to plan accordingly prior to import of data to the model.

When creating models from GIS, facility and network representations should be clearly defined before model construction begins. The utility's objectives should also be defined and understood before any model construction takes place. Defining how each field in the GIS maps to model elements will greatly assist in the data management plan and effort.

2.3. DATA SOURCES AND AVAILABILITY

The data required to create a model originate from a variety of sources, as summarized in the following list. The availability, consistency, quality, and quantity of data received, in combination with the relative size and complexity of each water system, vary significantly with each utility. Engineering judgment, water system experience, and technical software capabilities are essential ingredients to incorporating all the data into a model of the distribution system. Some data are required for a model and other data are supportive and not necessarily required. Creating a model requires staff to do the following when gathering data, with sampled data sets listed within each:

1. Identify and collect geographical data:
   - Land use and zoning for existing and future conditions;
   - Population and development projections from the local planning authority;
   - Aerial photographs (online imagery also available);

- Elevation data: digital terrain model (DTM), triangulated irregular network, contour maps, global positioning system (GPS), light detection and ranging (LIDAR), or other elevation data; and
- Street maps and parcel maps including street names, rights-of-way, and city and township boundaries.

2. Establish an inventory of existing facilities:

- Pipes: location, diameter, length, age, material, lining, and connectivity to other pipes;
- Storage facilities: location, capacity, dimensions, base and water surface elevations (minimum and maximum), accessibility, mixing systems, and system connections;
- Water supply facilities: location, elevation, hydraulic grade line (HGL), capacity, operational patterns, and control settings;
- Groundwater wells: location, elevation, capacity, pump curves, pumping water levels, recent pump tests, and control settings;
- Connections (e.g., import connections, emergency interties, etc.): location, elevation, capacity, hydraulic grade line, and purpose (take and/or give);
- Pumping and booster stations: location, elevation, capacity, typical suction and discharge pressure, and pump curve(s);
- Control valves: flow-control, throttle-control, pressure-regulating, and pressure-sustaining valves (location, elevation, capacities, settings, and operability);
- Other system valves: check valves, isolation valves, and other normally closed valves; and
- Boundaries: service area and pressure zones.

3. Review water consumption records:

- Water sales (billing records);
- Service area population;
- Land use;
- Number and type of service connections (current and prior years);
- Consumption by major customers;
- Nonrevenue water (NRW) and water loss records;
- Diurnal consumption patterns for various customer types, if available; and
- Consecutive systems, water hauler locals, and other bulk water consumption venues.

4. Review water production records:

- Peaking factors;
- System diurnal demand pattern, developed as a function of hourly inspection produced or transmitted to system. A diurnal demand pattern is a unitless profile that describes how demands change in a 24-hour period. The diurnal profile value at a particular hour is multiplied by the average daily demand to obtain an hourly demand; and
- Pressure zone–specific diurnal demand patterns, if sufficient data is available.

5. Develop water demands using existing water consumption, water production, population, and land use data. Establish current and projected water demands for the total system, pressure zones within the system, and major users on the following basis:

    - Annual average day demand (total water use in a year divided by the days in a year);
    - Maximum day demand (the highest demand that occurs in a single day during an entire year);
    - Peak hour demand (the highest hourly demand that could occur during the maximum day);
    - Fire flow demand requirement (typically based on land use, structure spacing, and fire authority);
    - Minimum hour demand or maximum storage replenishment rate (occurs during maximum day);
    - Other operationally critical demands within the water system's service area;
    - Land use classifications; and
    - CIS databases with customer addresses and water use records.

6. Review system operating records and interview staff:

    - Annual, seasonal, and diurnal water production records;
    - Service and booster station flows and pressures;
    - Clearwell and storage level variations for various demand scenarios;
    - System operation criteria and set points for pumping stations, storage facilities, and control valves;
    - Identification of any closed isolation valves;
    - System operating pressures determined from previous testing, monitoring, and fire flow testing;
    - Observed low-pressure areas that indicate system deficiencies;
    - Observed fire flow capabilities;
    - Measured roughness characterization; and
    - Known operations challenges.

2.3.1. Data Confidence

All data used for the model should be objectively evaluated to determine the importance and level of confidence in each data set provided for model creation, development, and calibration. The level of confidence in the data used, how the data are assigned to the model, how the data are documented, and what efforts are required to find more detailed or reliable information are all important factors in weighing the level of effort required, and the costs, to obtain qualified data given the model's objectives. Furthermore, the confidence level of the data directly affects the confidence level of the model calibration and results. Particular attention should be given to data that would have a larger impact on model results. For example, a 10-ft (3-m) error in ground elevations could have a much greater effect on resulting pressures than a 10-ft (3-m) error in pipe length.

2.3.2. Types of Data

Data used for model construction can generally be categorized into four main groups: geographical data, physical facilities data, demand data, and operational data. A detailed water system inventory can reveal much of this information. Certain analyses require additional information, depending on the type of study. Energy costs, infrastructure costs, water constituent or contaminant concentrations, and demand or customer growth projections are examples of additional data that may be required.

2.3.2.1. Geographical Data. Geographical data includes the geographical coordinate system, jurisdictional boundaries, census block boundaries and populations, street centerlines, aerial imagery, and other information useful to establish the physical location of the model.

Elevation or topographic data are used to assign an elevation to the model nodes or junctions. Distribution node elevations are frequently assigned as equal to the ground elevation. Ground elevation data can be obtained from a variety of sources including utility as-builts, planimetric data, LIDAR studies, or DTMs. For better accuracy, elevations at pumps, control valves, and storage facilities may be based on as-built drawings or field surveys. Elevations for junctions that represent pressure monitoring locations must be carefully documented and related to the actual elevation of the pressure-recording device. The model uses the calculated hydraulic grade and elevation to calculate pressure at a given location.

Use of digital data, such as DTMs, to overlay model junctions for elevation interpolation is typically a preferred method of elevation assignment. The opportunity to minimize human error and allow computer-based interpolations assists with efficient and accurate performance of this task. The quality of the DTM and confidence level should be evaluated before proceeding with automated elevation assignments. Quality assurance should be performed following the automation of this task to compare model junction contours to DTM contours.

2.3.2.2. Physical Facilities Data. Physical data include the geometric descriptions and hydraulic attributes of pipes, pumping facilities, storage facilities, and other control facilities. Physical facilities require proper geometric data and associated hydraulic data. In addition, proper elevation data are required.

Distribution system geometry includes the connectivity, length, and diameter of pipes within the distribution system and within specific facilities. Upon construction of the geometric representation of the water distribution system, hydraulic attributes must be assigned to both facilities and distribution system piping. Hydraulic data for input to the model can include pump type and curve, valve type and characteristics, minor loss characterization, and pipe roughness characterization. It is necessary to determine reasonable roughness coefficients, or *C*-factors (the friction factor in the Hazen–Williams head-loss equation most commonly used in the United States), in order to accurately represent the head loss that will occur within the system. Generating a reasonable roughness coefficient typically involves consideration of several factors including pipe diameter, pipe condition, material, year of installation, results of field hydrant testing, and industry standard roughness coefficients for given pipe characteristics. Engineering judgment should be applied to validate the methods, data, and assumptions used to develop this relatively variable hydraulic coefficient.

2.3.2.3. Demand Data. Demand data used for calibration and simulations of a model represent the total water supplied to the distribution system at any given time. This may include water supply components such as well production, treatment plant production, and wholesale and/or neighboring utility supply by interconnections. The supply total that is produced and delivered into the distribution system equals the total water consumption of the system, including system water losses.

In recent years, water consumption data (also referred to as *sales data* or *customer billing data*) have been used to obtain a base demand allocation across the distribution system. This demand is assigned to model junctions along the water system piping, using customer billing data addresses as a spatial reference. Total water supply recorded as flowing into the system is compared to the billing data, with the difference, commonly nonrevenue water, allocated throughout the system. There are other approaches for demand generation and allocation to model junctions, some of which are summarized in subsequent sections of this chapter.

2.3.2.4. Operational Data. Operational data include parameters such as flow rates, reservoir elevations, valve and pump controls and set points, and fixed pressures that establish boundary conditions in the model. Operational data applied to the model take into account the performance characteristics of the water system. Typically operational data include information supplied by the utility operator in connection with the utility SCADA system used to monitor and control system facilities. Operational data required for input to the model can include maximum and minimum tank levels, valve position or flow rate, on–off set points for pumps or valves, initial pump status, pump speed, and pump sequencing within the pump station.

2.3.3. Sources of Data

Historically, data on distribution systems were kept as paper records and maintained by the water system operators or, as in the case of a new distribution system, by the developer and/or their client. Today, many utilities have moved – or are currently moving – their water system records to computer-assisted design (CAD) or geographic information systems (GIS), so this should be the first place to look for water system data. A GIS provides the opportunity to more efficiently construct a new model or update an existing model.

For utilities without a GIS, development of a detailed model of the water system can be a first step in creating a GIS of their water system. Expanding the model data collection process to include creation of a GIS adds time to the model-building process but ultimately delivers substantial benefits to the water utility.

2.3.3.1. Paper Records. Paper record sources for distribution system data include the following:

- Water distribution mapping,
- Contour mapping,
- Aerial photography,
- As-built construction drawings,
- Operations and maintenance manuals,
- Zone boundary maps,
- Water master plans or studies on the water system,
- Annual reports, and
- Planning documents that contain information on the water system's age.

Today's modeling software packages have the ability to graphically present results of a distribution system analysis. These software packages also allow data entry using digitizing techniques, often with aerial photos or other types of a land base that aid in positioning spatial data. To do this, however, the input files must include the geographic coordinates (x–y) for each node. The coordinate system used should be verified for spatial compatibility. If GIS for the system is not available, the coordinate system of the paper maps or other asset data could be used. A standard coordinate system, such as the US Geological Survey State Plane or the Universal Transverse Mercator coordinate system, is preferred. If the model is on an established coordinate system, the model can be converted to a different coordinate system if the need arises. Using the same coordinate system as the rest of the utility's data simplifies model maintenance and enables the use of model results in conjunction with other utility information.

2.3.3.2. Electronic Records. With the creation of a model from electronic data sources, there is an expectation that future model updates will be done quickly in response to infrastructure upgrades or system replacements. Standard procedures should be developed to extract data from electronic sources to allow for timely updates.

The types of electronic data sources typically used as inputs to the model include the following:

- GIS,
- SCADA,
- DTM/DEM (digital terrain/digital elevation model),
- CMMS,
- CADD (computer-aided design and drafting) drawings, and
- CIS or other customer usage database systems.

Depending on the type of data structure chosen for model creation, a GIS generally contains more information than is required for building a model. As a result, GIS should be reviewed before being used as a basis for model updates. In addition, commercially available GIS utilities can be used to significantly automate the process of GIS data extraction. The functionality of these GIS utilities typically includes adding a new ID or modifying an existing ID, adding or modifying attributes, modifying topology to prepare it for import to the model, and logging changes in the GIS since the last export. Once the work flow is set up, tested, and proven, future model updates are more efficient and accurate.

SCADA represents the collection of equipment, communications, and software that allows the real-time control and monitoring of system operations. Typically, SCADA systems are used to control distribution system operations and to report on equipment behavior and on the distribution system itself. SCADA systems are designed so that pressures, flows, pump status, reservoir levels, disinfectant residuals, valve position, and other information are collected at regular intervals, can be stored in a database, and are usually presented graphically to the operators. The SCADA system therefore provides a wealth of information that is downloadable for use in the model or to compare against model results.

The supervisory control function can be manual or automatic. In manual mode, an operator can manually operate valves, pumps, and other devices via a software interface or at the facility. In automatic mode, a remote telemetry unit (RTU) located at the facility can be used to automate the control of a facility, or SCADA software may control the facilities remotely based on a preset operating strategy. In practice, most SCADA systems have a combination of manual and automatic controls, depending on the field device and system characteristics. It is recommended that the modeler develop close communication with the system operators to develop a realistic, detailed understanding of how the system operates and how manual system controls are tracked; this information is highly useful during model calibration.

Elevations can be obtained from a variety of electronic sources. A DTM is a type of DEM and is a three-dimensional representation of the ground surface. DTMs are used to assign elevations in a model and are usually used with CAD or GIS packages.

Utilities that use CMMS usually have a record of repairs, improvements, or other water distribution system changes. A CMMS is often a source of valuable information regarding the condition of the water distribution network and related facilities and is, therefore, a valuable source of information when the model is constructed and updated.

CAD systems are used to store drawings or maps of a water distribution system. Although more and more utilities use GIS to track utility assets, the CAD system is still a source of valuable asset information. Today's CAD systems can integrate with GIS formats as well.

Utilities often store information useful for model development in relational databases. Customer billing systems, for example, usually contain information about each customer's water use. Some utilities also store asset information, repair data, and other information in a database that may be useful to the modeler.

2.3.3.3. Physical Inspections. A site visit to water supply sources, pump stations, reservoirs, pressure-regulating valve stations, and monitoring points is very useful for understanding how the distribution system operates and helps verify that site drawings accurately represent current infrastructure.

The physical inspection of a facility can provide information on how the system operates as well as the physical in-ground plumbing (piping) around facilities; this will improve the modeler's perspective for model facility configurations. Field inspections should be conducted for each system facility to gain an understanding of how each facility functions. Ideally, site inspections are conducted with an operator who can explain how the facility is operated under varying system conditions. A complete understanding of the function of each facility should be obtained while on site. In addition, digital photos of the entire site, including pumps and valves, may prove valuable later when the model is constructed away from the facility. These field inspections should include collecting data for the following:

- Number, type, and condition of the water facilities;
- Types, materials, and ages of pipes and overall condition of the piping network;
- How system components are controlled and perform over the range of demands;
- How the facilities and individual components interface with each other;
- Limitations and deficiencies of the facility;
- Valve location and position (open or closed) and bypasses; and
- Facility site layout and available spacing relative to other potential future facilities that the model and planning efforts may determine.

Observed physical characteristics should be compared with the network and operations data to verify that the piping connections and element locations are accurately represented. The location of all pressure and flow monitors should also be confirmed. Elevations of pressure sensors should be surveyed, if possible, and compared with elevations used in the model. Using SCADA or field tests, control valves should be checked to determine whether or not they are at the correct set points and that they are fully operational and not stuck in one position. Noises indicating valve or pump cavitation should be noted; a pump may be operating considerably off its original design point because of improper sizing, impeller problems, or incorrect set points.

Where possible, pipes that have been replaced should be visually inspected to determine the interior condition and if the pipe's roughness has changed in time. This can be compared with the results of roughness coefficient testing programs (see chapter 3) to determine whether or not the values measured are consistent with observed conditions. A hydrant flow test should be conducted before and after pipeline replacements (and after installation of pipeline additions) to verify that all system valves are open and that the facilities are functioning as intended.

2.4. PHYSICAL FACILITIES DEVELOPMENT

2.4.1. Model Elements

Models use two types of elements to represent the various water facilities: node elements and link elements. Node elements are points in the model that include junctions, tanks, and reservoirs where flow can be added or removed from the system and where hydraulic grade lines (HGLs) and pressure reputes are calculated. Link elements are lines in the model that represent pipes for which flow results are calculated. Pumps and valves can be defined as either link or node elements in the model interface depending on the modeling software (and, in some cases, the software version). Hydraulic solvers represent flow elements, including pumps and valves, as links.

2.4.1.1. Nodes. The primary functions of a node include connecting links, storing elevation data, and allowing water to enter or leave the network. A node can also represent a boundary condition such as a fixed grade node, tank, pump, or valve. When a node is not a tank or other operational element such as a pump or valve, it may be called a *junction node* or simply a *junction,* especially when used to connect pipes. Nodes may be added to the model to represent hydrants or meter locations where demands could be placed, or demands may be assigned to nodes that are in the model for other reasons. Model results for nodes report HGL, pressure head (HGL minus elevation), fire flow results, water age, and flows (into or out of the network) at node locations.

At a minimum, nodes are required where two or more pipes connect and when pipe data changes, such as diameter, age, or material (which can impact pipe roughness), where flow enters or leaves the system, and where HGL is set or needs to be calculated. Models that have relatively long pipe runs between node locations and substantial changes in topography may not have nodes at the highest and lowest points in the distribution system. In these cases, additional nodes should be inserted in the network near these high and low points to represent these potential locations of extreme pressures. If nodes are not included in the model at hydrant locations, they may be added to the pipeline to represent hydrants used during field testing.

2.4.1.1.1. *Elevation Data.* Elevations used in the hydraulic analysis are separated into two categories: facility elevations and ground elevations.

Facility elevations should be input with a high level of confidence at locations critical to the model's calibration and operation, such as at boosters, tanks, and control valve stations. Field data may be gathered at these locations during the model calibration process to determine whether or not the model is performing as expected. Facility elevations, if determined as necessary, are often determined by surveying.

Ground elevations are typically used for the remaining model nodes. These elevations are not required for calculating HGL but are necessary for determining available delivery pressures in the water distribution system. Alternatively, when elevations are applied to nodes in the model, a standard bury depth for the pipe can be subtracted from the ground elevation to represent pipe elevations if pressure in the pipe is needed.

Depending on the application of a model, elevations should be as accurate as reasonably possible. An elevation difference of 5 ft (1.5 m) translates to a potential error in model output of approximately ±2 psi (13.8 kPa). This potential error should be kept in mind during both calibration and system analysis.

2.4.1.1.2. *Quality Assurance Tips for Assigning Elevations.* High-quality GPS equipment that is calibrated to known benchmarks may be used to provide accurate elevation data. As-built drawings are also used for elevations. Elevations from DTMs and DEMs vary widely in quality and should be used with caution. Even though the elevations may be printed out to several decimal places, they are frequently extracted from sources that are only accurate to 20 ft (6 m) and hence are only as good as their source. When automated elevation interpolation methods are used, the maps should be reviewed and any geographic break lines that limit the use of automation interpolation techniques should be identified.

Due to the significant impact of elevations on model results, it is recommended that the elevations be reviewed once they are in the model. This can be done by generating elevation contours from the model and comparing them directly to the source file, whether the source is a contour file or DTM. This side-by-side comparison may help identify errors that were introduced during the interpolation and assignment process, such as conflicting or inconsistent vertical datums.

2.4.1.2. Links. Pipes are critical elements of the model. The model uses links to represent flow conveyance elements, pipelines, or pipe segments. Pipe links are required to convey water from node to node and to account for the head losses between nodes. In some modeling software, links are used to represent pumps and valves in the interface and modeling database. These types of links require additional data not required for pipes.

2.4.1.2.1. *Pipe Data and Development.* The pipeline data required for model calculations consist of diameter, pipe length, and a roughness coefficient. Pipe roughness coefficients may be estimated based on the diameter, material, and/or pipe age but ideally should be based on observed head loss during field calibration. Estimates of minor water losses from valves, pipe tees and bends, and so on, may also be included, if not accounted for in the roughness factor (see 2.4.1.2.3).

The model considers the available flow area of a pipe for hydraulic calculations. During model construction, nominal pipe size is typically used, as this produces a reasonable result in the hydraulic loop equations for the purposes of modeling. However, the modeler should be aware of the difference between nominal diameter and true internal diameter when performing detailed calculations, such as for hydrant flow tests or when comparing the economics of different pipe materials.

For larger pipe sizes, the difference between nominal and actual diameter is significant because increased wall thicknesses are required in larger-diameter pipes to maintain sufficient pipe rigidity. For transmission main modeling, the actual internal diameter is recommended over the nominal diameter. When the actual internal diameter is not known, an appropriate roughness factor is determined if the transmission main is monitored during calibration.

The length entered into the model is the length of pipe between node locations, including fittings. Typically, this length is obtained from the GIS piping layer but can also be obtained from design drawings, scaled maps, or as-builts. In areas with substantial changes in ground elevation, care should be taken to enter the true total length of pipe and not just the horizontal distance between nodes.

2.4.1.2.2. *Pipe Roughness Coefficients.* Pipe roughness can be estimated from diameter, material, and/or age (or pipe condition). A testing program is recommended to determine the appropriate correlation of these variables to roughness factors for local conditions (see chapter 3). Pipe roughness should also be conservatively adjusted to include minor losses, if minor losses are not explicitly accounted for at each pipe, as described in Section 2.4.1.2.3.

Over the years, new pipe materials have been introduced and manufacturing methods have been largely improved. As a result, many mature water systems are composed of a wide variety of pipe materials with varying hydraulic properties.

Pipe materials found in water distribution systems include:

- Unlined cast iron,
- Lined cast iron,
- Ductile iron (cement lined),
- Steel,
- Prestressed concrete cylinder pipe,
- Asbestos cement (sometimes referred to as *Transite*),
- Polyvinyl chloride, and
- Polyethylene or high-density polyethylene.

Pipe manufacturers typically provide roughness coefficient information for new, relatively clean pipes. In general, pipes are initially very smooth inside and perform hydraulically as described in the literature for a few years. Over time, however, transformations may occur within the pipes through a number of mechanisms. These mechanisms include corrosion, tuberculation, sedimentation, and biofilm development. Corrosion and encrustation rates depend on pipe material and water chemistry. These processes effectively reduce the pipe's interior diameter, which is subsequently reflected in the model by adjusting the roughness coefficient or modeled pipe diameter.

Empirical and quantitative formulas have been developed to characterize water flow through pipes. Several of the older formulas were based on tests of pipes manufactured in the early 1900s. These earlier pipes were generally manufactured using less stringent quality controls than are used today and were connected with many fittings that caused significant disturbance in the flow path. Two formulas, whose derivations were based on smooth pipe, are commonly used in water supply to calculate head loss or volumetric flow rate. They are the Hazen–Williams formula (empirical) and the Darcy–Weisbach formula (quantitative). Both formulas are available for use within modeling software packages; however, the Hazen–Williams formula is used predominantly for model calculations in the United States and will be assumed as referenced in discussions unless otherwise indicated.

Table 2-1 lists general roughness coefficients used in initial model development. The smoother the interior wall of the pipe, the higher the *C*-factor. This translates into a smaller head loss and subsequent higher pipe carrying capacity. The inverse is true as well. Though Table 2-1 provides a general guideline for *C*-factors used, the local roughness factor must be tested to determine the extent of the pipes' hydraulic capacity deterioration, as hydraulic capacity is influenced by pipe age, material, and local water quality conditions.

Table 2-2 lists general roughness coefficients (ε) for the Darcy–Weisbach equation. With this equation, unlike with the Hazen–Williams equation, the smoother the interior wall of the pipe, the lower the roughness coefficient, which translates into a higher pipe carrying capacity. The roughness coefficients listed in Tables 2-1 and 2-2 are provided as general guidelines and do not take into consideration very low values (high values for Darcy–Weisbach) that may be the result of local conditions. As previously indicated, local testing is recommended to develop a better estimate of roughness.

Table 2-1 Pipe roughness Hazen–Williams *C*-factors for discrete pipe diameters

Type of Pipe	1.0 in. (2.5 cm)	3.0 in. (7.6 cm)	6.0 in. (15.2 cm)	12 in. (30 cm)	24 in. (61 cm)	48 in. (122 cm)
Uncoated cast iron—smooth and new,		121	125	130	132	134
30 years old						
Trend 1—slight impact		100	106	112	117	120
Trend 2—moderate impact		83	90	97	102	107
Trend 3—appreciable impact		59	70	78	83	89
Trend 4—severe impact		41	50	58	66	73
60 years old						
Trend 1—slight impact		90	97	102	107	112
Trend 2—moderate impact		69	79	85	92	96
Trend 3—appreciable impact		49	58	66	72	78
Trend 4—severe impact		30	39	48	56	62
100 years old						
Trend 1—slight impact		81	89	95	100	104
Trend 2—moderate impact		61	70	78	83	89
Trend 3—appreciable impact		40	49	57	64	71
Trend 4—severe impact		21	30	39	46	54
Miscellaneous						
Newly scraped mains		109	116	121	125	127
Newly brushed mains		97	104	108	112	115
Coated spun iron—smooth and new		137	142	145	148	148
Old—take as coated cast iron of same age						
Galvanized iron—smooth and new	120	129	133			
Wrought iron—smooth and new	129	137	142			
Coated steel—smooth and new	129	137	142	145	148	148
Uncoated steel—smooth and new	134	142	145	147	150	150
Coated asbestos cement—clean		147	149	150	152	
Uncoated asbestos cement—clean		142	145	147	150	
Spun cement-lined and spun bitumen-lined—clean		147	149	150	152	153
Smooth pipe (including lead, brass, copper, polyethylene, and polyvinyl chloride) —clean	140	147	149	150	152	153
Polyvinyl chloride wavy—clean	134	142	145	147	150	150
Concrete—Scobey's coefficient (C_s)						
Class 1—C_s = 0.27; clean		69	79	84	90	95
Class 2—C_s = 0.31; clean		95	102	106	110	113
Class 3—C_s = 0.345; clean		109	116	121	125	127
Class 4—C_s = 0.37; clean		121	125	130	132	134
Best—C_s = 0.40; clean		129	133	138	140	141
Tate relined pipes—clean		109	116	121	125	127
Prestressed concrete pipes—clean				147	150	150

Source: Lamont (1981).

Table 2-2 Equivalent sand grain roughness for various pipe materials for use in the Darcy–Weisbach equation

	Equivalent Sand Roughness, ε	
Material	(ft)	(mm)
Copper, brass	1×10^{-4}–3×10^{-3}	3.05×10^{-2}–0.9
Wrought iron, steel	1.5×10^{-4}–8×10^{-3}	4.6×10^{-2}–2.4
Asphalted cast iron	4×10^{-4}–7×10^{-3}	0.1–2.1
Galvanized iron	3.3×10^{-4}–1.5×10^{-2}	0.102–4.6
Cast iron	8×10^{-4}–1.8×10^{-2}	0.2–5.5
Concrete	10^{-3}–10^{-2}	0.3–3.0
Uncoated cast iron	7.4×10^{-4}	0.226
Coated cast iron	3.3×10^{-4}	0.102
Coated spun iron	1.8×10^{-4}	5.6×10^{-2}
Cement	1.3×10^{-3}–4×10^{-3}	0.4–1.2
Wrought iron	1.7×10^{-4}	5×10^{-2}
Uncoated steel	9.2×10^{-5}	2.8×10^{-2}
Coated steel	1.8×10^{-4}	5.8×10^{-2}
Wood stave	6×10^{-4}–3×10^{-3}	0.2–0.9
Polyvinyl chloride	5×10^{-6}	1.5×10^{-3}

Source: Compiled from Lamont (1981), Moody (1944), and Mays (1999).

Hazen–Williams Formula

English units: $$V = 1.318Cr^{0.63}S^{0.54} \quad (2\text{-}1)$$

Equivalent metric equation:

$$V = 0.849Cr^{0.63}S^{0.54} \quad (2\text{-}1\text{M})$$

The head loss h_f may be calculated from:

$$h_f = \frac{4.72Q^{1.852}L}{C^{1.852}D^{4.87}} \quad (2\text{-}2)$$

$$h_f = \frac{10.62Q^{1.852}L}{C^{1.852}D^{4.87}} \quad (2\text{-}2\text{M})$$

Where:

V = mean velocity, ft/s (m/s)
C = Hazen–Williams pipe roughness coefficient
r = hydraulic radius of pipe, ft (m)
S = slope of energy line
h_f = head loss [ft (m)] in pipe length L, ft (m)
Q = discharge, ft^3/s (m^3/s)
L = length of pipe, ft (m)
D = nominal inside diameter of pipe, in. (mm)

From a theoretical standpoint, a pipe's C-factor should vary with flow velocity under turbulent conditions. However, the difference is usually within the error range of the roughness estimate, and therefore the C-factor is assumed to be constant regardless of flow. If C-factor testing is done at very high velocities (>10 ft/s [3 m/s]), a significant error can result when the resulting C-factors are used to predict head loss at low velocities (Walski et al. 2007). At very high Reynolds numbers, the C-factor can actually decrease with diameter for smooth pipe.

Darcy–Weisbach Formula

$$H_L = f\left(\frac{L}{D}\right)\left(\frac{V^2}{2g}\right) \tag{2-3}$$

Equivalent metric equation:

$$H_L = f\left(\frac{L}{D}\right)\left(\frac{V^2}{2g}\right) \tag{2-3M}$$

Where:

H_L = head loss [ft (m)] in pipe length L, ft (m)
f = Darcy–Weisbach friction factor
L = length of pipe, ft (m)
V = mean velocity, ft/s (m/s)
D = nominal diameter of pipe, in. (mm)
g = acceleration of gravity, 32.2 ft/s^2 (9.81 m/s^2)

The friction factor, *f*, is determined from the Colebrook–White equation, which is represented graphically as the Moody diagram (Figure 2-2), knowing the pipe relative roughness (ε/D) and the Reynolds number (R), as described in AWWA Manual M11 (2004). Most modeling software packages require that pipe roughness be entered into the model; *f* is then calculated automatically.

2.4.1.2.3. *Minor Losses.* Minor losses occur at valves, tees, bends, and other appurtenances within the distribution system due to turbulence within the flow of the water. Minor losses are generally computed by multiplying a minor loss coefficient by the velocity head. At the low velocities observed during normal conditions – in the 1 to 2 ft/s (0.3 to 0.6 m/s) range – the velocity head and resulting minor loss are very low. Minor losses are not typically applied to all fittings in a distribution system model.

If a model is calibrated without explicitly accounting for minor losses in the distribution system, the roughness factors and pump curves that result from the calibration effectively incorporate the impact of minor losses. However, minor losses are often input at facility locations such as pump stations and control valves to improve accuracies and account for losses that result from the higher velocities at those locations and at hydrant service laterals (to represent hydrant bends and fittings).

A model that includes minor losses produces a more precise representation of the physical facilities within the model; however, field testing is recommended to confirm whether it is more accurate. Minor loss *K* coefficients are typically determined experimentally and are found in tables from various manufacturers and in textbooks. Often they are expressed in terms of pipe equivalence. Data are available for many different types of fittings and appurtenances. Examples of typical minor loss coefficients are shown in Table 2-3.

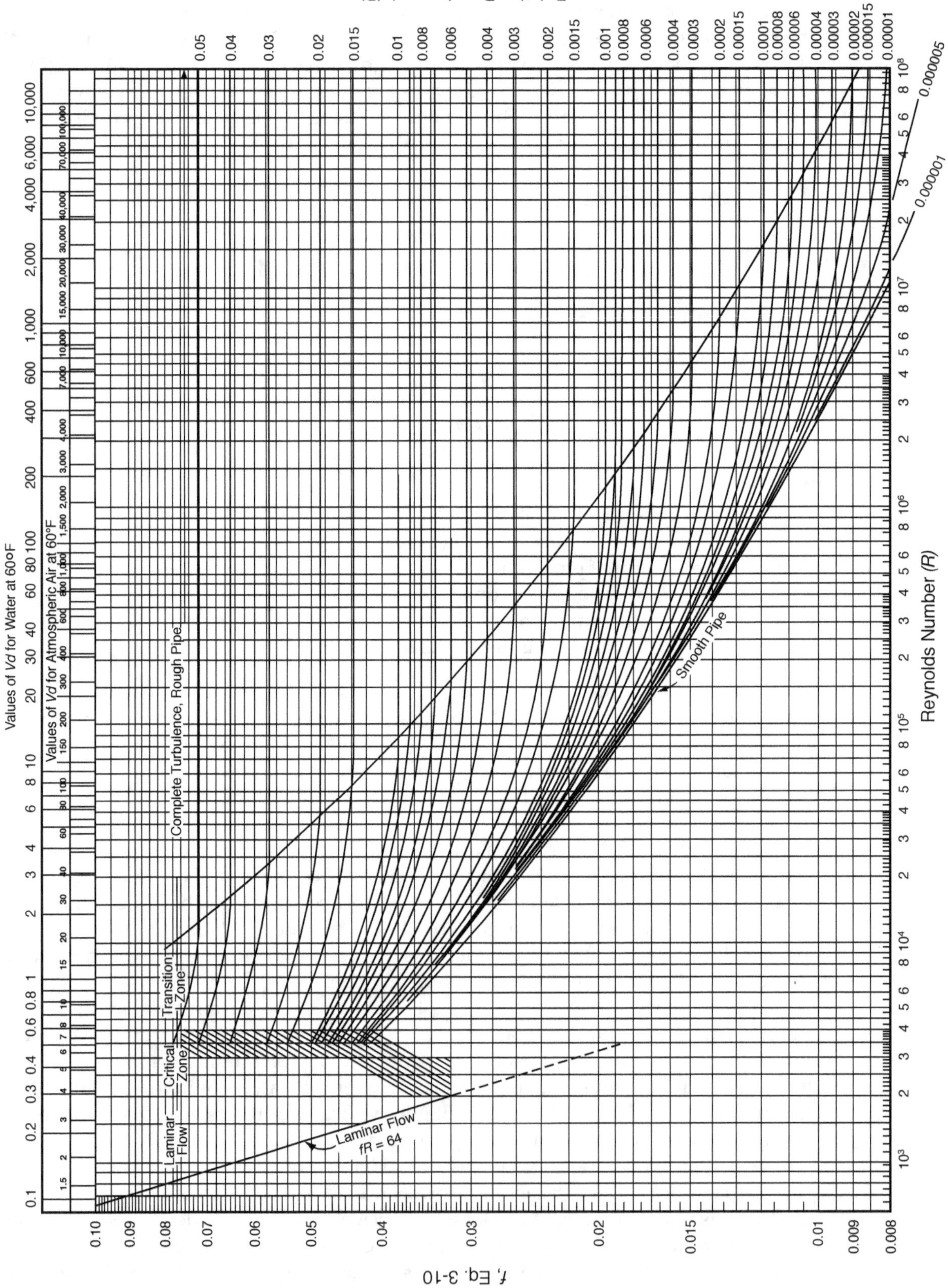

Figure 2-2 Moody diagram

Source: AWWA 2004

Table 2-3 Typical minor loss coefficients*

Type of Component or Fitting	Minor Loss Coefficient
Tee, line flow	0.2–0.4
Tee, branch flow	0.8–2.0
Elbow, regular 90°	0.3
Elbow, long radius 90°	0.7
Elbow, regular 45°	0.1
Elbow, long radius 45°	0.2
Globe valve, fully open	7–10
Gate valve, fully open	0.15–0.4
Gate valve, ¼ closed	0.3–1.5
Gate valve, ½ closed	2–4
Gate valve, ¾ closed	20–30
Swing check valve, forward flow	2–4
Ball valve, fully open	0.05
Butterfly valve, fully open	1.0–1.5

* Some software packages have an extensive library of minor loss coefficients.

It is particularly important to include minor losses for PRVs and other control valves when the model is used to simulate extreme situations, such as fire flow or main breaks, so as not to overestimate the capacity of these facilities. In high-flow events, if the control valve is fully open, the set point is no longer relevant, and flow is controlled solely by the minor loss of the valve. It is recommended that modelers obtain the actual loss curve from the control valve manufacturer. PRVs are typically globe valves, which have substantial minor losses, as shown in Table 2-3.

2.4.1.2.4. *Pipe Direction and Check Valves.* In general the pipe direction as defined in the model by the *from node* and the *to node* is not important. However, for pipes that contain check valves, pumps, and control valves, the direction must be correct and consistent with the direction of flow. Pipes that contain these elements must provide hydraulic representation that reflects the actual direction of the device. Pump and control valve elements are discussed in more detail later in this chapter.

Check valves permit flow through a pipe in one direction only. They are used occasionally in distribution systems to prevent flow from moving across a zone or other type of boundary. Check valves are located on the discharge side of many pumps, but these valves do not necessarily need to be included in the model because most modeling software does not allow reverse flow through pumps with a defined pump curve. However, check valves within the distribution system should be modeled.

2.4.2. Modeling of Non-Pipe Elements

Model facilities include reservoirs (water sources), tanks (water storage), pump stations, wells, connections to other systems, and control valves such as PRVs, PSVs, and similar elements. These elements may or may not exist within a GIS and may require modification or additional detail in the model. GIS facility elements are routinely oversimplified for mapping purposes and do not require model detail. For example, the GIS may represent an entire pump station with multiple pumps by a single point feature. This same facility requires significantly more detail within the model. Figure 2-3 illustrates a pump station represented in GIS that includes one point (PS) compared to the multiple elements used for the same facility in the model (P representing each pump within the pump station, including upstream and downstream piping).

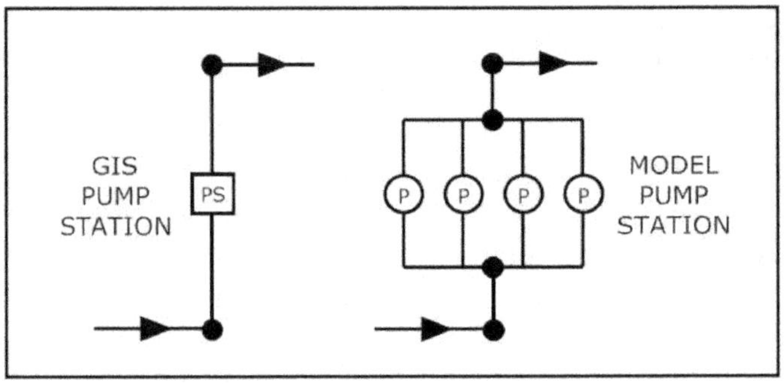

Figure 2-3 Geographic information system detail versus model detail

Abbreviation: GIS, geographic information system.

2.4.2.1. Water Storage Facilities. A water storage facility allows water to be stored from, or released to, the water system. Model elements that represent water storage facilities are either tanks or reservoirs. These elements are used to store and release water as demands vary throughout the day within the water system network.

Tank elements require geometric input including diameter, water levels, volume curves, and base elevation. The base elevation plus water levels serve to define the HGL, which may vary based on system demands. The water level that indicates an empty tank or the minimum level limit for available storage should also be determined. The tank would be modeled as empty when drained to this minimum level. The behavior of empty and full tanks can vary between different modeling software. In a steady-state simulation, initial tank levels are static and are used to define the boundary HGL and total tank volume. In EPSs, tank levels fluctuate in response to system demand; use of an accurate tank diameter or tank volume curve is a critical part of model validation. A tank can have more than one pipe connected to it, which could represent separate drain and fill lines. In a model, tank elements represent ground level and underground storage facilities, elevated water storage tanks, standpipes, or hydropneumatic tanks.

A reservoir represents a fixed hydraulic grade supply source, an unlimited supply source such as a large lake, a surface water treatment plant clearwell, or a groundwater basin. Reservoirs are typically modeled with only one pipe connecting them to the system. These characteristics allow groundwater supplies (such as aquifers) to be modeled as the source of supply for groundwater wells. When modeling a well using a reservoir, the reservoir level should be set to the pumping water level of the aquifer.

2.4.2.2. Pumping Stations. Pumps lift water to higher elevations and add energy into the water distribution system. Pumping requirements depend on system demands, network layout, ground topography, and system losses. Pumps increase head at a flow rate determined by downstream demands and storage hydraulics.

A pump station is typically represented as a single point in GIS for mapping purposes, regardless of how many pumps are located at the station. The use of a single pump element to model an entire pump station should be avoided whenever possible. It is preferred that each pump be modeled as its own pump element, with its own pump curve and controls. There is a continuing trend to accurately represent the true piping configuration at pump stations in the GIS.

To simplify identification of total flow through the station, modeled pumping stations may include one dedicated suction pipe into the pump station for each pressure zone that feeds the pump station and one dedicated discharge pipe leaving the pump station for each pressure zone that pump station serves. The dedicated suction and discharge pipes allow the modeler to show the entire flow into or out of the station from and to each pressure zone. Where the model is attempting to reflect the actual station geometry as accurately as possible, total station flow may be available only by adding the flow from two or more pipe elements.

2.4.2.2.1. *Pump Curves.* Most pumps used in distribution systems are centrifugal or turbine pumps with either a constant- or variable-speed drive.

A pump head characteristic curve is the unique representation of total dynamic head delivered (y-axis) versus flow rate (x-axis) for a particular pump, as illustrated in Figure 2-4. The size of the pump and the shape and size of the impeller determine the characteristic curve of the pump. Variable-speed pumps have a family of curves, where each curve characterizes the pump at different operating speeds. In most cases, pump characteristic curves are available from the pump manufacturer or supplier. Existing pumps should be tested periodically to determine their characteristic curves, as impeller wear over time affects pump performance.

Most modeling software allows pump performance information to be entered in multiple ways. Some allow the input of a single design point, head versus flow, and then the software assumes a curve fit intercepting the shutoff head, the design point, and some lower point representing maximum flow at zero head. The curve assumed by the model may not reflect the actual pump performance characteristics. Depending on the type of analysis, it may be best to input actual pump curve information using multiple points of head and flow versus a single point that represents an assumed pump characteristic curve.

When modeling a water distribution system and selecting pumps, the range of system operating conditions should be considered. A very flat pump curve results in larger changes in flow rate for smaller changes in pressure. In contrast, a very steep curve produces relatively small changes in flow rates over a relatively wide range of pressures.

It is recommended that the most accurate and detailed pump curve data be used in the model. Compiling the most comprehensive data available to create detailed pump curves is the preferred approach and often starts by locating the manufacturers' pump curves for all pumps and validating these curves through field pump testing. In order of preference and accuracy, the best information for developing pump curves is (1) recent field tests of head versus capacity, (2) performance tests conducted during pump manufacture for initial approval, (3) manufacturer's standard pump curve for that particular unit, and (4) reported pump capacity and head (design point).

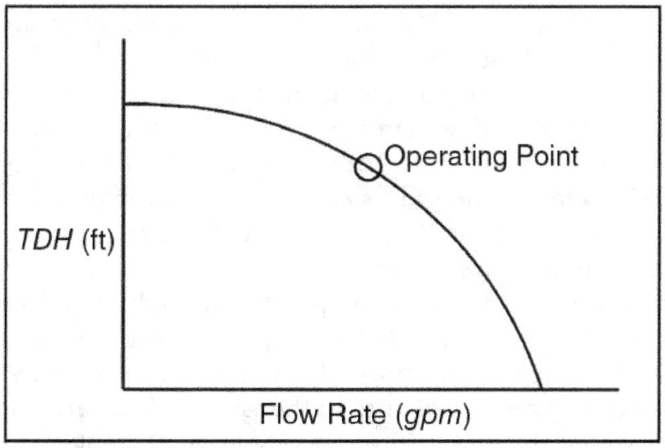

Figure 2-4 Pump head characteristic curve

Abbreviation: TDH, total dynamic head.

For pumps with variable-speed drives, most software requires input of the pump performance at a given speed (usually maximum speed). The software then calculates pump performance at reduced or increased speeds based on distribution system hydraulic characteristics and pump affinity laws. For systems with variable-speed drives, it is important to know the pump speed during the calibration period to achieve acceptable results. Pump speed is often not recorded in historical SCADA data and may need to be estimated based on recorded flow rates or discharge pressures.

2.4.2.3. Valves. The valves used in a water distribution system are broadly classified into three main categories: isolation valves, control valves and check valves, and pressure relief and vacuum valves. Isolation valves enable a utility to isolate portions of the system for maintenance or emergency response purposes. Isolation valves are also used in the manual closed position as zone valves to create pressure zone breaks to manage pressures within the system. Isolation valves are discussed further in section 2.4.2.3.1. Check valves allow flow in only one direction and are discussed in section 2.4.1.2.4. Control valves apply a hydraulic restriction or setting to the water system, impacting either hydraulic gradient or flow rate. Control valves are discussed further in section 2.4.2.3.2. Pressure relief and vacuum valves are not typically modeled. However, for transient surge model applications, the inclusion of these valves is desired.

2.4.2.3.1. *Isolation or Zone Valves.* Manual isolation valves are typically either gate valves or butterfly valves. They are located throughout the distribution system and within pumping facilities. In most cases, valves are left either in the fully open or fully closed position and only operated during maintenance or emergency response events. Most isolation valves in the distribution system are normally open, although some valves are fully closed to separate pressure zones. These valves are commonly referred to as *zone valves.*

2.4.2.3.2. *Control Valves.* There are six types of control valves typically included in models: pressure-reducing valves (PRVs), pressure-sustaining valves (PSVs), altitude valves, flow control valves (FCVs), throttle control valves (TCVs), and general-purpose valves (GPVs). Some proprietary software includes additional types of valves such as pressure breaker valves or vacuum breaker valves. In the field, these valves have a sensing component, usually a solenoid valve, that opens and closes the main valve based on pressures, flows, or water levels in the distribution system.

Check valves could be included as another type of control valve. However, they differ in function substantially from the other control valves identified in this chapter. Check valves permit flow through a pipe in one direction only, are not adjustable, and do not require a model setting – only a flow direction.

PRVs detect pressure on the downstream end of the valve and will throttle if necessary to ensure that the downstream pressure does not exceed a set value. They are used in the distribution system where there is a substantial drop in topographical elevation. Without the PRV, excessive pressures can develop in the distribution system. When multiple PRVs service a single area, the potential exists for the valves to work against each other as pressures oscillate between the valves within the zone to reach equilibrium between the valve settings. Setting the pressure points to a slightly different grade line, making one of the valves predominant, sometimes alleviates this problem.

PSVs measure the pressure on the upstream end of the valve and throttle to maintain a setback (i.e., upstream) pressure in the system. They are often used as control valves for reservoirs where flow to refill the reservoir, after depletion from pumping, is throttled. The reservoir feed is regulated to ensure that customers serviced along the main have adequate supply pressure.

Altitude valves are used in conjunction with storage facilities to automate the filling cycle of a reservoir while preventing tank overflow. A level sensor opens and closes the altitude valves based on two (high and low) level set points. Altitude valves are either fully open or fully closed, so they can impose a large demand on a water distribution network when they open. For this reason, they are often used in conjunction with PSVs that throttle flow. Some modeling software simulates altitude valves at each tank by not allowing any additional flow into a tank when the simulated water level is at the modeled maximum tank elevation.

FCVs are used to regulate the flow rate of water. Valve position is adjusted either automatically or by operator control through the SCADA system; this is based on imparting enough head loss to maintain the desired flow rate through the valve. Typical applications of FCVs in a water system include limiting the supply to a customer at a predefined or not-to-exceed flow rate and the delivery of water from one pressure zone to another at a rate that allows for adequate fluctuation of water level in storage facilities.

TCVs are used to induce head losses in the distribution system. In a strict sense, FCVs are TCVs because a valve is throttled to deliver the desired flow. However, for this discussion and for modeling purposes, TCVs are different from FCVs. TCVs may be used in the model to simulate a partially closed valve in the distribution system. A TCV may be used to restrict flow in one of a set of transmission mains, thereby forcing flow into one of the pipes and creating better hydraulic conditions within a pressure zone. Also, a TCV can be used to simulate a partially open valve between two areas in the distribution system where such a situation exists.

2.4.2.3.3. *Control Valve Modeling Tips.* In certain cases, the use of a single control valve element to model multiple valves provides acceptable hydraulic results while also increasing the model's stability. Two or more valves should not be directly connected in series. Instead, a pipe should be used to separate the two valves. It should be noted that this may cause the model to become unstable. A single valve should never have more than two pipes connected to it (one upstream pipe and one downstream pipe). Valve stations should be modeled with one suction pipe and one discharge pipe to make it easier to record the entire flow into or out of the station.

Many utilities keep a record of the PRV settings in their systems. During model construction, these may be taken at face value and entered directly into the model without the realization that the settings may be estimates, that the actual pressure setting may be significantly higher or lower than the originally documented number, or that some valves have been adjusted since the last time the records were updated.

The modeler should also keep in mind that minor losses are often input for throttle-control valves. Globe valves can have high head losses that result from the valve's geometry. Valves that are modeled as TCVs can have a variable position with a curve input to the model that defines the relationship between the minor loss coefficient and the valve position.

2.4.2.4. Hydrants. When planning a model, the modeler should establish how fire hydrants will be simulated in the model. There are several options available:

- Adding a representative junction along the water main at the hydrant lateral to represent the hydrant location;
- Including the detail of the hydrant lateral and the hydrant in the model; or
- Not breaking the water main and relying on the nearest junction to represent the hydrant.

The short- and long-term needs of the model should be considered when choosing among these options. For instance, if the model is developed primarily for the analysis of fire flows, the nearest-junction approach may not be suitable. In this case, if the water main is not split in the GIS at the hydrant lateral, it may be valuable to request that the GIS group add a fitting in the GIS to represent the hydrant lateral connection point and splitting the main at this point. This approach will ease the process of long-term model maintenance.

Hydrants are typically represented by junctions in the model, although some model software provides the ability to model a hydrant as its own element type. In the field, hydrants are located at the end points of hydrant service laterals. If a model includes these service laterals, the number of pipes in the model is much greater. As previously described in this chapter, it is recommended that these service laterals not be included in the model.

2.4.2.5. Turnouts. Turnouts, or interconnections, are connections to an adjacent system where water can be received or delivered. Most modeling software packages do not have a single model element to represent a turnout. Therefore, turnouts must be modeled using other elements that are available. Modeling a turnout connection that supplies a system can be defined using two elements: a reservoir and a flow-control valve. The reservoir is used to simulate the water source and the pressure by which the water is served. The flow-control valve regulates the amount of water that is provided to the system from the source. Used together, these elements can accurately represent a turnout as a supply delivery point into a system. For turnouts that represent flow out of a system, no special modeling techniques are required. It is likely sufficient to model this situation as a typical node with an assigned demand. Modeling bidirectional flow through turnouts is more complicated but can be accomplished using a combination of reservoirs, demands, and flow-control valves in conjunction with operational controls.

2.4.2.6. Wells. Most modeling software packages do not have a single element for a groundwater well. Therefore, wells must be represented using other elements that are available. Most typical approaches to model a well include the following:

- A reservoir and a pump best represent the actual operation of a well. The reservoir is used to simulate the groundwater basin from which the well is pumping. The pump is used to simulate the well pump and motor. Used together, these elements can accurately represent a well.
- A negative demand can be assigned to a junction that acts to input water into the system at the well location. However, this is not recommended as it provides a less accurate representation of the facility and can be confusing when reviewing demand summaries within the model.

When modeling the groundwater basin as a reservoir, it is important to set the HGL of the reservoir to the pumping (not static) water elevation for the well. Usually reported as pumping water level (this is the depth from the wellhead to the water surface while the pump is running), the HGL would be calculated as the wellhead elevation minus the pumping water level. Alternatively, the static water elevation minus drawdown would also provide the pumping water elevation. In some cases, the pumping water elevation may be negative, indicating that the water level is below sea level. Care should be taken to define the gradient of the reservoir that represents the groundwater pumping level. These water levels can change dramatically depending on the time of year and the groundwater influence of pumping from other tributary wells.

For some wells, the pumping water level may vary during the year or for different pumping rates. In this case, the modeler may wish to select the lower level, the typical level during maximum day demand, or an annual average. Selecting levels that are higher than the actual levels for a specific demand period can result in overestimating a well's production capacity, causing it to operate at a different point on its pump curve and affect discharge pressures, model calibration, and distribution system recommendations. When modeling a water treatment plant that delivers flow to the distribution system without pumping, a FCV is required in the model downstream of the reservoir to simulate variable delivery rates.

2.4.3. Quality Assurance Steps

2.4.3.1. Topology and Connectivity Review. Topology and connectivity review of GIS data imported into a model is necessary after an import from the GIS. The following issues are specific to nodes and pipes but can also be applied to tanks, pumps, valves, and reservoirs.

Many software packages have built-in tools to identify and resolve the topology issues discussed in this section. It is important to note and communicate the model building impacts of these topological issues to the supporting groups that provide system data layers. In the following connectivity scenarios, flow may or may not be delivered across the intended pipe segments. If this condition exists throughout the model, flow patterns, system losses, and resulting pressures would differ significantly in the model from actual field conditions. Certain types of issues may prevent a model from running until they are resolved. Field flow tests are an excellent way to validate network topology as well as locate closed valves that should be opened.

Nodes in Close Proximity: Nodes in close proximity designate nodes that are within a very close tolerance of each other but are not connected by a pipe. In most cases, they may appear to overlap, as shown in Figure 2-5. For a model to run properly and before demands are allocated, it is imperative to identify and remove unnecessary nodes and connect appropriate pipe segments.

Pipe-Split Candidates: Pipe-split candidates represent separate pipe segments that should be connected by a common node, as depicted in Figure 2-6. It is important to identify where these occur and to make sure that the pipes are split where necessary.

Intersecting Pipes: Crossing pipes are pipes that cross but do not intersect at a common node, as shown in Figure 2-7. These pipes should be identified and examined carefully to determine whether or not they should be split at intersections. It may be helpful to consult an atlas map of the system to assist in determining if pipes should be intersected.

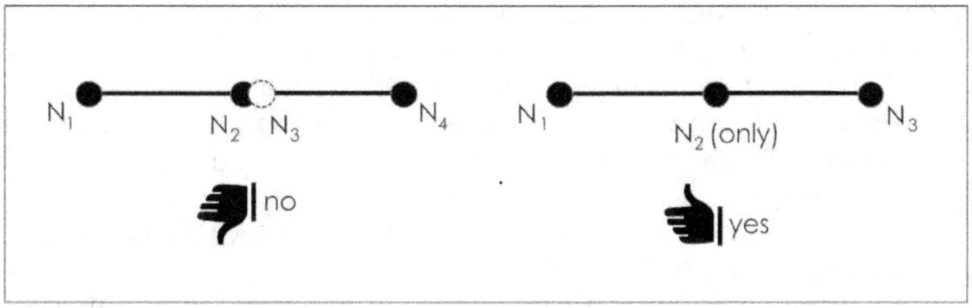

Figure 2-5 Nodes in close proximity

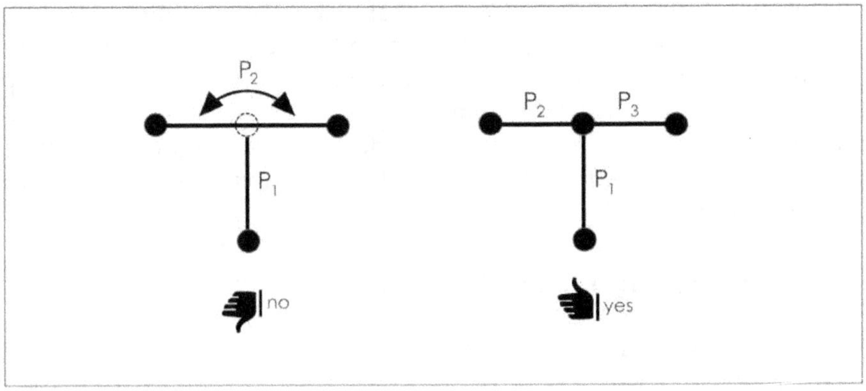

Figure 2-6 Pipe-split candidates

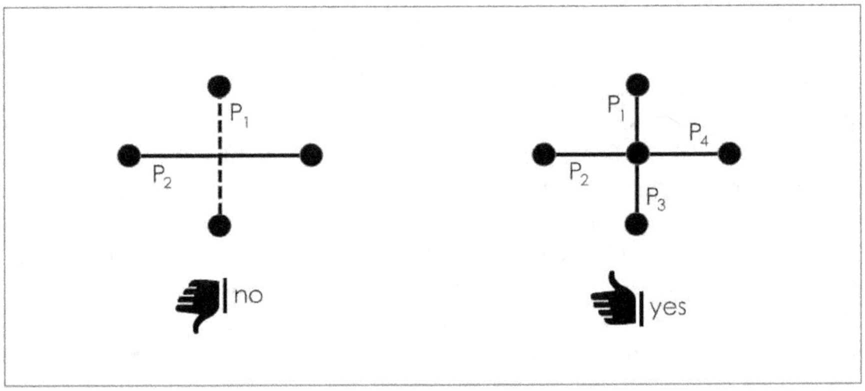

Figure 2-7 Intersecting pipes (Note: Confirm pipes are connected)

Disconnected Nodes: Disconnected, stray, or orphan nodes (i.e., nodes that are not connected to any pipe) are separated from the piping network, as shown in Figure 2-8. For a model to run properly, all disconnected nodes must be resolved by connecting to a pipe, deleting the node (if appropriate), making the node inactive (depending on software), or another edit. Sources of disconnected nodes include missing pipes, inaccurate node attribution, duplication of nodes, and hydrants without their leads. Disconnected nodes should be resolved prior to demand allocation.

Diameter Discrepancies: Diameter discrepancies are typically identified when pipes with significantly different diameters are connected in series. For example, a pipe segment identified as a 36-in. (91.4-cm), connected to a 3-in. (7.6-cm), connected to another 36-in. (91.4-cm) pipe is most likely an error. These diameter discrepancies should be identified, and an atlas map or record drawings should be reviewed to determine if their diameters should be modified.

Duplicate Pipes (Parallel Pipes): Duplicate pipes are pipes that share both the same *to node* and *from node*. These can be superimposed pipes that share the exact same shape or pipes that have the same connections but are not duplicates. When duplicate pipes have been identified, the user can determine if a parallel pipe actually exists and, if so, reroute (redraw) the pipe if needed, or the user can remove or delete it from the model if such a duplicate pipe does not actually exist, as shown in Figure 2-9.

Disconnected Pipes: Disconnected, or orphan, pipes are pipe segments that have no end (terminal) nodes and are therefore not connected to the distribution system. Once disconnected pipes have been identified, junctions should be added to their end points or the pipes should be deleted as appropriate. Many software packages allow the user to automatically add junction nodes at their end points, as shown in Figure 2-10.

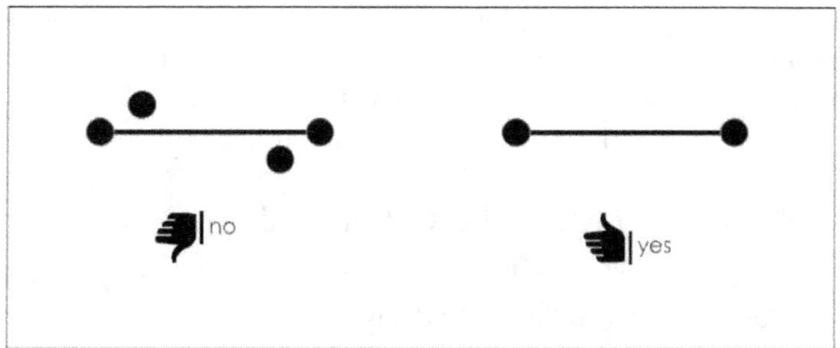

Figure 2-8 Disconnected or orphan nodes

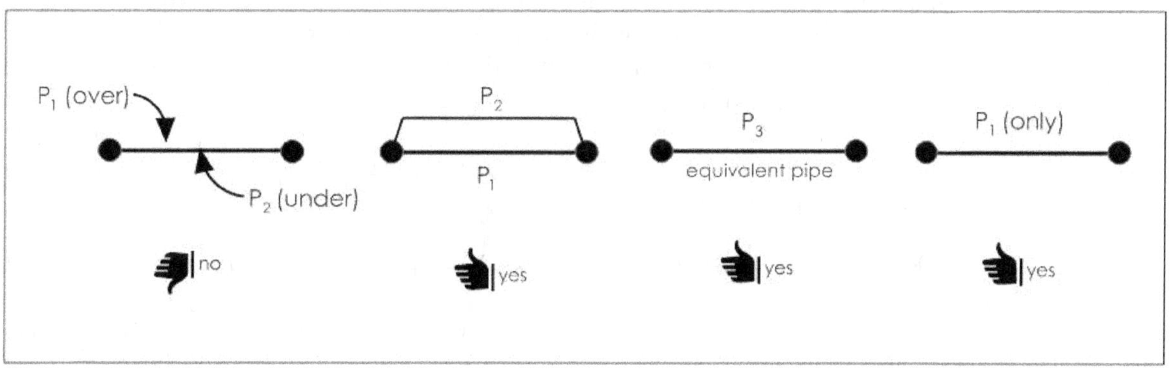

Figure 2-9 Parallel pipes

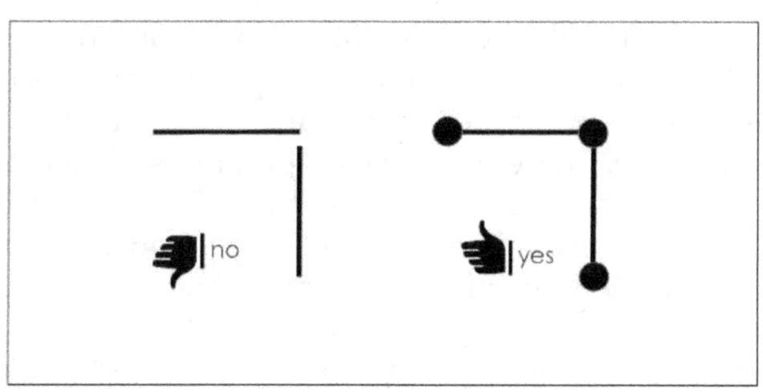

Figure 2-10 Disconnected pipes

2.5. DEMAND DEVELOPMENT

Demand data are added to the model after facility information has been attributed and distribution system topology has been established. There are four steps typically considered when allocating demands to the model: (1) determine the types of water demand, (2) determine the sources of water demand, (3) allocate base demands to the model, and (4) adjust base demands for planning, operational, or design scenarios. There are multiple methods for obtaining, reconciling, and determining demands and then loading these demands into the model. Several methods are presented in this section. The modeler needs to evaluate the modeling objectives, budget, and available data to select the most appropriate demand calculation and allocation method.

2.5.1. Determine Types of Water Demand

The first step in determining the types of demand within a given water system is to review unique characteristics and land use designations of that system. The information collected should include a listing of major customer types in the community. Irrigation practices and special industries should also be considered, as they may have unique usage rates and/or patterns that require special consideration in the model.

The five customer usage types or sources of water demand that are most often seen in hydraulic analysis include residential, commercial, industrial, and wholesale demands, and water loss. In some communities, additional customer classes such as multifamily, government, and institutional may be required. Large facilities such as military bases, prisons, universities, and hospitals may be kept separate from the other customer classes because they have unique consumption patterns. The modeler needs to select the customer classifications most appropriate for the water system being modeled and the data available to develop the demands.

2.5.1.1. Residential Demands. Residential demands consist of domestic and irrigation consumption.

Domestic consumption consists of indoor water used for flushing toilets, washing, cooking, and drinking. Conservation efforts as well as changing building codes have resulted in reduced average day consumption.

If using general per capita, parcel size, or land use methods to calculate consumption, the age of a neighborhood and the renovation rates in a community may be important factors to consider when developing demands.

Irrigation consumption consists of outdoor water used on lawns and gardens. This component of consumption is community specific and dependent on yard sizes and climate. A number of communities have water conservation programs that specifically target outside water use. These conservation programs often have a significant effect on the magnitude of irrigation demand. Again, this may be a consideration if using per capita, parcel size, or land use methods to develop demands. It is less of a consideration if the demands are developed directly from customer service information.

2.5.1.2. Commercial Demands. Commercial customer usage types typically consist of demands by stores, restaurants, gas stations, offices, and similar businesses. Commercial establishments may have fairly consistent daily usage rates and predictable fluctuations during any given day. The hours of operation and the type of building influence the demand pattern. Depending on the proportion of commercial developments in the community, it may be necessary to separate the commercial category into two groups, depending on general hours of operation. For example, delineate offices and shopping centers that may have different hours of operation from laundromats and car wash businesses.

Historical commercial flow data should be used with caution in forecasting future demands. A number of changes in the design of air conditioners, toilets, and other water fixtures have resulted in significant water cost savings for businesses. Again, the modeler should be aware of renovation changes in the commercial sector and conservation measures initiated by the local utility.

2.5.1.3. Industrial Demands. The industrial customer category consists of all larger water customers and is typically comprised of manufacturing plants. Because water service to these types of sites may be metered at a location far away from the actual site, the careful allocation of demands to the proper location on the distribution system is important. Each industrial site is likely to have a unique flow pattern based on hours of production and water use characteristics. Field flow measurements may be helpful when determining diurnal patterns of large customers. Seasonal variation is usually not an issue, but variability based on the day of the week may be significant.

A survey of the large customers in a community is generally recommended to determine their water use patterns and demand characteristics. Other departments within the city or utility may have reports or permits from these large industrial users that would be helpful in determining their average demand and diurnal patterns.

2.5.1.4. Wholesale Demands. Wholesale demands consist of water sold to another water utility. These demands may have unique diurnal patterns for the delivering system. In many cases, the wholesale customer system has its own storage facilities, resulting in a more constant demand on the delivering system over the course of a day. Wholesale customers may be contractually limited to the rate of water delivered on a daily or hourly basis, and delivery may be physically restricted through a pumping station or control valve. Field flow measurements may be necessary to determine the appropriate diurnal pattern to apply to the model.

2.5.1.5. Water Loss. System water loss is the difference between the total water supply to the system and the authorized billed and unbilled water consumption. Water loss in the distribution system must be added to demands in the model so that the total amount of water supplied equals the total water demand. A good reference for determining and managing water losses is AWWA Manual M36 (2016).

The amount of water loss in a distribution system may vary widely from system to system. Values ranging from 4 to 70 percent of the total water produced have been reported. Also, the amount can vary from year to year and from pressure zone to pressure zone within a system. The higher values are generally associated with communities that are not fully metered, have soil conditions that prevent surfacing of water from main breaks, or do not maintain an active meter testing/replacement program.

2.5.1.6. Sources of Water Loss. Water loss is always present in a water system. Water loss consists of two distinct components: real loss and apparent loss. Real loss is the physical loss of water from the distribution system, including leakage and tank overflows. Apparent losses include meter inaccuracy, billing error, and unauthorized use.

Water loss results from several factors. These factors include unidentified leaks in pipes, main breaks, fire hydrant flushing, reservoir drainage for maintenance, unauthorized use, unmetered services, inaccurate and nonfunctioning meters, and on-site water plant usage.

Water plant site usage consists of water used for filter backwashing, chemical mixing, rinsing tanks, and sanitation purposes. In certain cases, this water is not metered and can represent as much as 5 percent of the total water production.

2.5.1.7. Allocation of Water Loss. Depending on the significance and location of the water loss, there may be different model demand allocation approaches to consider. These approaches include, but are not limited to, (1) an even distribution across the entire network model to consider the water loss demand system-wide, or (2) zone-by-zone specific allocation, if the water loss can be calculated by pressure zone.

Water loss is usually divided equally among all nodes in the network because specific or isolated causes are difficult to determine, unless district zone metering has been created throughout the distribution system. System-wide district zone metering allows for a more accurate allocation of water loss. To increase allocation accuracy, water utilities use leakage tests in subareas of the distribution system and prorate the water loss to other areas that have similar characteristics such as pipe material, soil type, and age of water mains. Special considerations should be made for areas that lack sufficient demand representation (e.g., unavailable data), as these areas may warrant different allocation of water loss.

It is important to note that most steady-state analyses are based on the extreme conditions of peak hour or maximum day demand plus fire flow. Under these conditions, the influence of water loss allocation may be negligible when compared to total system demand conditions.

2.5.2. Determine the Sources of Water Demand

Water demand can be determined using multiple sources, each of which offers varying levels of quality, completeness, and accuracy. These sources should be evaluated carefully before being allocated to the model.

CISs provide information on water demand at the customer meter (commonly referred to as *water consumption data*). Water consumption data may benefit the model by providing both the amount and location of usage, which may improve the accuracy of subsequent water system analyses. Additional sources for estimating demand within the water system include water production by pressure zone, land use layers, and population data. Each data source may reasonably represent water system usage and be acceptable for the model while providing varying levels of accuracy. Each data source can be used independently or as a support to validate the primary data source used.

There are two general approaches to demand loading: a top-down approach where demands are known by system/source production or land use and spread out to model nodes; and a bottom-up approach where loading is based on individual customer billing data, which can be assigned to nodes and then corrected by an allowance for non-metered water.

2.5.2.1. Customer Information System. CISs store records of water sales at the customer meter. A CIS is a database of customer water consumption that provides an excellent source of data when determining and allocating water demands to the model. The CIS provides two unique advantages for customer water usage compared to other data sources. First, it represents actual usage over a given time period and, second, it provides the spatial location of where water consumption occurred. These are strengths of a CIS that make it a primary and preferred source for water demand information for the model. If metering data does not have an *x*-*y* coordinate in the same coordinate system as the model, it needs to be georeferenced to determine the coordinates. Georeferencing is not always straightforward. An AMI can provide useful customer usage data, often at a higher data measurement frequency than the utility's billing cycle.

CIS typically tracks the following information for each customer:

- Customer name,
- Account number,

- Address (service location and billing location),
- Meter-read water usage (actual water consumption data),
- Customer type (customer/demand classes),
- Meter number (meter number may also be available in GIS, providing an easy way to develop the demand distribution),
- Date period for each meter reading,
- Average water usage (average water consumption), and
- Meter information (e.g., make, size).

Most billing systems contain monthly, bimonthly, or quarterly water consumption information, water use categories, meter route numbers, and customer addresses. Certain billing systems contain additional information such as number of residences for a multi-family account or business type for a commercial account.

Water consumption data from a billing system may be collected as 12 consecutive months of data for the purpose of generating an average day demand. Consumption data may also be collected for a particular season or period of interest for the purpose of calculating a higher or lower demand and peaking factor relative to the average day demand. Some meters are read on a monthly basis, while others are read bimonthly or quarterly. This meter reading frequency should be confirmed and accounted for prior to calculating the average demand for each customer.

2.5.2.2. Other Sources of Water Demands. If metered consumption data are not available from the CIS, other techniques can be used to estimate and assign demands to the model. The three most common demand calculation techniques are based on production data, land use classifications, and population counts.

Often, all three techniques are used in conjunction to develop estimated demand allocation. These techniques are also used to model future growth scenarios.

2.5.2.3. Water Production Data by Pressure Zone. Water utilities typically maintain records of total daily production supplied to the distribution system. Depending on the availability and accuracy of production records for each facility, water production may be allocated across the model or by pressure zone. If allocated across the model system wide, it is useful to reference population and land use data sources for spatial allocation.

To calculate demands by pressure zone, a record of the flows that enter and leave the pressure zone is required. Flowmeters and changes in reservoir levels are used to estimate the total volumes consumed in each pressure zone. This is done by performing a mass balance calculation for each hour of the day within the pressure zone. Care must be taken to identify any unmetered automatic control valves or check valves that could move water across an adjacent pressure zone boundary. Additional information such as population counts and land use characteristics can further validate and refine the calculated water demand by pressure zone.

2.5.2.4. Land Use Classifications. Land use classifications can be used to estimate future consumption for a variety of land use types or to estimate current consumption if actual CIS or billing data are not available. The most common land uses are residential, commercial, and industrial. Land use classifications are usually represented on a map as polygons. Land uses typically follow the local zoning regulations; in most cases, there are more land use classifications than required for water use classification. Consideration should be given to the utility's community plan for future land use (zoning) when planning for growth and considering water demand projections.

Development of water consumption using land use typically involves a water-duty factor that is associated to a land use classification. The water-duty factor is a ratio of water usage per unit area per land use category. It uses dimensions of gallons/day/acre (cubic meters/day/hectare) per land use type. These water-duty factors can be calculated using population data or customer water consumption within a specific land use area. It is common to create preexisting water-duty factors for a given service area or for a neighboring service area, with similar characteristics for use, reference, or comparison.

Some communities have developed a GIS to track land use and zoning. GIS adds value in its ability to efficiently and accurately calculate the area of any land use (or zoning) polygon designation (in acres), apply the water-duty factor to the land use classification area (as gallon/day/acre or cubic meters/day/hectare), and generate a total estimated water demand (as gallons/day or cubic meters/day) for land use types. The total demand generated for a land use classification may then be assigned to appropriate model junctions contained within the land use polygon.

Water-duty factors per land use category are also used to estimate future water demands for a given service area in order to evaluate master planning scenarios where historical data may not be available to perform demand calculations. Water-duty factors may be used to support water demand projections and to validate future water demand. There is also an emerging trend of using multivariate regression demand models to predict future water demands because of the lack of historical information regarding new conservation measures or the effects of climate change.

2.5.2.5. Population and Employment Estimates. Population and employment estimates can be used to estimate future consumption in residential, commercial, and mixed-use areas, and can be used to estimate current consumption if actual CIS or billing data are unavailable. Generally, population and employment information is obtained from government census data. In some communities, the local transportation department maintains traffic count data by neighborhood for use in designing collector and arterial road systems.

The modeler assigns a population and/or employment value to each node and uses separate per capita multipliers to convert the population and employment numbers to consumption values. Typically, modeling software allows for multiple types of demands at each node. Per capita water consumption can vary from community to community based on lawn irrigation needs and the area's age. Historical water consumption should be reviewed to provide a basis for estimating per capita consumption. Future water demand is also evaluated by multiplying changes in population densities or new service area populations by per capita consumption rate.

2.5.3. Demand Allocation Process

Demand allocation is the process of assigning water demands to the appropriate junctions in the model. The base demand may be considered as the water demand allocated to the model that serves as the basis for adjustment for alternate demand conditions. The model needs realistic water demand within the network, located at or near the location where it occurs, to predict the flow distribution within the network and calculate the resulting hydraulics. Therefore, the objective of allocating base demands by distributing the most appropriate data source of water demands across the model is met by establishing a representative distribution of flow within the water system network for subsequent demand adjustments or scaling.

Allocating base demand can be a highly automated effort, a semiautomated effort, or a manual effort, depending on the base data source being used, the supporting data available, and the objective of the hydraulic model application. The following section discusses the general approach to using CIS water consumption data as the source for allocating base demands to the model.

2.5.3.1. Water Consumption Data. Allocating base demand using water consumption data is effective for modeling because the actual water consumption, as well as a spatial representation of where that consumption occurred, can be applied. Assuming the utility has customer meters as a GIS point layer and the meters have been linked to the CIS, water consumption from these meters can be allocated to the nearest pipe or node using GIS spatial analysis tools or demand allocation tools provided by modeling software. Thiessen polygons may be created around strategic nodes in the GIS or model. Water consumption within each polygon can then be aggregated to the node within that polygon.

If meter locations are not available in the GIS, customer addresses may be used to establish a point location for each meter. This approach cross-references GIS street centerline addresses with the CIS customer billing address and approximates the location of the CIS meter with its associated consumption data. Utilities that use GIS to manage city tax parcels can also use parcel addresses to create a spatially enabled customer meter point file from the CIS. The modeler may use this information to link the land parcel to customer consumption data in the CIS. With the meter layer created, the water demand allocation methods previously described would apply. The modeler should ensure that the coordinate system used by the street map conforms to the coordinate system of the model nodes.

Large industrial and wholesale customers should be assigned separately to ensure that large customer demands are properly allocated. Care should be taken to ensure that consumption data are allocated to the correct pressure zone, especially near pressure zone boundaries. Thiessen polygons, created to follow pressure zone boundaries, can be useful to ensure that consumption data from one zone do not get improperly assigned to junctions in another zone. Quality assurance is recommended during and following the customer meter creation process, as CIS customer addresses may not always be representative of the actual location of water consumption.

If a high-quality street centerline shape file or other GIS data source is not available for geocoding, a postal code boundary map may be used. The modeler should ensure that the addresses used are service addresses, not billing addresses, as these may be different.

2.5.3.1.1. *Water Data Allocation Challenges.* Although geolocating addresses can be a highly accurate form of demand allocation, problems can still exist. It is not uncommon for a significant portion of the CIS data to not be geocoded correctly. When this occurs, the address is typically geocoded to the centroid of the area (zip code) and must be manually located or assigned. Alternatively, the demands from these points can be spread across and added to the other demands in the model.

2.5.3.2. Service and Nonservice Nodes. When base demands have been accurately quantified, it is important to consider how base demands will be applied to the model. Prior to importing base demands to the model, it is recommended that the modeler identify service junctions and nonservice junctions. Nonservice junctions are those junctions that should not be assigned a demand value. Nonservice junctions include junctions at the suction and discharge points of pump stations, junctions immediately downstream of water storage tanks, junctions along a large transmission main, or mains from one pressure zone that cross but do not serve customers in another pressure zone. Once service and nonservice junctions have been identified and attributed appropriately in the model, the service junctions should be isolated and assigned base demands.

2.5.4. Adjusting Base Demands

After base demands have been allocated to the model, it is common to adjust or scale these demands to reflect different demand conditions for calibration and analysis purposes. Depending on the data source, the demand condition of the data source (e.g., average day, maximum day, minimum day), and the intended application of the model, different demand adjustments may be required.

Typically, base demand allocation using water consumption is reflective of an average day demand and scaled to reflect water production for the desired demand condition. Due to the potentially comprehensive nature of CIS, however, it may be possible to acquire water consumption data specifically for historical maximum months, which could allow for average maximum day water consumption to be used as a base demand allocation. In this case, it would be consistent to adjust the maximum day demand water consumption to reflect a maximum day water production value.

It is recommended that base demands allocated to the model be reflective of average day demand as a consistent frame of reference for subsequent demand adjustments. While average day data are usually used as the basis for a model, demands vary continuously over time based on time of day, day of week, season, and special events, for example.

2.5.4.1. Geographic Variations. Just as the demand varies by the type of customer, it also varies depending on the neighborhood. The average size of the dwelling, average number of water use fixtures, age of the neighborhood and accompanying efficiency of the fixtures, and the size of lawns can result in large variations in average residential demand between older and newer neighborhoods. If the demand development and allocation process has been based on actual customer billing data, base demand allocation should account for these variations.

2.5.4.2. Calibration. For calibration purposes, the goal is to have the model perform in a way that is similar to water system operations. As a result, base demands using water consumption data, land use data, population data, or other sources outside of production values should be scaled to reflect the total water production in the system during the calibration period of interest. Adjusting base demands to reflect water production also generally considers water loss experienced by the distribution system, which also should be accounted for in the network. It may be challenging to accurately adjust water demands to represent the locations of water loss. However, if negligible, a uniform scaling of base demands to reflect water production may be acceptable to account for the water loss component. Large customer demands may be adjusted separately to ensure accuracy of flow distribution.

2.5.4.3. Analysis. For analysis purposes, the goal of having the model perform in a way that is similar to water system operations remains, as does the goal of maintaining a reasonable level of conservativeness when identifying deficiencies and making recommendations. Therefore, demand adjustments for analyses may reflect water production for a given demand condition and may also consider an adjustment to account for added conservativeness, and to consider maximum day demand, peak hour demand, and minimum day demand (depending on the intended analyses). The conservativeness of a demand factor is based on system experience, the level of confidence in the data used, other agencies with similar characteristics, the intent of the model application, and the engineering judgment of the utility and engineer.

Demands may also be adjusted using scaling factors to reflect future demands as supported by land use and/or water demand projections for analysis of future conditions. Depending on the scenario evaluated, it may also be necessary to determine seasonal variations or hourly variations in demand.

2.5.4.4. Seasonal Variations. Seasonal variation is sometimes significant in distribution systems. Typically, seasonal variation is directly related to irrigation. However, communities with large educational, military, institutional, industrial, or tourist demands should consider the potential seasonal effects of these demands and account for them when developing a model. Systems that use an aquifer storage and recovery strategy will have large seasonal "demands" as they pump into the aquifer.

Local climate variations determine the irrigation and air conditioning requirements. The modeler should consider the period of extended hot weather to determine when the associated peaking factor adjustment would be applied. Systems that experience extreme seasonal variations may be analyzed with unique considerations, resulting in a customized approach to demand development and base demand adjustments within the model for analysis purposes.

2.5.4.5. Diurnal Curves. Initially, base demands are allocated to the model as steady-state demands, that is, they represent an instant in time or snapshot of a certain demand condition.

2.5.4.6. Demand Variations Over Time. Representing these variations in the model requires a diurnal curve, which is a set of factors, usually on an hourly basis, that represents the ratio of hourly demand to the average base demand during the selected time period.

Field testing and/or analysis of system operating data are required to determine the specific diurnal demand curve for a community. Depending on the data available for diurnal curve calculations, a curve may be developed for the entire water system, for major geographic areas, by pressure zone, by customer, by land use classification, or by season.

To develop a diurnal demand curve for the overall system, water supplied from treatment plants, pump stations, and groundwater wells must be recorded, in addition to storage inflow and outflow. Ideally, this information is readily obtained from the utility's distribution SCADA system. A diurnal curve is developed by summing the inflows and outflows in a discrete area for each hour of the day to calculate the demand throughout the day. Demands are then normalized to yield an average demand of 1.0. This same mass balance technique is used to develop curves for hydraulically isolated pressure zones.

Diurnal demand curves for specific customer types, such as low-density residential, high-density residential, and commercial, may be directly measurable if the utility has installed service meters that log hourly demand data or if metered facilities (such as a pump station) feed an area known to have a single customer type. If none of these diurnal data sources are available, diurnal patterns can be estimated by isolating a homogenous land use area and monitoring the flow over a specified time period, typically 24 hours. Strap-on portable meters can be used in specific areas to measure flow such that the diurnal curve can be calculated for specific customers or groups of customers. The diurnal demand curve for the service area under consideration is derived by summing all flows into and out of the area, measured by meters that surround the area. Figure 2-11 shows an example diurnal curve.

Diurnal patterns also vary significantly by season. The modeler should consider this factor when developing a sampling program to ensure that the data collected are suitable for the scenario modeled. For example, a winter sampling program is not useful for a model assessing transmission main requirements to meet peak hour summer demands.

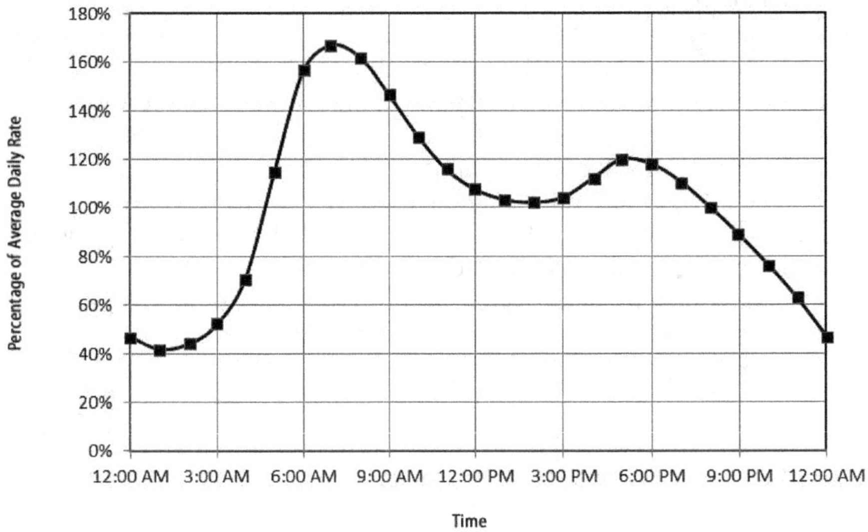

Figure 2-11 An example diurnal curve

2.6. OPERATIONAL DATA

Operational data consist of facility status changes involved in the day-to-day and seasonal running of a water system to successfully deliver water to utility customers. Understanding, appreciating, and applying operational data are vital to model calibration and analyses.

Operational data are generally classified as the following two main types: automated controls and manual controls. Automated controls are programmed into the utility SCADA system for automatic water system operations and consist of pump and valve control logic and set points for pumps, control valves, and altitude valves. Most modeling software packages include a control scheme that can replicate the behavior of a SCADA system. The model automatically changes valve position or pump status in response to changes in pressure, flow, water level, or time at determined points in the model.

Manual controls are typically used to represent operator intervention, which overrides the automated controls programmed into the SCADA system. It is often challenging to develop model control logic for manual controls because the set point to trigger a change in facility status is not predefined, and operators may modify the way the system is operated from day to day. Typically, the modeler replicates the manual control using time-based controls that are based on when the SCADA system showed the change of facility status occurring. This approach to emulate manual controls may be suitable for model calibration but it is typically not recommended for analysis.

Operations may change on a seasonal basis. For instance, a utility may change seasonally between surface water and groundwater supplies due to water quality or water rights issues or may take tanks out of service due to lack of winter turnover. If the regional network is not modeled for utilities that purchase water from a regional supplier, seasonal hydraulic grade line changes in the regional supply should be accounted for.

Typical operational data required for models are listed in Table 2-4. The items are often interrelated, meaning model input changes may automatically result in a recalculation of other model parameters. Operations data are used extensively in the model calibration process so that model inputs and outputs agree with actual measured values within a defined calibration tolerance. There are four main sources of operations data: operations staff, written records, charts/loggers, and SCADA.

Table 2-4 Operational data required by facility or equipment type

Facility/Equipment	Operational Data Required
Pumps	Pump status (on–off) Pump controlling tank, pressure, flow, or other set points for pump control Pump speed or frequency settings (for pumps with variable-frequency drives)
Pressure-regulating valves	Upstream/downstream pressure, valve position, and setting Valve controlling tank, pressure, flow, or other set points for valve
Flow-control valves	Valve position, flow setting Valve controlling tank, pressure, flow, or other set points for valve
Node pressures	Any pressure data collected at pressure monitoring stations in the distribution system
Pipe flows	Any in-line flow data measured in the distribution system (includes wholesale supply locations or zone boundary meters)

2.6.1. Operations Staff

The most important source of modeling information is the utility operations staff. The modeler should develop a good working relationship with this group while developing models. Operations staff can inform the modeler of how the actual system operates as opposed to how it was designed to operate. Operators can identify specific areas to study in order to determine why the actual operation varies from the design, such as finding closed valves, determining incorrect set points, and identifying distribution system deficiencies. During model calibration, input from operations staff helps in interpreting data and identifying necessary model adjustments.

Operations staff involvement in model development greatly enhances the acceptance of analysis results and may motivate more efficient operation of the water distribution system. For these reasons, involving the operations staff in model development results in a more accurate, usable, and sustainable model for the utility. It can also help to build staff confidence in the model results.

2.6.2. Written Records

Written records or standard operating procedures consist of the reports and documents that explain how the distribution system is theoretically supposed to operate. Utilities may have an operations manual that provides a schematic for each facility and a narrative of how each facility should operate based on varying conditions. The modeler should verify these types of reports against actual operating conditions, which may be different because of constraints and conditions in the distribution system that may not have been present or recognized when the reports were written.

Typical written reports that should be reviewed include the following:

- Operations and maintenance manuals for each facility;
- Design reports and record drawings;
- Operator training manuals;
- Emergency manuals (to identify how things are expected to operate in an emergency);
- Historical records of deficiencies in the system (e.g., low-pressure complaints, water quality complaints);
- Maintenance records for all facilities (e.g., pump stations, reservoirs, valves, pipes);
- Valve status reports;

- Hydrant flushing reports; and
- Pump performance tests.

Each utility will have a different combination of written records. In some cases, the quality of the reports is not consistent.

2.6.3. Charts, Loggers, and SCADA

Charts are historical records that show time-relative data. Traditionally, these have been paper records and are generally in a circular or strip format. Digital data-logging equipment has reduced the need for paper charts, but software is still required to graphically represent and interpret the collected digital data. Care should be taken to ensure that charts and digital data are calibrated.

Software packages are available from data-logger equipment suppliers that help in analysis of the data. These packages may be worth the cost to reduce the effort required when using standard databases and spreadsheets for analysis.

2.6.4. Supervisory Control and Data Acquisition

SCADA systems are a powerful tool for collecting and maintaining historical information on the real-time or historical operation of a water system. The SCADA system database may include flows, pressures, alarms, and tank levels and equipment information such as the on–off status for pumps, valves, tank levels, and more.

Data are collected via SCADA software and stored in database files for evaluation. The amount of data collected is determined by the polling frequency of the SCADA system. It is also possible to determine total system demand based on a mass balance of flows from the treatment plants and flow in and out of the reservoirs or change in reservoir levels.

2.6.4.1. Importance of SCADA to Modeling. Historically, only the most critical operating points were monitored via SCADA due to the cost of additional field devices and communication lines. These costs have since declined, allowing utilities to increase the number of monitoring points in the system. If the budget allows, the modeler should review the data being collected and possibly coordinate the installation of additional monitoring points at key locations.

SCADA provides a wealth of data on the actual operating conditions of the water distribution system. Differences between the model results and SCADA readings are used to calibrate the model or identify locations where field crew checks can be conducted to locate closed or partially closed valves.

SCADA can generate data files for model calibration and allows the modeler to confirm the accuracy of the model on a regular basis. SCADA software can usually extract facility records to ASCII, delimited text, spreadsheet, or other database formats to facilitate comparison with model results. Models are typically calibrated once they are constructed, then every few years thereafter, depending on the frequency and types of changes in the system. Some model software packages generate log sheets that identify where the model and the system differ.

Development of common interfaces and linkages between the model and the SCADA system improves the modeler's ability to consider operational scenarios. This is accomplished by entering current operating conditions from SCADA into the model and allowing the operator to perform "what-if" analyses to test different scenarios. For example, different pump and valve combinations and the results of those combinations can be tested from the same operational starting point.

2.6.4.2. Data Requirements from SCADA. In general, SCADA data can be used in two ways when modeling:

- *Model input:* These include boundary condition data and system control data. Boundary conditions include information on which pumps are operating, flow and pressure control points, valve status, reservoir levels, water production, and demand loading (if available). This information is required to perform steady-state analyses and to set the initial conditions for EPSs. System control data are related to the changes in boundary conditions that occur over time and include information on the change in pump or valve status and the variation in water levels or pressures at system boundaries. This information is required to perform EPS analyses. See chapter 6 for more information on how SCADA data are used in EPS models.
- *Verification or reference data:* These include information collected by flow, pressure, and level monitors and may be presented in a spreadsheet alongside the model output to confirm that the model is sufficiently representing actual operations. These data may be used to provide guidance on areas of the model that should be adjusted for calibration purposes but are not used directly in the hydraulic calculations. These data are used for both steady-state and extended-period calibration.

It is important to note that SCADA data can include transient or surge data (transients are rapid changes in pressure and flow) that can provide a misleading perspective of the network's average steady-state conditions. Transient data are typically generated during pump startup and shutdown or during changes in valve status. The use of average data instead of instantaneous data values will help ensure that the data placed into the model are a true representation of the pressures and flows in the system. A review of data values in adjacent time periods is also useful to identify anomalies in the data.

Data errors also occur during a SCADA system's communication failure with the facility monitoring point. These should be flagged in SCADA. When assembling data for the model, it may be necessary to interpolate values if some measured values appear to be missing or are not readily available. Sometimes it is better to use a running average of adjacent data points to avoid representation of data measurement or transmission anomalies.

2.6.4.3. SCADA Issues. The following issues should be addressed when using a SCADA system in coordination with a model for setup and/or calibration purposes:

- *Data format:* Most field instruments record data in binary format and the data is converted to digital format in the local RTU or the programmable logic controller. When exporting data from a SCADA system, it is helpful to generate this data in a format that is suitable for use in spreadsheet or database programs rather than as a simple text file. Depending on the software used, this may necessitate use of an ASCII file, Microsoft (MS) Excel file, or MS Access database as a common data interchange format. Delimiters between columns of data, such as tabs or commas, are helpful.
- *Units:* Often SCADA data are presented in reports without units. The user must take care to know what units are being reported and make the appropriate conversions when using these data for modeling. In addition, graphical scales are often different for different tanks on the same graph.
- *Measured versus calculated values:* It is important to note whether the data being used are measured directly or are calculated. In many instances, flow rates from multiple flowmeters (such as in a pumping station) are summed to provide a

single data point. In these cases, the degree of error in the data could be higher and must be noted when comparing with modeling results.

- *Links to model IDs:* By far the most time-consuming aspect of relating SCADA data to models is the relationship between IDs in the SCADA system (often referred to as *tags*) and those used in the model. Many SCADA points have no direct relation to models, and these systems have distinct nomenclatures with unique tag numbers. The IDs tracked in the SCADA system are usually not the same as the IDs used in the model. Models may have simpler naming conventions but will store data in multiple tables. It is helpful to develop separate tables that cross-reference the SCADA points used in the models and the model data.
- *Accuracy:* SCADA may report pressures to more figures than the accuracy of the device. For example, the SCADA output may say the pressure is 61.78698 pounds per square inch (psi) when the value is only accurate to ± 2 psi.

2.7. REFERENCES

American Water Works Association (AWWA). 2016. Manual M36, *Water Audits and Loss Control Programs*. Denver, CO: AWWA.

2004. Manual M11, *Steel Pipe—A Guide for Design and Installation*. Denver, CO: AWWA.

Lamont, P.A. 1981, Common Pipe Flow Formulas Compared With the Theory of Roughness, *Journal AWWA*, 73(5): 274.

Mays, Larry W., ed. 1999. *Hydraulic Design Handbook*. New York: McGraw-Hill.

Moody, L.F. 1944, Friction Factors for Pipe Flow. *Transactions of the ASME*, 66(8):671–684.

USEPA. 2006. *Initial Distribution System Evaluation Guidance Manual for the Final Stage 2 Disinfectants and Disinfection Byproducts Rule, EPA/815/B-06/002*. Washington, DC: US Environmental Protection Agency.

Walski, T.M., Chase, D.V., Savic, D.A., Grayman, W.M., Beckwith, S., & Koelle, E. 2007. *Advanced Water Distribution Modeling and Management*. Exton, PA: Bentley Systems.

2.8. ADDITIONAL RESOURCES

Bonema, S.R., Elain, C., Mercier, M., & Tiburce, V. 1995. Linking SCADA With Network Simulation. In *Proceedings of the AWWA Annual Conference*. Denver, CO: AWWA.

Cesario, A.L. 1995. *Modeling, Analysis and Design of Water Distribution Systems*. Denver, CO: AWWA.

Chase, D., & Jones, G.L. 1994. Linking Hydraulic Network Models and SCADA Systems: The ABCs of System Integration. In *Proceedings of the AWWA Computer Conference*. Denver, CO: AWWA.

Deagle, G., & Ancel, S.P. 2002. Development and Maintenance of Hydraulic Models. In *Proceedings of the Information Management and Technology Conference.* Denver, CO: AWWA.

Edwards, J.A. 2002. Water Demand Allocation Using GIS. In *ESRI User Conference Proceedings.* Denver, Colo. August.

Edwards, J.A., Koval, E., Lendt, B., & Ginther, P. 2009. GIS and Hydraulic Model Integration: Implementing Cost-Effective Sustainable Modeling Solutions. *Journal of the American Water Works Association,* 101(11):34–42.

Green, D., & Montgomery, C. 1998. SCADA Communication Experience at the Detroit Water and Sewage Department. *Proceedings of the AWWA Annual Conference.* Denver, CO: AWWA.

Fortune, D., Campbell, P., & Cavor, R. 2000. Cost of Ownership of Hydraulic Models. In *Proceedings of the Infrastructure Conference.* Denver, CO: AWWA.

Hutchison, W. 1991. Operational Control of Water Distribution Systems. In *Proceedings of the AWWA Seminar on Computers in the Water Industry.* Denver, CO: AWWA.

Japan Water Works Association and Awwa Research Foundation. 1994. *Instrumentation and Computer Integration of Water Utility Operations.* Denver, CO: AWWA and AwwaRF.

Maidment, D., ed. 2000. *Hydrologic and Hydraulic Modeling Support in GIS.* Los Angeles, CA: ESRI Press.

Mau, R.E., Boulos, P., Heath, E., & Brennan, W.J. 1996. Advanced Network Modeling Applications: Dynamic Design and SCADA Interface. In *Proceedings of the AWWA Engineering and Construction Conference.* Denver, CO: AWWA.

Moore, P.B., Harrington, D.A., Hauffen, P.M., Ray, R., & Tejada, R.W. 2008. Avoid the GIS-Model Disconnect: Identifying the Major Hurdles Associated With Hydraulic Model Creation and Updating From GIS. *AWWA/Distribution Systems Symposium Proceedings.*

Ray, R., Moore, P.B., Harrington, D.A., & Hauffen, P.M. 2008. The Achilles' Heel of GIS-Built Hydraulic Models: Maintaining/Updating a Model From GIS Data. In *AWWA Conference Proceedings.* Denver, CO: AWWA.

Rehnstrom, D.J., & Butler, C.L. 2001. Maximizing the Use of Your Hydraulic Model. In *Proceedings of the Information Management and Technology Conference.* Denver, CO: AWWA.

Schulte, A.M., & Bonema, S.R. 1995. SCADA Data for Predictive, Training and On-Line Water Distribution System Simulation. In *Proceedings, International Water Supply Association, Specialty.* Denver, CO: AWWA.

Shamsi, U. 2002. *GIS Tools for Water, Wastewater, and Stormwater Systems.* Reston, VA: ASCE.

——. 2005. *GIS Applications for Water, Wastewater, and Stormwater Professionals.* Florence, KY: CRC Press.

USEPA. 2005. *Water Distribution System Analysis: Field Studies, Modeling and Management, A Reference Guide for Utilities.* EPA/600/R-06/028. Cincinnati, OH: US Environmental Protection Agency.

Water Environment Federation. 2004. *GIS Implementation for Water and Wastewater Treatment Facilities.* Alexandria, VA: Water Environment Federation.

Wilson, L.L., Moshavegh, F., & Bolze, M.A. 2001. Constructing, Maintaining, and Utilizing a Large Water Model. In *Proceedings of the Information Management and Technology Conference.* Denver, CO: AWWA.

Wood, P. 1994. Application Integration for Improved Utility Operations. In *Proceedings of the AWWA Computer Conference.* Denver, CO: AWWA.

Chapter 3

Hydraulic Tests and Measurements

3.1. INTRODUCTION

A primary objective of computer modeling is to develop a tool that will predict the performance of an existing water distribution system and simulate the impact of changes to the system. Hydraulic tests and measurements (field tests) are important in developing such a model. Hydraulic tests are used to obtain and verify system information and are an integral part of model calibration. Tests for model calibration are commonly planned to obtain normal system operational information and to stress the system hydraulics. A test that imparts stress on a portion of the system provides valuable information to aid in model calibration.

At times, information is not available without taking measurements in the distribution system. Data such as pump curves, pipe roughness coefficients, flow rates, system demands, and storage levels can be obtained from field tests when no other source of this information is available. There are also times when information provided from other sources is incorrect or not up to date, the model does not match the data provided, or a large variation exists between a theoretical estimate and an actual measurement. In these cases, hydraulic tests and measures can be used to verify and understand such discrepancies in order to determine why the model does not behave as expected. During the model calibration process, field testing is critical in identifying closed valves, incorrect data records, severely tuberculated pipes, and other challenges within a distribution system. By using field testing to calibrate the model, the accuracy of simulation model results increases, confidence in the model is reinforced, and the model evolves as an efficient planning, design, and operational tool. This chapter covers the following areas related to hydraulic tests and measures:

- Planning and preparing for field tests,
- Pressure and flow rate measurements,
- Meter calibration,

- Pipe roughness coefficient testing,
- Hydraulic gradient tests, and
- Fire flow tests.

3.2. FIELD TESTING—PLANNING AND PREPARATION

Field testing is necessary to develop a calibrated model that can be used with confidence. To be effective, field testing should be well planned. See the sample field preparation checklist below. There are a number of different parameters to test for and measure depending on the condition and/or purpose of the model. The potential for failure is high when planning is insufficient, parameters that affect the test are poorly understood, incomplete data are collected, and/or test equipment does not function properly. When planning a field test, testing objectives must first be established. These objectives include calibrating all or a portion of a model, measuring or estimating pipe roughness coefficients (commonly the Hazen–Williams C-factor), checking meter calibration, verifying pump performance or control-valve settings, identifying the cause of anomalies in the distribution system, and determining available fire flow for proposed developments. Personnel experienced with the distribution system operations, such as field personnel from the utility and local fire department staff, should be consulted when planning field tests. Their understanding of and agreement to testing plans are critical.

Additional benefits of the test may be captured through test execution and data collection. Observations during the test, such as asset condition and operability, and general water quality characteristics can be recorded and communicated for appropriate follow-up action. Tests such as measuring fire hydrant flow can also serve as an opportunity for customer interaction. An informational flyer that explains the purpose and benefits of the testing or project can be provided to answer customers' questions. Use of a flyer can allow crews to stay on schedule.

Prior to gathering any field data, all the equipment used must be calibrated. Any SCADA equipment or permanent meters should also be calibrated and verified if the data are to be used to calibrate the model. Calibrating equipment prior to field testing will ensure confidence in the data collected. Poorly calibrated equipment can result in gross inaccuracies that can cause problems later with model calibration. Proper planning requires time and effort. If test results do not align with generally expected values, variables such as equipment calibration can be excluded in the investigation as a benefit of proper test planning.

► FIELD TESTING PREPARATION CHECKLIST

- Define test objectives
- Select test locations
- Plan test day/time to minimize system disruption and impact to public
- Understand hydraulic conditions and available SCADA data
- Identify and inspect equipment, traffic control devices, and other personal protective equipment
- Prepare data collection forms
- Notify applicable stakeholders

3.2.1. Site Selection

Hydrant flow tests are very useful for calibration of systems that do not have large transmission mains. Multiple perspectives for selection of test locations should be considered. Initially, site selection should focus on spatially diverse geographic areas and hydraulically dissimilar areas within the system. Once these areas are identified, a number of site-specific factors should be considered when selecting hydrants, pumps, and other appurtenances. If testing is conducted for system-wide model calibration, each pressure zone should be represented to verify local hydraulic grade lines (HGLs) and supply availability. For large pressure zones, multiple flow tests may be needed to provide a geographically diverse set of data. If evaluating fire flow capacity or pipe roughness, consideration must be made for how many tests are necessary to capture piping variations within the system. Data such as pipe installation year and material can be used to group similar areas of the system for the purpose of identifying test locations and applying test results to the model. The same thought process can be used to recalibrate only a portion of the model.

If hydrant flow tests are to be performed, they should occur at locations in the system expected to yield significant pressure drop in order to provide the most meaningful results. Although tests that yield relatively little pressure drop do provide some information and may confirm intended system operation, they may not be useful for model calibration if the pressure drop is less than the acceptable margin of error (typically, 5 to 10 psi [34 to 69 kPa]; see section 4.2.4). Considering potential pressure drop during test planning will help to avoid wasted time on unproductive tests.

External factors should also be considered, in addition to system hydraulics and operation. The optimal test site would yield the most valuable calibration data that can be reasonably gathered while balancing impacts to the community or the environment. For instance, the site expected to yield the best calibration data might be unacceptable due to downstream drainage conditions, extensive traffic control requirements, or public perception. It is good practice to plan for an alternate test location in case of unexpected circumstances. Sometimes a hydrant may be found to not operate properly, and planning for an alternate hydrant in advance saves time. For residual pressure monitoring, it is beneficial to select a location that is as close to the flowing hydrant as possible, preferably upstream, to capture the most pressure drop for analysis and to monitor for excessive drop during testing. However, it is also beneficial to monitor pressure elsewhere in the system to understand flow direction(s) and system set points. For example, multiple pressure monitoring points within a zone can assist in diagnosing improperly closed valves. Positioning a data logger on the upstream side of a normally closed secondary or backup source from another zone (such as a pressure-reducing valve [PRV] with low setting or a check valve) can be useful in determining whether the facility opened or functioned as expected during the test. It may also be desirable to monitor pressure at a known low-pressure point.

3.2.2. Maps and Test Plan Documents

A particularly valuable tool for field testing is a map of the water distribution network that identifies zone boundaries, key transmission mains (including size), elevations, hydrant locations, and major system facilities. It is common practice for a map to be created from a utility's geographic information system (GIS). This map should be annotated with all the important elements necessary for planning and implementing the field testing. A system-wide or regional map is beneficial for defining the area of the system and showing the locations of all planned tests. A site-specific map can be more detailed with an aerial image, rights-of-way and landowner details, and ground elevation contours.

The site-specific map, for example, can include locations of fire hydrants to be tested, valves that need to be closed, traffic control plans, estimated discharge flow routing to storm drains or open ditches, and/or any system operational changes needed for a particular test. The map can be used to locate potentially impacted customers (for notification regarding the aesthetics of the water) and system high points that limit allowable flow due to low pressure (minimum pressure during testing is not necessarily at the flow location). The map can also be used to indicate installation points for a temporary pressure logger (recorder) or flow monitoring equipment, as shown in Figure 3-1. The map should also identify existing system data collection monitoring points and any other information useful for the planning process.

While paper maps are still commonly used, application of mobile device technology is increasingly being applied to water distribution system operations, planning, and testing. Utilities with advanced GIS capabilities may use system mapping, available on portable computers or tablets, to plan the test and to capture the test results. Results such as global positioning system (GPS) location of test equipment and photos of test setup can be used to complete test reports.

When multiple tests are conducted, standardized test forms are recommended. Standard forms provide consistent formatting to clearly document the auxiliary data to be collected, the work performed, field conditions, and results. This documentation is particularly useful for personnel not present at field testing and for information retrieval long after the test is conducted. Although forms require some extra effort during test preparation and shortly after each test, they do improve efficiency later during model calibration. Figures 3-2 and 3-3 show sample test forms.

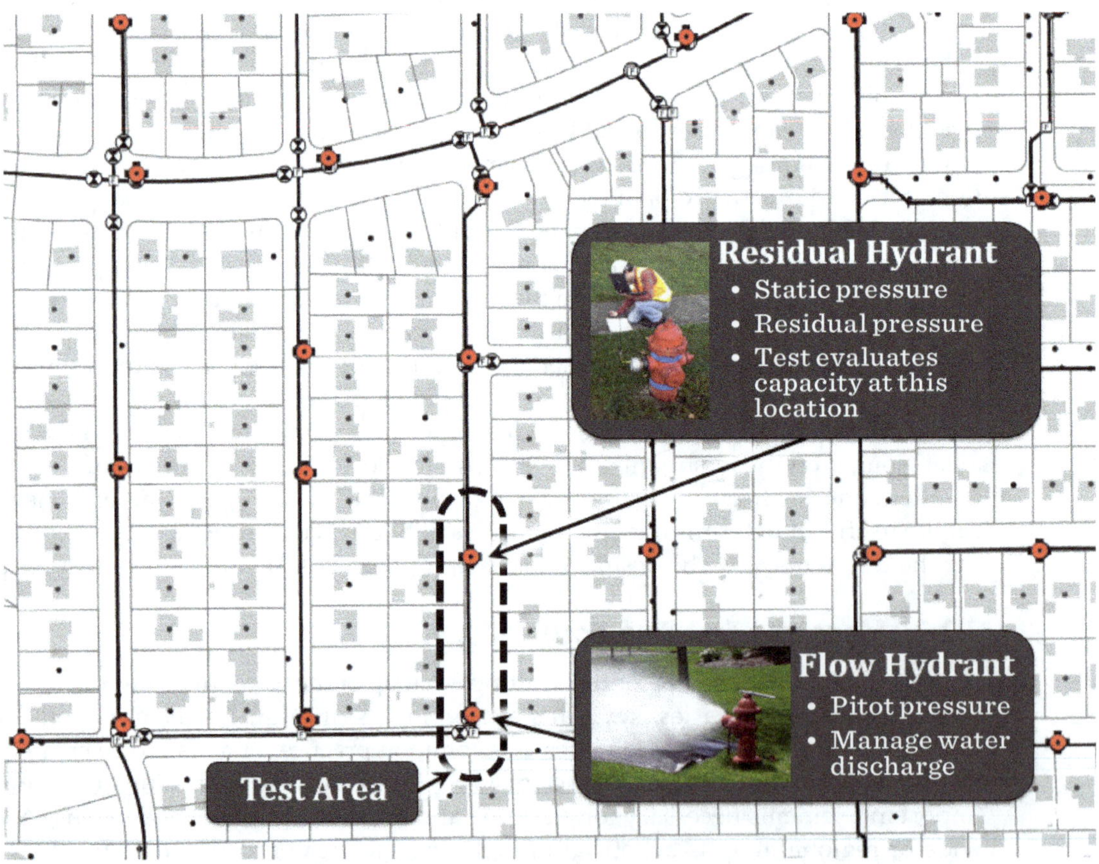

Figure 3-1 Site-specific map of planned fire flow test

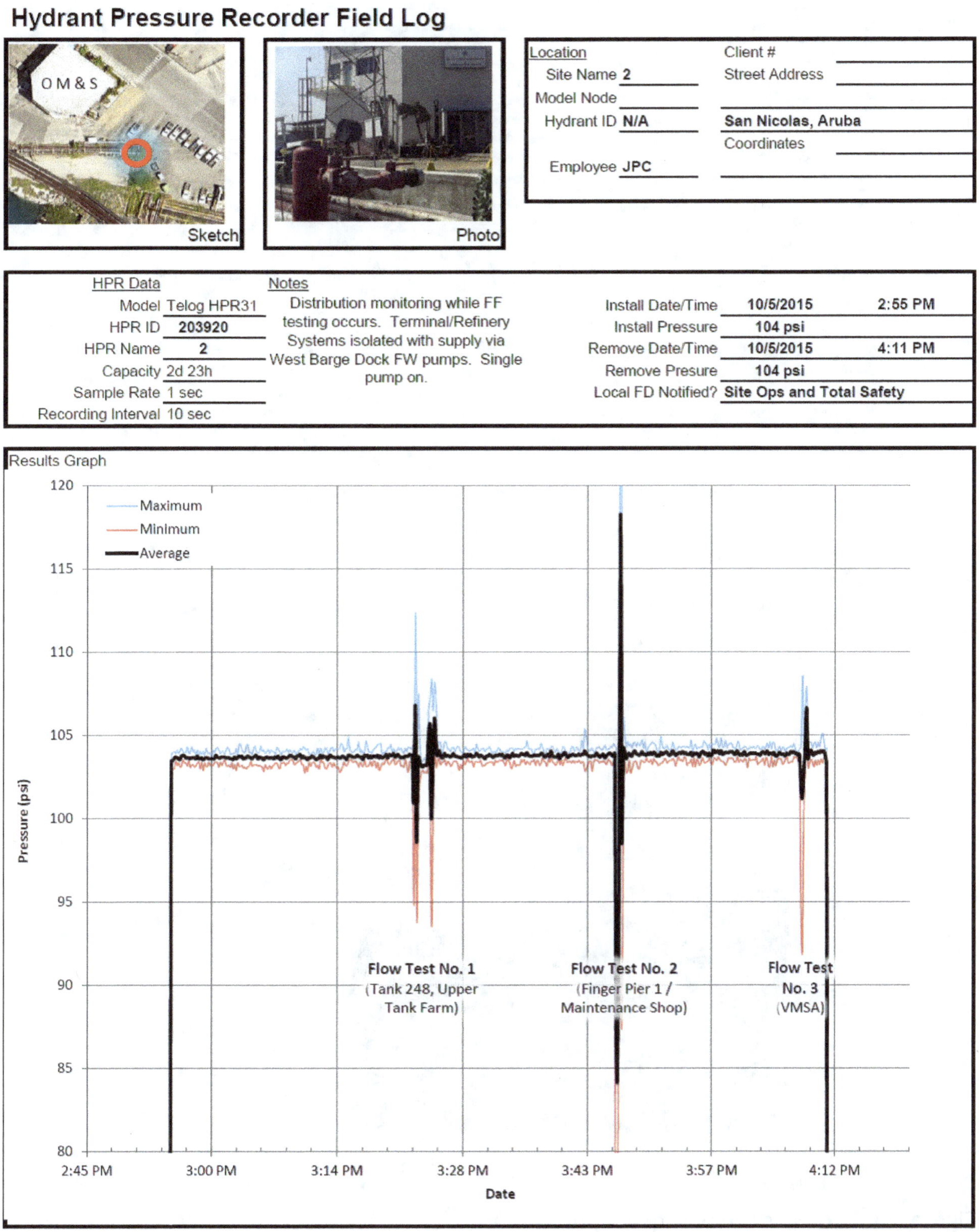

Courtesy of Jim Cooper, Arcadis.

Figure 3-2 Field log of pressure recorder during a flow test

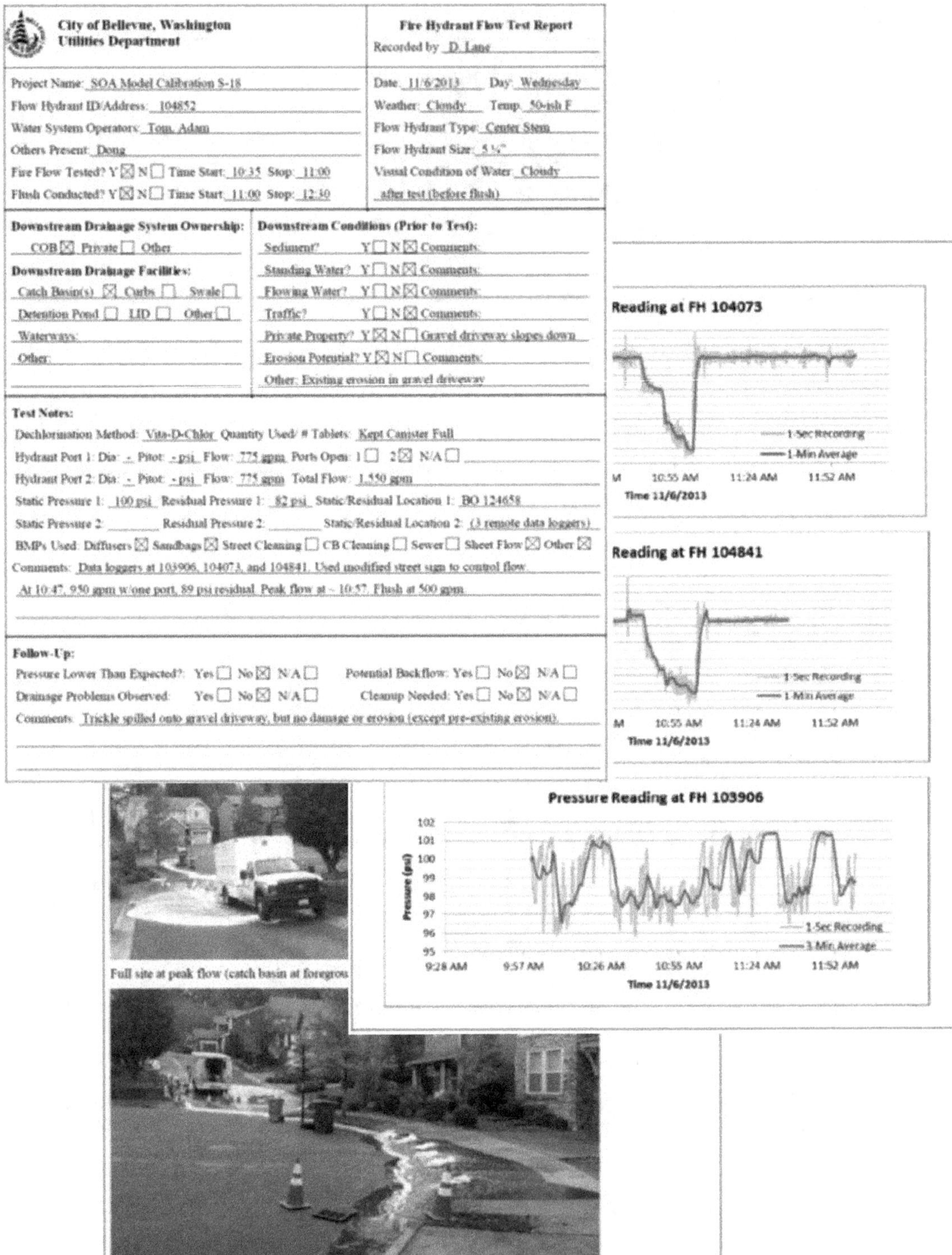

City of Bellevue, Washington
Utilities Department

Fire Hydrant Flow Test Report
Recorded by D. Lane

Project Name: SOA Model Calibration S-18
Flow Hydrant ID/Address: 104852
Water System Operators: Tom, Adam
Others Present: Dong
Fire Flow Tested? Y ☒ N ☐ Time Start: 10:35 Stop: 11:00
Flush Conducted? Y ☒ N ☐ Time Start: 11:00 Stop: 12:30

Date: 11/6/2013 Day: Wednesday
Weather: Cloudy Temp: 50-ish F
Flow Hydrant Type: Center Stem
Flow Hydrant Size: 5 ¼"
Visual Condition of Water: Cloudy after test (before flush)

Downstream Drainage System Ownership:
COB ☒ Private ☐ Other
Downstream Drainage Facilities:
Catch Basin(s) ☒ Curbs ☐ Swale ☐
Detention Pond ☐ LID ☐ Other ☐
Waterways:
Other:

Downstream Conditions (Prior to Test):
Sediment? Y ☐ N ☒ Comments:
Standing Water? Y ☐ N ☒ Comments:
Flowing Water? Y ☐ N ☒ Comments:
Traffic? Y ☐ N ☒ Comments:
Private Property? Y ☒ N ☐ Gravel driveway slopes down
Erosion Potential? Y ☒ N ☐ Comments:
Other: Existing erosion in gravel driveway

Test Notes:
Dechlorination Method: Vita-D-Chlor Quantity Used/ # Tablets: Kept Canister Full
Hydrant Port 1: Dia: - Pitot: - psi Flow: 775 gpm Ports Open: 1 ☐ 2 ☒ N/A ☐
Hydrant Port 2: Dia: - Pitot: - psi Flow: 775 gpm Total Flow: 1,550 gpm
Static Pressure 1: 100 psi Residual Pressure 1: 82 psi Static/Residual Location 1: BO 124658
Static Pressure 2: Residual Pressure 2: Static/Residual Location 2: (3 remote data loggers)
BMPs Used: Diffusers ☒ Sandbags ☒ Street Cleaning ☐ CB Cleaning ☐ Sewer ☐ Sheet Flow ☒ Other ☒
Comments: Data loggers at 103906, 104073, and 104841. Used modified street sign to control flow. At 10:47, 950 gpm w/one port, 89 psi residual. Peak flow at ~ 10:57. Flush at 500 gpm.

Follow-Up:
Pressure Lower Than Expected?: Yes ☐ No ☒ N/A ☐ Potential Backflow: Yes ☐ No ☒ N/A ☐
Drainage Problems Observed: Yes ☐ No ☒ N/A ☐ Cleanup Needed: Yes ☐ No ☒ N/A ☐
Comments: Trickle spilled onto gravel driveway, but no damage or erosion (except pre-existing erosion).

Courtesy of City of Bellevue Utilities.

Figure 3-3 Example fire flow test report

Flow test forms include general information such as the date, start and finish times, weather conditions, names of personnel at the test, test location, and test photos (inserted after the test). Information relevant to calibration should be recorded, such as identification (e.g., asset number) of flow hydrant(s) and residual hydrant(s), Pitot gauge pressure (or flow, for manufacturer-calibrated pressure gauges), hydrant port diameter, test flow rate, residual pressure(s), telemetry data (added after the test), and space to record general comments. Relevant system conditions such as tank levels and pump on–off status should also be documented. Each utility may also choose to record additional information that suits its specific needs. For instance, utilities with water use restrictions may want to document conservation measures, and utilities with stringent stormwater regulations may choose to document dechlorination method and erosion control measures.

3.2.3. Hydraulic Conditions During Hydrant Flow Test

Results obtained from a field test for the purpose of model calibration require additional system hydraulic data to ensure proper mass–flow balance and calibration. These can include storage tank water levels, control-valve settings, pressure zone isolation valves status, real-time water demand (if available), and flow rates of water entering and exiting the area where the test occurred. This includes specific operational schemes that occur during the test, such as seasonal variations. Any applicable hydraulic boundary conditions within the test area must also be recorded. This additional data may be available for viewing on a utility's supervisory control and data acquisition (SCADA) interface. As part of test planning, it is recommended that the modeler confirm that the desired data is trending and stored in a historian database prior to field data collection such that it can be retrieved following the test for use in model calibration. Additional discussion on data requirements in general is provided in chapter 2, and information specific to model purposes is provided throughout this manual.

Where possible, flow should be monitored at points where water is entering and exiting the system or pressure zones, such that the modeler has the information necessary to calculate a mass balance of the distribution area. If not already equipped, flow monitoring devices can be placed at distribution system entry points – for example, the discharge point of water treatment facilities or connections with imported water supplies, pump stations, and/or PRV stations. Typically, flow is monitored where water lines are easily accessible. However, it may be necessary to excavate a water line in order to monitor flow. If continuous flow monitoring is performed, diurnal demand patterns can be obtained from the data.

3.2.4. Equipment Preparation

After field testing requirements are established and site locations are selected, an assessment of equipment necessary to carry out the tests should be made. Some utilities have equipment such as electronic pressure recorders available to mount on fire hydrants. However, equipment can also be rented or purchased, depending on need. It is important to verify that all equipment has been calibrated prior to use in the field. Preliminary testing to ensure all equipment is appropriate and in proper working order is critical to ensure data obtained are valid for the intended purpose. Batteries should be fresh or fully charged, and digital memory storage must be adequate for the tests being performed. Programming of electronic devices must be applicable for the test being performed, such as frequency of measurements and data storage.

3.2.5. Other Considerations

The time of day and duration of the tests should be considered to ensure customers are not adversely impacted during testing and to ensure that the equipment and the system will not be adversely impacted by weather conditions. Testing for periods of several days or weeks might also be needed. If operational problems occur, the entire test may not necessarily be invalidated. Data from the days without operational anomalies may still be used, depending on the nature of the anomaly. The more data that are collected about the test conditions, including hydraulic, atmospheric, and system boundary conditions, the easier it is to understand anomalies found in test results. Fire flow testing is typically performed on weekdays during daytime hours. However, depending on the situation, it may be necessary to perform system testing on weekends and/or at night. It is important that no flow testing be performed near active roadways and pedestrian areas during periods of potential low temperatures to prevent freezing and ice on these surfaces.

Flow testing at high velocity typically results in aesthetic impacts such as discolored or cloudy water. If impacted customers can be identified in advance of testing and if testing date(s) are known, the utility can proactively notify customers of the impacts in advance (Figure 3-4). This will prevent customer service calls and manage public perception and can be accomplished via mail, use of the reverse 911 system, use of door handle tags or signage, through social media, or by other means.

Downstream drainage of flushed water should be evaluated to prevent environmental and property damage. Locations with the potential for erosion or flooding (e.g., downhill driveways) may need to be avoided, or mitigating measures may need to be put in place. Sediment and debris that accumulate on roadways or in catch basins should be removed to prevent the discharge of contaminants into local streams. Private drainage facilities may need to be avoided. Local regulations regarding dechlorination of discharged water should be followed.

It is also best to ensure that utility field personnel or local fire department staff are available to assist with testing and that system operators are aware that system testing is occurring. It is crucial that communication channels are kept open during this time to ensure that the utility can adequately continue to serve its customers during the testing period and that firefighting operations are not obstructed. It may be necessary to involve local building owners because a drop in pressure could trip alarms in some types of fire suppression systems. Without proper coordination, flow testing may cause false alarms that could result in needless fire department response and evacuation of residential and commercial buildings.

When planning fire flow testing, it is also necessary to make sure that sufficient head losses can be achieved during testing. Head loss in the mains must be greater than the accuracy and precision parameters of the instrumentation. Otherwise, the field test might fail to yield meaningful results. Generally, the greater the head loss, the more accurate the test results will be. To ensure sufficient head losses are achieved on large mains, two or more hydrants may need to be opened simultaneously during the test.

Measures should be taken to prevent or minimize theft of monitoring devices that will be attached to hydrants for many days. Each device should be labeled with an identification mark and contact information. Devices attached to hydrants should be fitted with locks to prevent tampering, as shown in Figure 3-5. It is important that the local fire department is aware the hydrant is locked and has the authority to cut off the lock if necessary during an emergency.

Personal and public safety during field testing for model calibration is also important. It is common that field data for model calibration is collected by engineers, modelers, and other office staff who do not typically perform field activities. If a hydrant is pressurized with a temporary pressure recorder, the local fire department must be notified of the pressurized hydrant, and the hydrant must be identified with field markings. Proper safety planning for field testing should, at a minimum, consider traffic control, public impact of the discharged water stream, flow direction, and erosion. Some authorities require dechlorination; however, little to no free chlorine remains in the water shortly after being discharged from the hydrant.

Courtesy of Jim Cooper, Arcadis.

Figure 3-4 Hydrant testing public notification sign

Courtesy of Jim Cooper, Arcadis.

Figure 3-5 Secured pressure-recording device on a hydrant

3.3. WATER DISTRIBUTION SYSTEM MEASUREMENTS

3.3.1. Pressure Measurements

Measuring pressures in a distribution system is key to understanding the system's performance. Pressure is typically measured in pounds per square inch (psi) or in metric kilopascals (kPa) with 1 psi (6.89 kPa) being equivalent to 2.31 ft (0.704 m) of water head. This is based on the characteristics of potable water typically found within a distribution system. Pressure can further be defined as absolute pressure or gauge pressure. Absolute pressure is the pressure measured with absolute zero (a perfect vacuum) as the datum, and gauge pressure is the pressure measured with atmospheric pressure as the datum (Walski et al. 2007). Because gauge pressure is generally used for water distribution modeling applications, all pressures indicated in this manual are gauge pressures. For this reason, all pressure measurements must be taken at a known elevation above a common datum.

Two types of devices are typically used for pressure measurement, the mechanical pressure gauge and the pressure logger. The mechanical pressure gauge has been used for years but can be difficult to read at times, which can introduce measurement error. Pressure fluctuations in many water systems, caused by flow transients and dynamic system changes, can make determining pressures difficult, especially during fire flow testing. In any case, the user must make sure gauges have the proper range and accuracy for expected pressures. Unfortunately, pressure gauges that are capable of reading higher pressures typically are less precise. Mechanical pressure gauges are less commonly used today for model calibration.

Pressure loggers provide the advantage of recording pressure over an extended period of time and can fully capture the system's response to changes in flow. Once recorded, pressure data are collected and downloaded. Software is available to graph system pressures (Figure 3-6), allowing the user to perform statistical analyses, such as calculating average values to eliminate noisy readings and easily obtain usable pressures. Regardless of the type of device used, the user must ensure that the elevation of the pressure gauge is accurate. Every 2.31 ft (0.704 m) of error in the gauge's elevation equates to a 1-psi (approximately 7-kPa) error in recorded pressure. Similarly, differences between the pressure gauge elevation and the corresponding model node elevation also introduce errors.

Typically, pressure measurements are taken at hydrants. However, pressures can also be taken from other easily accessible locations within the system. These include pump stations, tanks, meter vaults, air release valves, and blowoffs. Regardless of where measurements are taken, it is necessary to ensure that the location is not affected by local conditions. For example, highly variable flow conditions may result in severe pressure fluctuations.

Readings taken at a customer meter location where usage is sporadic can result in pressure fluctuations due to head loss or turbulence in the service line that result when flows change. If other options are available, pressure should not be measured at meters. However, in the absence of other alternatives, the user must make sure that there are no customer PRVs or check valves on the service line upstream of the gauge. Gathering pressures at such a location will almost certainly introduce errors in the results. If pressures are measured at such locations, it is important to make sure that the customer is not using water.

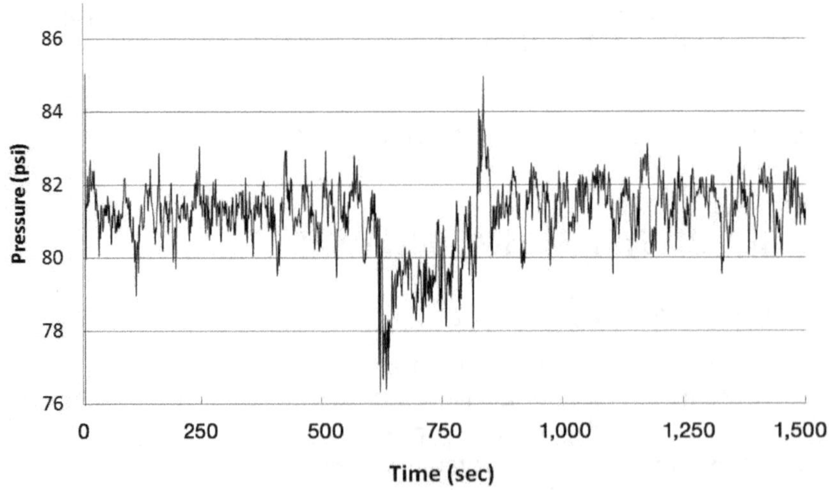

Figure 3-6 Chart of pressure logger system pressures

3.3.2. Flow Measurements

Flow measurements are an integral part of numerous water distribution system tests that can be performed, including pipe roughness coefficient tests, diurnal demand measurements, pump head-flow relationship tests, and fire flow capacity tests. Flow-measuring equipment in water supply includes devices such as hydrant Pitot gauges, Pitot tubes, strap-on meters (those that do not require a tap into the flow stream of the pipe), and master meters (such as Venturi meters).

3.3.2.1. Hydrant Pitot Gauges. Hydrant Pitot gauges (Figure 3-7) measure the discharge indirectly by measuring the velocity head with a pressure gauge. Figure 3-8 shows a hand-held Pitot gauge in use. The gauge measures pressure at the hydrant discharge point where the water is no longer under pressure. The flow equation for a Pitot gauge is

$$Q = 29.83\, p^{0.5} \times d^2\, C_H \tag{3-1}$$

Where:

Q = rate of flow, in gallons per minute
p = gauge pressure of the flow stream, in pounds per square inch
d = diameter of hydrant discharge port, in inches
C_H = hydrant discharge coefficient
Conversion: $Q_{,gpm} \times (6.3 \times 10^{-5}) = Q_{,m^3/s}$

The magnitude of a hydrant discharge coefficient, C_H, depends on the physical transition from the hydrant barrel to the discharge port, which is typically 0.9 for round and smooth, 0.8 for sharp and square, and 0.7 for ports that protrude into the hydrant barrel, as shown in Figure 3-9. See AWWA Manual M17 (2016a) for a more detailed discussion on field procedures for hydrant flow testing.

Hydrant Pitot gauges should be calibrated using a dead-weight tester. The gauge resolution must be within 0.5 psi (approximately 3.45 kPa) to accurately measure pressure heads below 10 psi (approximately 70 kPa). Flowing hydrant pressure is used to measure and calculate the flow rate as previously described. However, a flowing hydrant cannot be used for a system pressure measurement because of friction loss in the hydrant valve and barrel. Therefore, pressure measurements should be taken from the hydrant upstream from the flowing hydrant; this is described in more detail in the following sections.

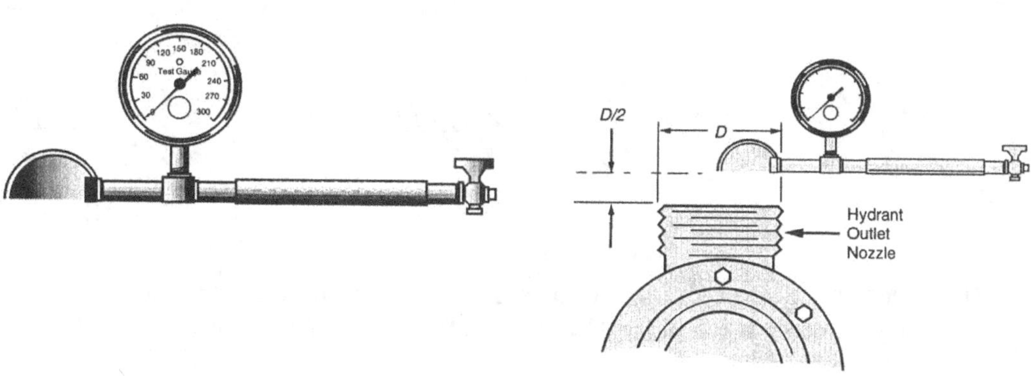

Courtesy of Tom Walski, Bentley Systems.

Figure 3-7 Hand-held Pitot gauge

Figure 3-8 Hand-held Pitot gauge in use

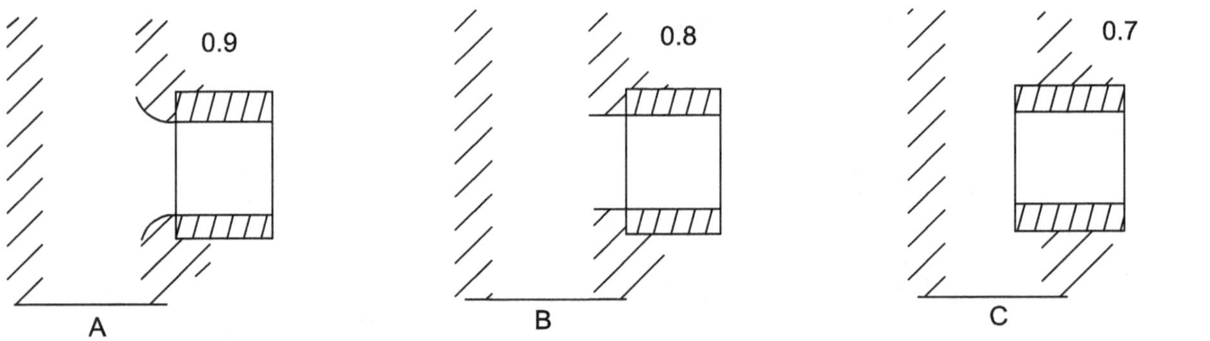

Figure 3-9 Three general types of hydrant outlets: (A) 0.9 for round and smooth, (B) 0.8 for sharp and square, and (C) 0.7 for ports that protrude into the hydrant barrel

Other Pitot gauges are designed to work with diffusers and/or dechlorinators. Diffusers have been developed to minimize erosion, remove chlorine from water discharged to the environment, and reduce other damage that can result from flow discharging directly from a hydrant nozzle. If diffusers are used (Figures 3-10 and 3-11), it is recommended that care be taken when using the discharge measurement data because the Pitot could be measuring discharge flow that is still under pressure. This may result in inaccurate discharge measurements if traditional discharge coefficients are directly applied to a diffuser (Walski and Lutes 1990). Engineers are encouraged to have the device manufacturer provide the necessary information such that discharge measurements can be adjusted accordingly. Diffuser manufacturers offer gauges designed specifically for use with their diffuser equipment and that also convert pressure to flow on the face of the gauge.

Courtesy of D. Lane, City of Bellevue Utilities.

Figure 3-10 Dechlorinating diffuser with manufacturer-supplied gauge

Courtesy of Jill Zimmerman, Albemarle County Service Authority.

Figure 3-11 Diffuser with pressure logger

3.3.2.2. Pitot Tubes. Pitot tubes measure velocity heads within a pipe. Velocity heads are typically small for nominal flows in water mains. For a flow velocity of 0.5 ft/s (0.15 m/s), the equivalent velocity head is only a 0.05-in. (0.0013-m) column of water, which is difficult to measure. To accurately test head loss, generally higher-than-normal flow velocity should be induced. Theoretical velocities are adjusted with an instrument coefficient obtained from laboratory calibrations. For meter and pump tests, a variety of flows should be planned, and the flow measurement device chosen appropriately for the task.

Flow rate in a pressurized pipe is computed by multiplying the pipe's cross-sectional area by the average velocity of the flowing water. Inside diameter is used to compute the cross-sectional area, the value of which is adjusted for obstruction of the Pitot tube as well as any corrosion or tuberculation inside the pipe. The average velocity may be determined from a velocity profile.

Figure 3-12 illustrates a method for determining velocity profiles in which a pipe is divided into five rings of equal area. Average velocity is the mathematical average of 10 measurements, 5 on each side of the pipe center. The ratio of the average velocity to the center velocity is referred to as the *velocity factor* (VF), which is unique for each gauging point. Figure 3-13 shows typical profiles at similar flow rates for two gauging points.

Insertion points for Pitot tubes are typically at air release valves and 1-in. (25.4-mm) corporations. However, larger corporations can be used with proper fittings depending on the testing equipment available. The best location to measure stable velocity profiles is along straight pipe upstream of the gauging point for a distance of at least 10, preferably 20, pipe diameters.

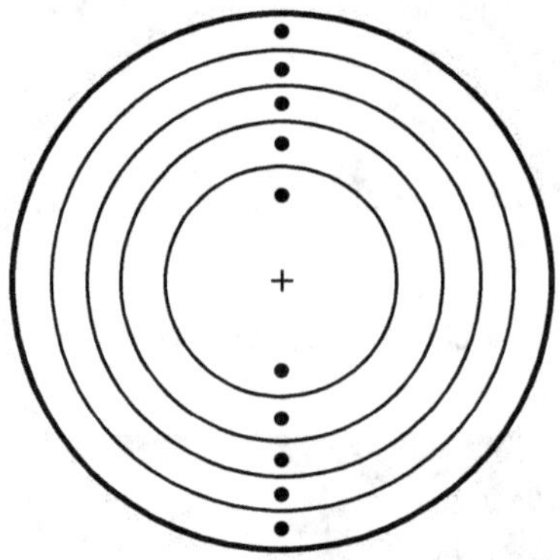

Figure 3-12 Traverse positions within a pipe with varying velocities

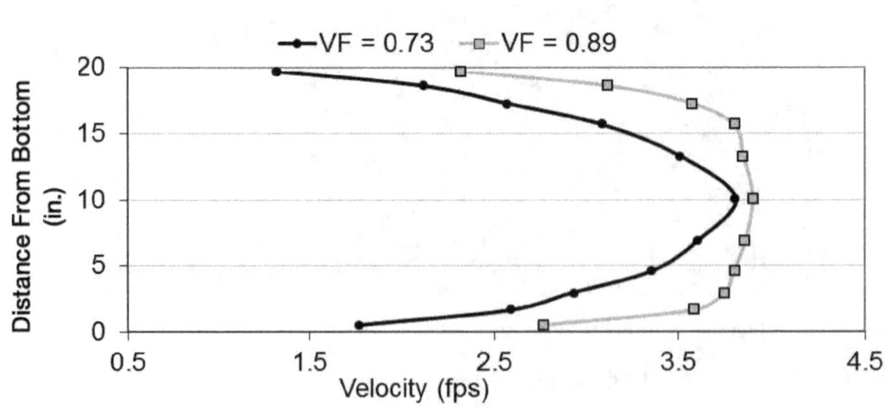

Figure 3-13 Typical velocity profiles at two gauging points

Abbreviation: VF, velocity factor.

3.3.2.3. Strap-on Meters. Strap-on, or clamp-on, portable meters are becoming more popular as technology advances, allowing for more accurate readings of flow within potable water systems (Figure 3-14). These meters do not require insertion of a device into the pipe for measurement; flow is measured by simply strapping the meter onto the outside of the pipe. Using ultrasonic or Doppler technology, the meter detects the velocity of fluid within the pipe and determines the amount of flow passing by the meter. When using these meters, their limitations must be considered, such as having the necessary length of pipe upstream and downstream to obtain a uniform flow for accurate readings. Often, the necessary straight length of pipe is unavailable in exposed piping (e.g., pump stations) such that excavation and installation on a buried pipe may be necessary to produce accurate data. In some instances, another type of meter should be used. The user must also know the type, condition, and thickness of pipe, which are critical to the meter's accuracy, and should carefully follow the manufacturer's instructions. A waterproof meter should be used if the application involves excavated below-grade pipe, due to the risk of submergence in the excavation. Very low flows can also impact meter accuracy. Generally, the higher the velocity, the higher the meter accuracy.

3.3.2.4. Master Meters. Master meters are flowmeters that are usually installed permanently in a pipeline to measure relatively large flows of water supplied or transferred. Master meters are typically present at water treatment plant discharges, larger transfer pump stations, purchased supply points, major wholesale customers, emergency supply locations, and district metering points. District metering is typically used to measure flow within a smaller area of the system and often to identify water loss within the system. AWWA Manual M36 (2016) includes more detailed discussion concerning district metering. Meters typically used for master metering include propeller/turbine meters, Venturi meters, and magnetic meters, all of which are discussed further in the following sections. More details about these and other types of meters can be found in AWWA Manuals M6 (2012) and M33 (2006).

3.3.2.4.1. *Propeller/Turbine Meters.* Propeller meters have been used as master meters in the past but are less common today (Figure 3-15). They consist of a propeller mounted axially in the line of flow within the pipe. As water passes through the meter, the flow induces a rotation of the propeller that is calibrated to measure the flow rate.

Turbine meters are typically not used for very large flows at water treatment plants but are often used for metering at pump stations and for large users and wholesale customers. Turbine meters translate the mechanical motion of a rotating turbine in water flow around an axis into a volumetric rate of flow. The turbine wheel is set in the pipeline. The flowing water pushes the turbine blades, imparting a force to the blade surface and setting the rotor in motion. When a steady rotational speed has been reached, the speed is proportional to fluid velocity.

3.3.2.4.2. *Venturi Meters.* The Venturi meter is widely used as a master meter (Figure 3-16). It consists of a Venturi tube (also called a converging–diverging nozzle) inserted directly into a pipeline. Because the diameter changes are gradual, there is very little friction loss. Static pressure measurements are taken at the throat and upstream of the diameter change. These measurements are traditionally made using a manometer. A typical Venturi tube with manometer is shown in Figure 3-17.

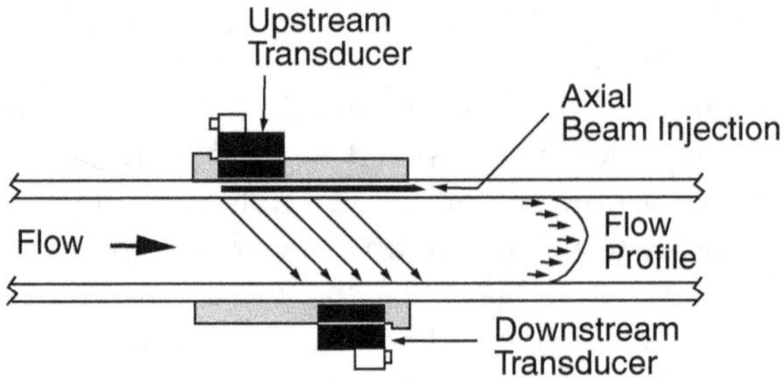

Figure 3-14 Schematic of a strap-on flowmeter

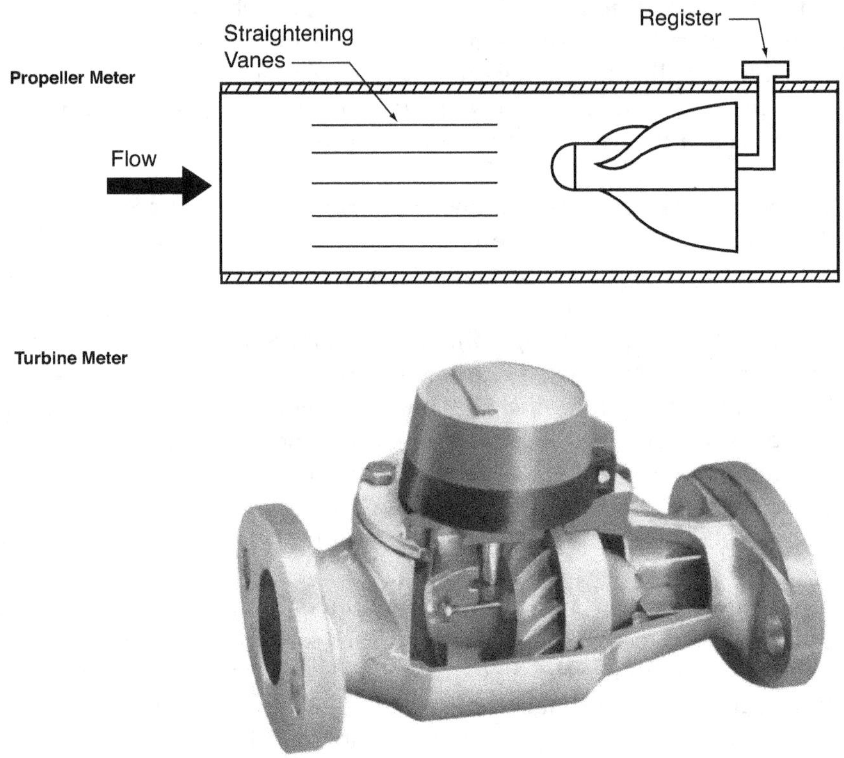

Figure 3-15 Schematic of propeller flowmeter and picture of turbine flowmeter

Courtesy of Jerry Higgins, Blacksburg Christiansburg VPI Water Authority.

Figure 3-16 Venturi tube in service

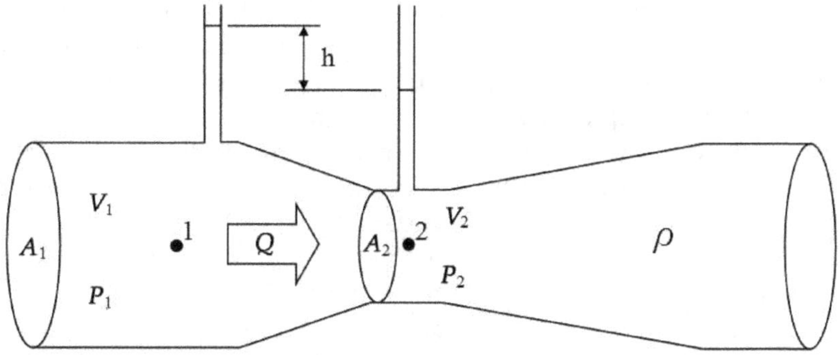

Figure 3-17 Typical Venturi tube with manometer

The calculation of flow using a Venturi meter assumes a horizontal orientation and frictionless, incompressible, and turbulent flow. Bernoulli's equation is applied for points 1 and 2.

Because

$$Q = V_1A_1 = V_2A_2$$
$$P_1 - P_2 = (\rho/2) \times (V_2^2 - V_1^2)$$

Then:

$$Q = A_1\sqrt{\frac{2(P_1 - P_2)}{\rho\left[\left(\frac{A_1}{A_2}\right)^2 - 1\right]}} = A_2\sqrt{\frac{2(P_1 - P_2)}{\rho\left[1 - \left(\frac{A_2}{A_1}\right)^2\right]}} \quad (3\text{-}2)$$

Where:

Q = flow
A = area
P = pressure
ρ = density of the fluid
V_1 = (slower) fluid velocity where the pipe is wider (A_1 area)
V_2 = (faster) fluid velocity where the pipe is narrower (A_2 area) as seen in the Figure 3-18
$(P_1 - P_2)$ = differential head

It should be noted that the equation shown assumes no friction loss through the meter. Typically, a *Cv* factor is applied to account for friction losses.

3.3.2.4.3. *Magnetic Meters.* Magnetic meters are often used as master meters and consist of a coil of wire around the pipe through which a current is induced as ions in the water pass through the pipe (Figure 3-18). These meters are calibrated so that the induced electrical current correlates to the flow rate in the pipeline. Magnetic meters are usually installed permanently, although temporary installations are also possible, and some magnetic meters are available as strap-on meters.

3.3.2.5. Meter Calibration Test. A meter calibration test compares master meter registration to flow rates measured by another instrument such as a test meter or an inserted probe. Meter calibration tests are used to verify production records, conduct pump tests, and calibrate computer models.

Meter calibration tests compare meter registration to the measured flow in a pipe, rather than simply checking electronic calibration of a transducer. Many meters include a primary device such as a Venturi tube and a secondary element such as a transducer to convert physical output to an electronic signal. A malfunctioning primary device causes meter error even if the secondary element is calibrated correctly. For example, a Venturi meter may provide inaccurate readings because of leakage between the Venturi tube and transducer, even though the transducer is functioning correctly.

Meter calibration tests are conducted for a range of flow rates. The range should cover what is expected to flow through the meter, including high and low flows. At high flow rates, a meter may provide readings that are sufficiently accurate; however, the same meter may provide inaccurate readings at lower flows that are more typical of normal operating conditions. Conditions in a shop test may not duplicate influences of upstream piping, turbulence, or other field conditions.

3.3.3. Supervisory Control and Data Acquisition

Operational SCADA information should also be collected during testing in order to gain an understanding of the system's hydraulic conditions during field testing. The type and availability of the data should be understood before field testing commences in order to best plan the test locations. If available, information on the following should be collected: flows, pressures, tank levels, and pump and valve status. Such information will be useful during model calibration and existing system analysis.

Courtesy of Robbie Cornett, Washington County Service Authority.

Figure 3-18 Magnetic meter

► COLLECT AVAILABLE SCADA DATA

- Boundary flow rates
- Pump status (on/off), flow rates, suction/discharge pressure
- Tank levels
- System pressures
- Valve status and settings
- Automatic meter readings (AMR)/automatic meter interface (AMI)

Abbreviation: SCADA, supervisory control and data acquisition.

SCADA data-logging frequency (also referred to as the *polling interval*) should be considered and adjusted if appropriate. If measurements are not taken frequently enough, the SCADA system may "miss" the flow test and fail to provide useful data. Conversely, a SCADA sampling period that is too frequent may produce an unreasonable amount of data that could create problems such as overwriting data on the measuring device or limiting hardware memory or storage capacity.

In recent years many utilities have installed pressure monitoring stations within their distribution systems, including upstream and downstream of major PRV stations. Data collected at these stations may be transmitted to SCADA and made available for use during field testing. Very few utilities collect flow data at PRV stations, but systems have been developed to calculate flow based on upstream and downstream pressures and the percentage of PRV opening. PRV manufacturers provide curves to calculate PRV flow based on these parameters.

3.4. WATER DISTRIBUTION SYSTEM TESTING

3.4.1. Fire Flow Tests

Fire flow tests can be used to calibrate friction factors in pipes and to verify settings and performance of control valves and pumps. They are generally used for model calibration or localized system investigation such as a suspected closed valve. These tests consist of simultaneously measuring flow and pressure at selected locations.

First, static pressure measurements on the test hydrant(s) are measured under normal system flow conditions. If possible, it is recommended that pressure be recorded at multiple hydrants or other locations (such as blowoffs) to provide more complete calibration data for looped flow fed from multiple directions. Next, the flow is measured on a nearby hydrant (flow hydrant), while dynamic pressure (residual pressure) measurements are measured on the test hydrant under the condition of steady flow (Figure 3-19). In larger networks or where larger mains serve the hydrant, two or more hydrants are opened to provide the necessary flows and resulting head losses. The flow rate is calculated from hydrant Pitot gauge readings and the hydrant discharge port size. A flowing hydrant (flow hydrant) should never be used for a residual pressure measurement due to the friction loss in the hydrant valve and barrel. Therefore, pressure measurements should only be taken from separate pressure monitoring locations (test hydrants) near the flow hydrant.

To achieve sufficiently accurate test results, there should be a minimum 10-psi (approximately 70-kPa) pressure drop (static pressure minus residual pressure). Ideally, the pressure gauge(s) or pressure logger(s) should have a minimum 0.5-psi (3.5-kPa) resolution to provide good accuracy. Gauges should be calibrated with a dead-weight tester, and pressure loggers should be calibrated before every set of tests.

In addition to helping model calibration by comparing measured flows and pressures from the fire flow test to modeled flow and pressures, fire flow tests can also be used to determine the available fire flow at a certain pressure. Derived from the Hazen–Williams formula, the results are often expressed as flow rates available at 20 psi (approximately 140 kPa):

$$Q_A = Q_M \times \{(S - 20)/(S - R)\}^{0.54} \tag{3-3}$$

Where:

Q_A = the rate of flow available at 20 psi
Q_M = the rate of flow measured during the test
S = the static pressure, in psi
R = the residual pressure, in psi
gpm × (6.3 × 10^{-5}) = m^3/s

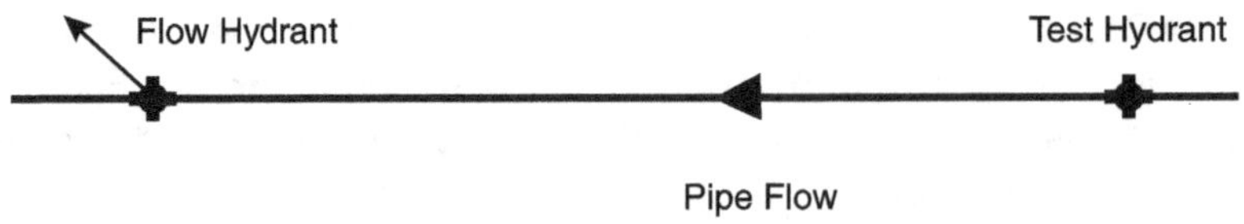

Figure 3-19 Simplified fire flow test configuration

The minimum system pressure during fire flow testing is 20 psi (approximately 140 kPa), as recommended by the National Fire Protection Association and by many water system regulatory agencies. A minimum pressure of 20 psi (140 kPa) prevents backflow or groundwater contamination and provides head for overcoming friction losses in the hydrant branch, hydrant, and suction hoses to pumper trucks during a fire. Flow during testing may need to be restricted to avoid pressures below 20 psi (approximately 140 kPa) at all locations, including high points and low-pressure spots (elsewhere in the system) as applicable, in addition to the test location.

Models can also be used to calculate available fire flows. Comparing the calculated available fire flow at 20 psi (140 kPa) from field tests can also help calibrate the model. However, comparing model results to measured flow and pressure is preferable to using the calculated available flow at 20 psi (140 kPa) because of possible complications from operational changes temporarily made during a fire, such as check valves opening, flow reversals, or pump changes. The model would need to reflect these temporary changes when simulating available fire flows if they are used for calibration purposes.

For fire flow tests to be useful for calibration, critical boundary conditions, pump status, and reservoir levels need to be known during the time of the test so that the model can be run under the same conditions for comparison. Critical boundary conditions include PRV settings. PRVs are sometimes used on pressure zone boundaries, instead of closed valves, so that low pressures will trigger the valves to open under fire flow conditions (low pressures). Under normal conditions, the PRVs are closed to separate pressure zones. The date and time should also be recorded as system demand conditions change over the course of a day and seasonally.

Fire flow tests in large systems have only limited value in calibrating models. Because most of the head loss occurs in the area near the flowing hydrant, measurements from fire flow tests allow only limited calibration in the region of the hydrant, and this does not provide a general indication of model accuracy.

Another enhanced approach to fire flow testing, at least with respect to time and labor efficiency, is to use continuous-recording pressure loggers while simultaneously releasing water from multiple hydrants (sometimes called *system stress testing*). When compared to normal protocols for conducting flow tests, this procedure can be accomplished by smaller crews and completed faster, and can result in more data for use in model calibration (Grayman et al. 2006). However, care must be taken when placing pressure loggers and selecting hydrants to flow so that an area can be properly calibrated in the model without creating a surplus of data that may provide little value in achieving model calibration.

3.4.2. Pipe Roughness Testing

Roughness testing can be performed to determine the C-factor for the Hazen–Williams equation or the roughness factor for the Darcy–Weisbach equation. Discussion here is limited to the Hazen–Williams equation. C-factors are friction coefficients used in the Hazen–Williams formula, which is an approximation of pipe roughness. Hydraulic models use the Hazen–Williams formula or other head loss equations, relating flow to head loss in each pipe element within the model. Like Moody diagrams and other equations, the Hazen–Williams formula, although empirical, provides sufficient accuracy within the normal range of velocities and temperatures in a water distribution system. As mentioned in chapter 2, if C-factor testing is done at very high velocities (velocities >10 ft/s [3 m/s]), a significant error can result when the resulting C-factors are used to predict head loss at low velocities (Walski et al. 2007). Therefore, velocities should be kept under this threshold during testing.

Measuring C-factors for every pipe in the distribution system is not practical; therefore, assumptions are made based on a sample of C-factor measurements. The sampling should include all combinations of pipe sizes and materials, as well as old and new pipes. Larger pipes tend to have higher coefficients (less roughness). Some utilities measure C-factors annually in various areas of the distribution system and develop a database of C-factors based on pipeline diameter, vintage, and material type. Regression analyses on this database provide a good prediction of C-factors in the distribution system during model development.

Careful consideration should be given to pipe installed prior to 1950. Unlined cast-iron pipes should be included in the sample, as C-factors vary widely from less than 25 to more than 100, depending on pipe age and water quality. If it is unknown whether pipes are lined or unlined, maps from the 1940s can be used to identify unlined pipes because the use of unlined cast-iron pipes virtually ended in the 1950s. It is also important to measure C-factors close to the treatment plant where treatment processes may have coated the insides of pipes.

Usually, the C-factors used in models account for losses in bends, valves, and other fittings. Accounting for losses in bends, valves, and other fittings when assigning C-factors simplifies model building and maintenance. Losses at these fittings may be small at normal velocities in water distribution systems but result in lower C-factors when included with friction losses.

When conducting field C-factor tests, nominal diameter, flow, head loss, and length of test section are determined and then the Hazen–Williams formula for the C-factor coefficient is solved. The formula for computing the C-factor of a pressurized circular pipe is

$$C = 3.551 \times Q \times D^{-2.63} \times L^{0.54} \times H^{-0.54} \tag{3-4}$$

Where:

C = pipe roughness (C-factor) coefficient
Q = rate of flow in test section, in gpm
D = diameter of pipe, in in.
L = length of pipe over which head loss is measured, in ft
H = head loss, in ft

Field results are verified by measuring head losses at two flow rates and comparing resulting C-factors. C-factors should agree within 10 percent. Another check consists of running two test trials on the same pipe but using different equipment for each trial.

Usually, flow must be artificially induced to produce velocities and head losses high enough to achieve accurate measurements. In pipes 12 in. (300 mm) and smaller, high flow rates are induced by opening one or more hydrants. In larger pipes, it is necessary to open multiple hydrants or use other methods such as opening blowoffs, throttling or closing off parallel mains or other pipes, operating pumps, or filling tanks to increase pipe flows.

Pipe diameters used in C-factor calculations are the same diameters used in the model, which are typically nominal pipe diameters. The length of pipe used to calculate a C-factor is the distance along the pipe between the pressure measurement points. Tests that include more than one pipe diameter between pressure sensors should be avoided. It is also preferable to select a location with minimal demand between the test points to ensure consistent flow through the pipe section.

Head loss measurements are equivalent to the drop in the HGL between the inlet and outlet of a test section. HGLs are taken at hydrants, air valves, corporations, or other taps. Measurement techniques depend on field conditions and the amount of head loss measured.

3.4.2.1. Parallel Hose Method. The parallel hose method measures head loss using a hose or pipe connecting the ends of a test section to a differential pressure transducer or manometer, as shown in Figure 3-20. This method measures head loss directly without knowledge of elevation. Portable test equipment includes reels of ⅛-in. (3.2-mm) or ¼-in. (6.4-mm) flexible tubing or rubber hose.

The parallel hose method generally requires at least 2 ft (0.6 m) of head loss to achieve accurate test results, provided differential pressure measurements are accurate within ±0.1 ft (0.3 kPa). Pipe lengths and flow rates that produce measurable velocities and losses without exceeding the range of the transducer should be selected.

The parallel hose method is not suitable for testing some pipes, as the practical limit for the length of hose is approximately 3,000 ft (1,000 m). Some short test sections require extremely high flow rates to produce measurable head loss. Having to deal with such long hose lengths can make this method impractical.

Sources of error include leaks and air in the hose. If possible, a valve should be closed immediately downstream of the test section, and the section should be checked for zero head loss at zero flow.

3.4.2.2. Two-Gauge Method. The two-gauge method uses pressure gauges at the inlet and outlet of a test section. Pressure measurements along with known elevations define the HGL. Head loss is determined by subtracting the HGL at the outlet from the HGL at the inlet, as shown in Figure 3-21. An accurate test requires a loss of at least 10 ft (3.1 m) using pressure gauges and elevations accurate to within ±0.5 ft (0.2 psi; 1.4 kPa) of head. Test results are extrapolated when applied to a model; a minimal observed head loss can result in significant extrapolation error. This level of accuracy requires calibrating the gauge with a dead-weight tester and accounting for the heights of instruments from benchmarks.

An advantage of the two-gauge method is that test lengths are increased to achieve additional head loss. However, isolating long test sections requires closing many valves. Another disadvantage is the need to know the elevations at which the pressure gauge measurements were taken. If elevation data are not readily available, it is possible to close a valve downstream of the test section and record pressures at zero flow. The difference in static pressure is converted to feet (meters) of water and used as the elevation of the higher gauge, assuming the elevation of the lower gauge is zero or datum.

3.4.2.3. Other Methods. Other methods of measuring head loss or pressure can be used under certain field conditions. For instance, reservoir levels define HGLs in some cases. The height to which water rises in clear hoses connected to pipe taps indicates the HGL directly; however, this method is only practical when pressures are relatively low. Parallel pipes with no flow are used to transmit the HGL from one end of a test section to the other. Some methods of calculation use pressure measurements at several flow rates to eliminate the need to know elevations.

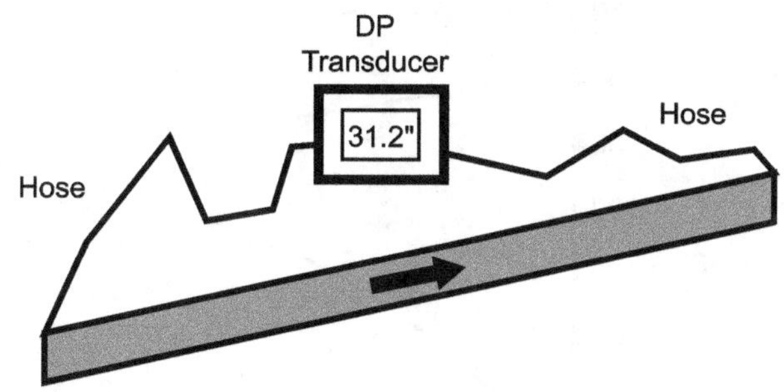

NOTE: in. × 25.4 = mm

Figure 3-20 Parallel hose method for measuring head loss

Abbreviation: DP, differential pressure.

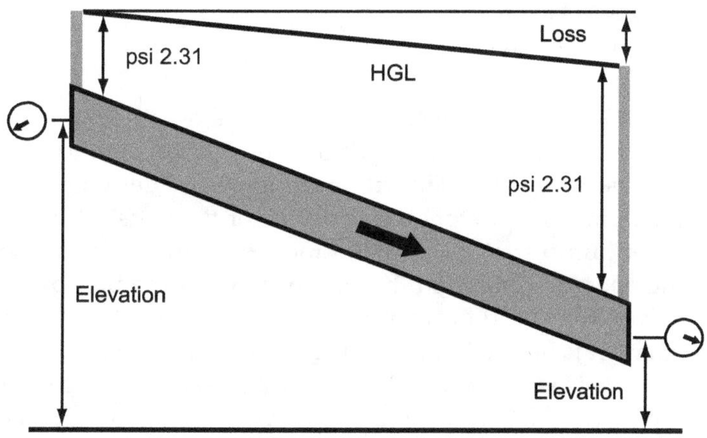

NOTE: psi × 6.895 = kPa

Figure 3-21 Two-gauge method for measuring head loss

Abbreviation: HGL, hydraulic grade line.

3.4.3. Pump Tests

Flow and head data collected from pump tests are entered into the model to simulate hydraulic performance and to calculate energy costs. Pump tests are used to determine the actual pump curve of a pump in the field. Assuming that a pump's design curve, or shop-tested pump curve from the manufacturer, accurately represents its field performance may introduce significant error into the model. A pump's performance in the field is sometimes quite different from design parameters. Differences between field performance and design parameters can be caused by equipment deterioration (e.g., worn impellers), undocumented equipment changes, or other field conditions.

As shown in Figure 3-22, pump tests can reveal how total dynamic head (TDH) and efficiency truly vary with flow. Most models use pump curves that are based on at least three operating points, each associated with a unique flow rate, TDH, and efficiency. Tests usually include normal conditions, design conditions, induced flow, throttled flow, and shutoff head.

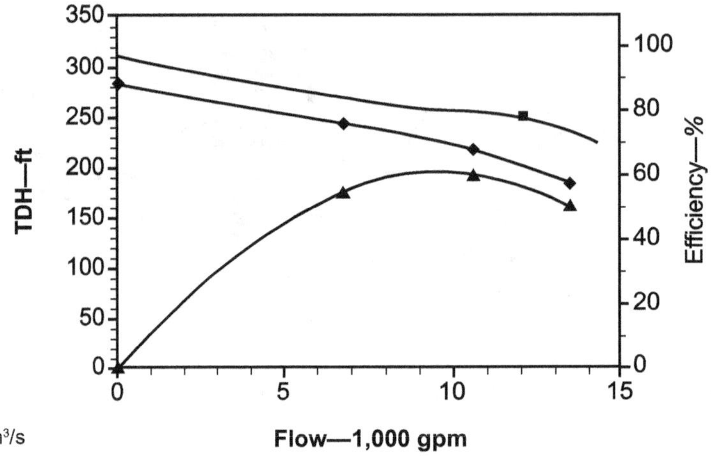

NOTE: ft × 2.989 = kPa
gpm × (6.3 × 10^{-5}) = m^3/s

Figure 3-22 Pump test results

Abbreviations: TDH, total dynamic head.

During pump tests, master meters or temporary flowmeters are often used to measure flow. The throttling of flow and shutoff conditions are often achieved by using the downstream valve. Throttling of flow provides data to the left of the operating point on the pump curve. Opening hydrants or blowoffs, filling a tank, or adjusting operations at other pump stations achieves induced flow. Inducing flow provides data to the right of the operating point on the pump curve. If the pumps are equipped with variable-speed drives, the speed of the pumps can be changed to determine points on the system head curve. It is best to hold speed constant while developing a curve.

TDH is defined as the energy added by a pump per pound of water flowing through the pump. TDH measured in foot-pounds per pound is dimensionally equivalent to feet of water. The equation for the TDH added by a pump is

$$\text{TDH} = (P + V + Z)_{OUT} - (P + V + Z)_{IN} \tag{3-5}$$

Where:

TDH = total dynamic head, in ft (× 2.989 = kPa)
P = pressure head, in ft
V = velocity head, in ft
Z = elevation of pressure sensor, in ft
$_{\text{OUT}}$ refers to the pump discharge
$_{\text{IN}}$ refers to the pump suction

Velocity head is computed by squaring velocity (in feet per second) and dividing by 64.4 ft/s^2 (two times the acceleration of gravity). Velocities are calculated by dividing measured flow rate by the pipe's cross-sectional area. Velocity heads can be significant, especially for induced flows.

Efficiency is calculated by dividing the pump's output hydraulic power by the input electrical power to the drive motor. The equation is

$$E = 0.01885 \times Q \times \text{TDH}/KW \tag{3-6}$$

Where:

E = efficiency, in percent
Q = rate of flow, in gpm
TDH = total dynamic head, in ft
KW = electrical power, in kW

Electrical power is measured directly using portable test equipment. Electrical power is calculated from measurements of voltage, amperage, and power factor. Power factors must be measured to verify accuracy. Electrical power may also be calculated from the rotational speed of power meter disks using the watt-hour per revolution factor (*KH*) and multipliers, which are stamped on most meters.

3.4.4. Hydraulic Gradient Tests

Hydraulic gradient tests reveal how HGLs vary along the length of a pipeline. Data points along the HGL are computed by adding elevation to each measurement of pressure head in feet. Hydraulic gradient tests provide data for calibrating computer models. Often, in combination with a computer model, the information can be used to locate closed or partially closed valves and under- or overloaded mains. The information can also help in determining where pipeline improvements may be warranted.

Hydraulic gradient tests consist of taking simultaneous flow and pressure measurements at intervals along a pipe path under steady-flow conditions. As shown in Figure 3-23, pressure measurements are taken along large trunk mains located between a plant and a tank. Also, pressure measurements are taken at the water treatment plant or supply point, at key facilities (major water users), on the inlet and outlet of pump stations, at tanks and major pipe intersections, and where pipe diameters transition. At key pipe sections, the flow rate is measured.

Pressure measurements in units of psi are converted to feet of water and added to pipe elevations to obtain HGLs. HGLs are then plotted against distance, as shown in Figure 3-24. The plots show the accumulation of head loss along pipe sections. The slope of the HGL becomes steeper in sections with significant hydraulic restrictions where head losses are greatest.

To be truly useful, hydraulic gradient tests should be conducted under conditions of high flow when head loss is significant or when high flow can be induced. In small systems, a simple way to induce high flow is by opening hydrants. In large systems, it may not be sufficient to open hydrants. In such cases, the test can be conducted when demand is high, when a tank is being filled, or by turning on pumps at appropriate locations.

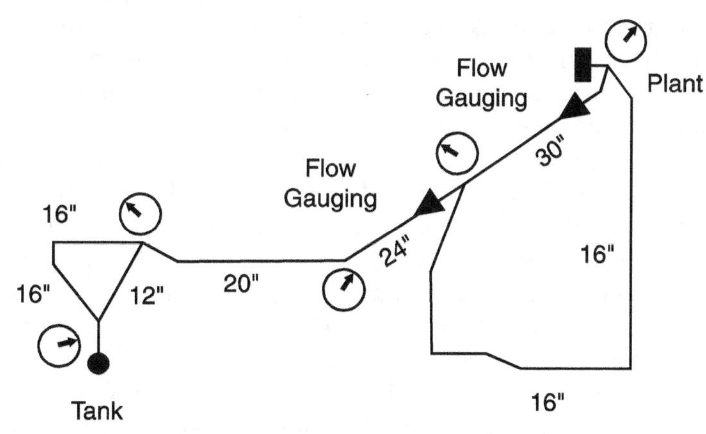

Note: in. × 25.4 = mm

Figure 3-23 Hydraulic gradient layout

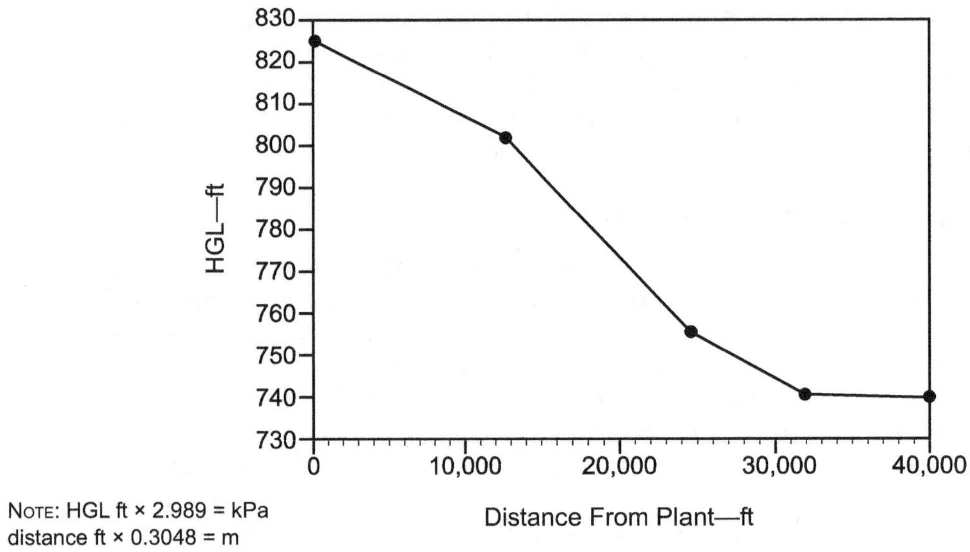

Figure 3-24 Hydraulic gradient test results

Abbreviation: HGL, hydraulic grade line.

3.4.5. Other Tests

Some models require that other types of tests be conducted. Calibration of PRVs requires accurate pressure measurements at known elevations, downstream (to confirm settings), and possibly also upstream to verify operation (if the PRV is closed under normal conditions). Pressure and flow measurements are useful in modeling other types of control valves. Pressure and tank level measurements are necessary for checking the accuracy of data from water plant charts and SCADA systems.

3.4.6. Testing Considerations

Any time a test that stresses the water system is performed, the overall effect on the water system must be taken into consideration and customers who will be affected should be notified. When such tests are performed, the system experiences decreases in water pressure. Due to higher velocities, sediment can be stirred up into the system, causing discoloration of water. This can be alarming to customers if they are not notified to expect these changes in the system during the testing process. When hydrants and valves are operated during the testing process, the utility should anticipate problems and be prepared to perform maintenance where needed. Especially where pressure data are measured, the data should be collected to capture several days or weeks, where possible, to ensure sufficient coverage of usable data across multiple locations. Where equipment and resources are limited, data should be gathered from the most critical locations or facilities that will best meet the objectives of the testing program.

The utility should also be prepared to provide traffic control, clean sediment and debris from roads and catch basins (to avoid impacts to streams), and repair damaged landscapes caused by the high volume of water from the hydrants. During the planning process, these issues need to be taken into consideration.

To avoid impacts to downstream property and limit the area affected by discharged water, flow diversion measures such as sandbags should be used. Utility vehicles or other improvised objects may also be used to effectively channel water away from traffic or property and into the drainage system (Figure 3-25).

Courtesy of D. Lane, City of Bellevue Utilities.

Figure 3-25 Modified surplus street sign, directing flow into a catch basin

The safety of water utility staff and the general public should be a primary concern during testing. Flow tests could involve risks of low visibility (particularly on streets during night testing), vehicle traffic, falling hazards (e.g., if catch basins or manhole covers are removed to receive flow), hypothermia, or other hazards, which should be anticipated and managed as appropriate. Testing during freezing weather should be avoided, due to the potential to create ice and hazardous downstream conditions.

Water conservation should also be considered during fire flow testing. Test duration should be minimized to avoid excessive water loss and to manage public perception. Testing during water usage curtailment periods should be avoided, if possible. Some water utilities implement a "catch and cart" procedure to recycle hydrant flow test water during periods of drought (Thomas 2009) to enhance public awareness of water conservation policies enacted by local authorities. The catch and cart procedure involves capturing water into a portable tank truck or trailer for transport to another location (e.g., park or public garden) for irrigation or other uses. Catch and cart does require additional setup costs as well as support from local authorities. However, the extra effort is likely to be worthwhile for utilities in locations subject to drought or water shortages.

3.5. DATA QUALITY

Data collected for model calibration should be more precise than data collected for routine operation and control. Calibration data are used to adjust pipe roughness and demands. However, the head loss in water distribution systems under normal operating conditions is usually very small, and errors in the measurement of head loss are of the same order of magnitude as the head loss itself (Walski 2000; Walski et al. 2007). Such data are helpful in checking boundary heads and ground elevations but are useless for adjusting pipe roughness because at such low flow rates, head losses cannot be measured accurately.

The key to data collection for calibration is to ensure that head loss is much greater than the error in measuring head loss. This is achieved by minimizing errors in head loss measurements or by increasing head loss. Head loss is maximized by taking measurements during peak demand periods or by conducting fire hydrant flow tests. If valves are known to be throttled, they should be opened prior to testing. Throttled valves can create high head losses and also introduce unknown minor losses.

Head measurements consist of both pressure and elevation measurements. Pressures are read with a gauge calibrated to 0.5 psi (3.5 kPa) or better. Elevation data should be at least as good (Walski et al. 2007). While reading elevations from a US Geological Survey topographic map with 20-ft (6-m) contour intervals may be acceptable for model building, the elevation of calibration data points should be significantly more accurate. Surveying the elevation of the gauge (rather than the ground elevation) is usually the most accurate way to determine elevation if maps with 2- or 3-ft (0.6- or 0.9-m) accuracy are not available. A GPS can be used to determine the elevation of pressure sensors but it will need to have a submeter or better vertical accuracy. The GPS should also be calibrated to nearby known benchmarks.

Adjustment of demands and pipe roughness in models alters the slope of the HGL. Therefore, it is essential to know the boundary heads when calibration data are collected. For example, knowing that the pressure at a point in the system is 40 psi (275.8 kPa) without knowing the water level in nearby tanks, the operation and discharge head of nearby pumps, or the setting and status of nearby pressure valves renders such data useless for making calibration model adjustments. Relying on the SCADA system for this information helps if changes are gradual, such as changes in tank water level, but is not useful when pumps can turn on and off and PRVs can open and close during a flow test because the polling interval of a SCADA system may miss a hydrant flow test. This requires stationing personnel at key points in the system during tests or installing data loggers to capture test results.

Readings through all flowmeters should be taken at a time that corresponds to pressure measurements. It is very helpful to know the flow rate through meters during a hydrant flow test if it is suspected that flow rates change during the test.

At times, historical flow and pressure measurement data are available and provided by the utility. The date, time, and account of procedures undertaken and conditions at the time are needed in order to use these data to build or calibrate the water distribution system model.

3.6. REFERENCES

American Water Works Association (AWWA), 2006. Manual M33, *Flowmeters in Water Supply*. Denver, CO: AWWA.

AWWA, 2012. Manual M6, *Water Meters—Selection, Installation, Testing, and Maintenance.* Denver, CO: AWWA.

AWWA, 2016a. Manual M17, *Installation, Field Testing, and Maintenance of Fire Hydrants.* Denver, CO: AWWA.

AWWA, 2016b. Manual M36, *Water Audits and Loss Control Programs.* Denver, CO: AWWA.

Grayman, W.M., Maslia, M.L., & Sautner, J.B., 2006. Calibrating Distribution System Models With Fire-Flow Tests. *Opflow,* 32:5.

Thomas, Roger, 2009. Water Conservation and Sustainable Use in Fire Suppression Systems. *Suppression, Detection and Signaling Research and Applications Symposium.*

Walski, T.M., 2000. Model Calibration Data—The Good, The Bad, The Useless. *Journal of the American Water Works Association,* 92(1):94–99.

Walski, T.M., and Lutes, T.L., 1990. Accuracy of Hydrant Flow Tests Using a Pitot Diffuser. *Journal of the American Water Works Association*, 90(7):58–61.

Walski, T.M., Chase, D.V., Savic, D.A., Grayman, W.M., Beckwith, S., & Koelle, E., 2007. *Advanced Water Distribution Modeling and Management.* Exton, PA: Bentley Systems.

This page intentionally blank.

Chapter 4

Hydraulic Calibration

4.1. INTRODUCTION

Models are only as good as the accuracy of their predictions of actual water system performance. The best way to ensure that a model provides results that closely match actual system conditions over a wide range of operating conditions is to perform model calibration. This chapter provides guidance on the steps required to calibrate a model, describes the various types of calibration, and discusses the typical issues and pitfalls that arise whenever calibrating a model.

Hydraulic models are essentially mathematical representations of water system hydraulics. To accurately replicate system performance, a model must include sufficiently accurate data that defines water usage in the system, the physical characteristics of the piping and facilities in the system, and the control settings of equipment and devices. A model may complete a simulation based on the information provided. However, if this information does not accurately reflect the system being analyzed, the result may have limited benefit and may actually lead to erroneous conclusions. Calibration can be tedious and time consuming. However, the potential benefits of a well-calibrated model for making accurate engineering decisions far outweigh the potential cost of mistakes caused by the use of incorrect model results. Once the effort-intensive process of calibration is complete, model maintenance becomes important to keeping the model relevant. Refer to chapter 9 for additional discussion on model maintenance.

4.2. WHAT IS CALIBRATION?

Calibration is the process of comparing the results of model simulations to actual field data and making corrections and adjustments to the model parameters in order to achieve the desired agreement between model-predicted values and field measurements. Typical comparison values include pressures, flow rates, and reservoir water levels. A well-calibrated model will stay in desired agreement with field measurements over a wide range of operating conditions.

Discrepancies between model results and actual system data can be due to many factors including incorrect model information, errors in reported field data, unknown constraints or controls in the water distribution system, and inaccurate assignment of model variables. Some of the parameters related to physical facilities that may require correction during calibration include system connectivity, pipeline diameters, node elevations, control-valve settings, check-valve direction, and pump characteristics. More variable or uncertain data that may require adjustment include model parameters such as minor losses, pipe roughness coefficients, and demand peaking factors.

It is important to remember that field data include a certain level of uncertainty as well and could be inaccurate. It should not be immediately assumed that there is something wrong with the model if it is not able to match the observed data. All inaccuracies, in the model and observed data, should be investigated carefully.

4.2.1. Calibration, Reconciliation, and Validation

Some people deem the term *model calibration* as a misnomer because it implies that the model will be adjusted to match the actual system. However, in many cases, there are conditions in the system that the model cannot simulate – for example, an unknown closed valve. For this reason, some suggest that the term *model reconciliation* is a better term than model calibration. In this manual, the generally accepted historical term *calibration* is used. It is acknowledged that the degree of calibration is highly dependent on the accuracy of available data and potential unknowns in the distribution system. The term *reconciliation* acknowledges that when model results do not match recorded data, the recorded field data may be incorrect or there may be unknown constraints in the distribution system, such as closed valves.

Understanding that a distribution system model has inherent limitations can prevent inappropriate use of the model. Adjusting input data to achieve a higher degree of calibration, or *best fit*, between modeled and measured results, without sufficient justification, will not overcome the inherent limitations in the data. The frequent use of a model, along with a high degree of familiarity with the system, is the best method for establishing model validity and will help the user to understand the calibrated model's limitations. Ongoing model verification is further discussed in chapter 9.

The term *validation* is often used to refer to the process of checking the results of a model following a model update in order to understand model accuracy. The term may also be used to refer to comparison of a model to a different set of field data than that for which the model was calibrated – for example, calibrating the model for average day conditions on August 12 and then validating the model against conditions on August 14. Some consider validation to be not as formal as the calibration process and that validation may involve a check of results to verify that pressures and flows are within expected and accepted ranges. The term *validation* is used in many ways by different modelers and is not as well defined a term as calibration.

4.2.2. Reconciling Discrepancies

Model results and measured values should be compiled into tables and graphs to facilitate comparison. The difference between measured and predicted values should be analyzed to see how closely the model reflects the actual system. If many of the differences are great, something may be wrong with the model setup or there may be errors in model connectivity. When only a few differences are significant, the modeler should concentrate on the areas where large differences exist. Reconciling differences at this time requires patience and ingenuity. Sometimes tweaking the model will bring simulated and measured results closer; sometimes talking with operators will reveal additional information that was not included in the original model. Calibration is an iterative process during which the modeler must carefully make assumptions; calibration is not about minimizing differences but instead about getting the model to correctly represent the system.

Unknown conditions in the distribution system often result in calibration inaccuracies. Any model assumptions made in the adjustment process should be the result of clear and logical findings in the system. For example, a large leak or a mistakenly closed valve may cause pressure discrepancies, which should be corrected in the field and replicated in the adjusted model.

A key point of model calibration is to fully understand why the model and field data do not agree. Once these differences are understood, adjustments are easier to identify and apply.

4.2.3. Types of Calibration

Model calibration can be categorized in several ways, as summarized here:

- Hydraulic versus water quality simulation,
- Steady-state versus extended-period simulation (EPS), and
- Macrocalibration versus microcalibration.

Currently, most models are calibrated only to hydraulic parameters in order to verify that pressures, flows, and tank water levels can be accurately simulated. However, calibration to water quality parameters is becoming more common. Two common water quality calibration parameters are chlorine residuals and disinfection by-products (DBPs). Water quality calibration requires additional data to be input into the model; for example, chlorine calibration requires input of initial chlorine concentrations, chlorine bulk decay curves, and wall decay coefficients. Further discussion of water quality calibration can be found in chapter 7.

Steady-state calibrations are hydraulic calibrations. They simulate distribution system hydraulic performance at a given moment in time, usually a high-demand condition such as peak hour demands or hydrant flow conditions. EPS calibrations can be conducted for either hydraulic or water quality. EPS analyses are a series of steady-state simulations linked together to approximate the behavior of a system over a period of time. EPS analyses represent changes in demands, operational controls for pumps and valves, and storage facility water levels. EPS analyses are required for water quality simulations and should be conducted using a model that has been hydraulically calibrated under EPS conditions.

Macrocalibration generally refers to calibrating an entire system or an entire pressure zone to recorded data captured by a SCADA system. Macrocalibration focuses on large discrepancies between observed and predicted values and helps identify gross model errors such as wrong boundary conditions, erroneous model connectivity, nodal elevations, or operational settings.

Microcalibration generally refers to analyses that verify that localized conditions in a specific area are being accurately simulated. Microcalibration can be equated to fine-tuning the model in a specific area. A typical microcalibration involves simulating field-recorded fire hydrant flows to verify the correctness of localized piping and valve closures. In smaller systems and in systems with minimal historical data capture, fire hydrant test results may be used as the basis for macrocalibration of the model.

An overall macrocalibration is usually performed when a model is first constructed or updated. However, if a model is used for a specific design study, it is often desirable to investigate the area in question and to conduct a microcalibration with a fire hydrant flow test as a check of model accuracies in the study area.

4.2.4. Calibration Goals

The goal of calibration is guided by the intended use of the model. A model that is sufficiently calibrated for master planning, for instance, may not be sufficiently calibrated for water quality analysis. The ultimate guideline for determining if a model is calibrated appropriately is that the model results should be capable of supporting the decisions to be made. Calibration should be stopped when the cost of additional calibration exceeds the benefits for supporting decision making or when the calibration goals are reached.

The required level and focus of calibration depend on the intended use of the model, model quality, and system variability. If the model's intended use is for pipe sizing, fire flow comparisons are critical. If it is used for water quality studies, tank mixing and chlorine decay are much more important.

There are no established standards for hydraulic calibration in the United States. Calibration criteria have, however, been published by several entities as described in the following sections. It is important to note that these sources present suggested guidelines, not regulated standards. It is also important to recognize that these goals were initially developed for steady-state hydraulic calibrations.

All models are approximations of actual represented systems. In a network model, both the mathematical equations used in the model and the specific model parameters are only numerical approximations. The goal in calibration is to reduce uncertainty in model parameters to a level at which the accuracy of the model is commensurate with the decisions to be made based on model results.

There cannot be any universal standards for model calibration because the required quality of calibration depends on the expected use of the model, the quality of field data, and resources available to perform the work.

4.2.4.1. ECAC Draft Calibration Guidelines. In 1999, the AWWA Engineering and Computer Applications Committee (now the Engineering Modeling Applications Committee) released the draft report "Calibration Guidelines for Water Distribution System Modeling," which presents a set of model calibration criteria. These criteria have been reproduced in other publications; it is important to recognize that these criteria were never approved by the committee because of their controversial nature. For that reason, they are not reintroduced here but are mentioned only because they may be encountered in other publications.

4.2.4.2. General Guidelines. The following are a commonly referenced set of guidelines:

- HGL predicted by the model should be within 5 to 10 ft (2.2 to 4.3 psi) of that recorded in the field.
- Water level fluctuations predicted by the model should be within 3 to 6 ft (10.9 to 1.8 m) of those recorded in the field.

The following applications have calibration guidelines based on these two criteria in some form:

- Master planning for smaller systems,
- Master planning for larger systems,
- Pipeline sizing,
- Fire flow analysis,
- Subdivision design,
- Distribution system rehabilitation study,

- Rural water system with no fire protection, and
- Emergency planning.

The last two applications have a suggested accuracy guideline of 10 to 20 ft (3 to 6 m) for HGL predicted by the model, or twice the allowable range for the others.

The higher-accuracy guideline for HGL of ±5 ft (1.5 m; pressure of 2.2 psi [15 kPa]) would typically be applied to models used for design and operations evaluations. The lower-accuracy guideline of ±10 ft (3 m; pressure of 4.3 psi [approximately 30 kPa]) would typically be applied to models used for long-range planning or for what is referred to in this manual as *master planning.*

4.2.4.3. United Kingdom Guidelines. Steady-state calibration criteria were established in 1989 in the United Kingdom by the Water Authorities Association and independent research company WRc (formerly World Research Centre). Steady-state performance criteria in the United Kingdom are summarized in the following list:

- Modeled flows agree to within
  - 5 percent of recorded flow when flows are greater than 10 percent of total demand (for transmission mains generally 16-in. [400-mm] diameter and larger with no service connections) and
  - 10 percent of recorded flow when flows are less than 10 percent of total demand (for distribution mains generally 12-in. [300-mm] diameter and smaller with service connections).
- Modeled field test pressures agree to within
  - The greater of 1.6 ft (0.5 m) or 5 percent of recorded head loss for 85 percent of test measurements,
  - The greater of 2.31 ft (0.7 m) or 7.5 percent of recorded head loss for 95 percent of test measurements, and
  - The greater of 6.2 ft (2 m) or 15 percent of recorded head loss for 100 percent of test measurements.

4.2.4.3.1. *Automated Calibration.* The process of adjusting model parameters so that the model reproduces results as measured in the field is usually an iterative, or trial-and-error, process that involves a significant amount of time. Automated calibration methods are being made available in commercial modeling packages, and their use may increase in the future. However, automated procedures may mask inaccuracies that might be uncovered with a thorough trial-and-error method. If automated procedures are used, care should be taken that such a situation does not occur. Automated calibration can work well once the source of the errors is understood. If it is known that errors in assigned pipe roughness are causing calibration inaccuracies, automated calibration procedures can greatly help refine the assigned roughness values. However, if the error is due to a closed valve, inaccurate pressure-reducing valve (PRV) setting, or pump curve discrepancy, automated calibration procedures cannot find these discrepancies and may produce significantly erroneous pipe roughness coefficients that perpetuate other errors. The results must be interpreted critically to avoid assignment of unreasonable values. Worse yet, adjusting the wrong parameter (e.g., adjusting roughness when the error is due to an incorrect tank level) can result in deterioration in model quality even though the results may look good; this is sometimes referred to as *calibration by compensating errors.*

The calibration accuracies of an iteratively calibrated model can be improved with the application of automated calibration procedures. Some modeling software has this built-in capability, and so it should be considered for use as part of a careful iterative calibration process. However, care should be taken to verify that system anomalies are adequately identified prior to using automated calibration procedures, and automatic calibration results should be checked for reasonableness.

4.2.4.4. Calibration Accuracy Reporting. Increasing regulatory focus on water quality in the distribution system has increased the value of a model that is able to run water age and source trace analyses. The Stage 2 Disinfectants and Disinfection Byproducts Rule (Stage 2 DBP Rule), by allowing modeling results to be used for the required initial distribution system evaluation, acknowledges that an EPS model is an effective tool for evaluating distribution system water quality. The absence of quantitative calibration criteria in the Stage 2 DBP Rule is significant because, by the absence of specific criteria, the guidance manual acknowledges the difficulties in defining an acceptable level of calibration for quantitative purposes.

Calibration accuracy data can be presented in several ways that can hinder direct comparison of models. All statistical analysis results should be documented sufficiently to identify the statistical methods and assumptions made and should include a discussion of deviations. Forced data such as using flow control valves or wells modeled as negative demands should not be included in the evaluation of calibration accuracies.

It is equally important that modeling engineers and utilities recognize the limitations of the available data. Calibration accuracies can be improved with additional field testing and the identification of anomalies. The costs for obtaining additional and more precise field data and the costs for additional efforts to improve model calibration accuracies must be considered and weighed against the benefit of improved calibration accuracies.

4.2.5. Data Sources and Errors

There are three general categories of data required to build and hydraulically calibrate a distribution system model. These are physical facilities data, demand data, and operational data, which are also described in chapter 2. Physical facilities data generally includes system maps and pipeline construction drawings; topographical (elevation) data; drawings of pumping, control valve, and storage facilities; and pump curve information. Demand information generally includes metered sales data, nonrevenue water data, and calculated demands based on recorded flows and storage facility water levels. Operational data generally includes information on pump on–off times and controls, storage facility fill rates as appropriate, and control valve settings. Supplemental data may be available that would include field tests to evaluate pipe roughness factors (physical data) or recorded demands for large uses (demand data).

The sources of data for model calibration vary widely from one utility to another. Some utilities have well-organized electronic data sources and sophisticated SCADA systems or other real-time monitoring equipment. Other utilities may have little information and/or less-reliable equipment. Utilities will find future calibration efforts easier by developing strategies to enhance their ability to procure accurate calibration data. Potential data sources are presented in the following sections along with a discussion of associated best practices and cautions.

4.2.5.1. System Maps and Asset Databases. It is important to have accurate maps that correctly identify pipeline sizes and connectivity. In the past, most system maps were in a simple drawing format with no underlying database. In some locations, especially in older systems, system mapping was unreliable and incomplete. Many utilities now incorporate system mapping within a geographic information system (GIS), which may include information such as pipeline diameter, installation year, material, and valve locations, to name just a few. Most commercially available modeling software is designed to make model construction from GIS mapping an easier process than it was in the past. It is important to recognize, however, that electronic information stored in GIS is often based on older mapping and may be unreliable and incomplete in certain areas.

4.2.5.2. Elevation Data. To obtain accurate pressures, appropriate elevations must be assigned to each node in the distribution system. Elevation data in electronic format are readily available and can be used within the modeling software to assign elevations to model nodes. However, the accuracies and limitations of available elevation data must be known. Elevations at pumping stations, control valves, storage facilities, and all locations where pressure is recorded for comparison should be based on more detailed information such as construction drawings or actual surveys.

It is important to understand the accuracy of the elevation data source. Topographic maps and digital elevation models based on those maps can have significant inaccuracies. A GPS can be used to obtain data, but it is necessary to understand its levels of accuracy because those values can be misleading. Elevation data based on LIDAR imaging are generally very accurate. LIDAR is a method of aerial surveying used to develop topographic maps.

During model calibration to observed pressure data, it is critical that the elevation of the model element that represents the observed pressure matches the elevation of the field pressure gauge. Alternatively, the comparison between model results and field observations should be done using HGL, which combines the elevation and pressure information into a single number. For example, pressures measured at a hydrant head will be about 1 psi (7 kPa) lower than the pressure at the ground (typical hydrant height of 2 ft to 3 ft (0.6 m to 0.9 m). When converted to water column pressure, this translates to about 1 psi (7 kPa), which is typically what is reported in the model. Similar care should be taken with pressure-regulating valve (PRV) settings. The elevation of control node in the model should match the elevation of the controlling pressure gauge in the system.

4.2.5.3. Supervisory Control and Data Acquisition. The preferred source of operational data is a SCADA system. The system can usually provide data on pump flows, pump discharge pressures, reservoir elevations, and, in some cases, even PRV flows. These data are very useful when setting up the model for the specific calibration scenario because the data for a snapshot in time are usually available. Data from SCADA systems can also aid in the development of a diurnal demand factor. This is done by conducting water balance analyses based on inflow into the system and changes in storage. It is important to recognize that SCADA systems can report data to higher precision than is justified. The accuracy of the sensor and any analog-to-digital conversion must be understood as well. When SCADA data are not available, operating information may be available from operators' hourly logs or on ink pen recorders. When using SCADA data, care should be taken to investigate whether the system adjusts the time during daylight saving time. Many SCADA systems use universal time and do not adjust the time when daylight saving time is applied.

4.2.5.4. Pressure Recording Devices. Pressure data may be available from chart recorders or may be electronically collected and stored through a SCADA system. For model calibration purposes, it is important to have pressure measurement devices that are accurately calibrated. Electronic pressure gauges can be out of calibration due to inaccuracies in the recording device and due to improper setup of the electronic signal that converts the mechanical pressure to an electronic signal that can be interpreted by the SCADA system. It is important to recognize that the SCADA system can be programmed to report pressure at any given elevation other than the actual pressure recorded. For example, the pressure measured in the field is relayed to the SCADA system, where a constant value can be added to the measured pressure before the pressure is written to a data record. This is sometimes done to record the approximate pressure at a different elevation than the recorded location.

The elevation of any pressure recording device must be known, and its level of accuracy should be commensurate with the accuracy of the pressure reading itself. For example, a pressure recorded to an accuracy of 1 psi (7 kPa; 2.3 ft [0.7 m]) combined with an elevation recorded to an accuracy of only 20 ft (6 m) results in hydraulic grade that is accurate to no greater than 20 ft (6 m). Model elements that correspond to the pressure gauges for model calibration should have the same elevation as the pressure gauge or a calibration comparison should be done using the HGL.

Temporary pressure recording devices have advanced from the old pressure ink chart recorders. Today's electronic pressure loggers (recorders) can be attached to hydrants or to other accessible points in the distribution system such as valve manholes or vaults and will collect data at any interval defined by the user. These devices usually sample pressure every few seconds and record maximum, minimum, and average pressures for the selected interval. These are useful for EPS calibration because they log pressure over time. They are also useful for steady-state calibration since they record static and residual pressure at additional monitoring locations without the need for additional staff. Due to costs, data loggers may need to be supplemented by less expensive pressure gauges, but they can provide labor savings if used repeatedly. Rather than comparing pressures, it is preferable to convert all measured pressures to HGL before making comparisons. See chapter 3 for additional information on system pressure testing.

4.2.5.5. Flow Recording Devices. Flow data may also be available from chart recorders or may be electronically collected and stored through a SCADA system. Many pumping stations incorporate flow recording devices. It should be noted, however, that flow recording devices may be out of calibration due to inaccuracies in the recording device and/or improper setup of electronic signaling interpreted by the SCADA system.

Flow recording devices are not as versatile as pressure recording devices for the temporary monitoring needed for calibration purposes. In many cases, flow rates can only be estimated using insertion instruments, which are not always easy to install. Ultrasonic flowmeters can be more accurate for large-diameter pipes, but installation requires a certain amount of space and the equipment can be expensive. The accuracy of ultrasonic flowmeters can be questionable for smaller-diameter pipes, especially when used to measure lower flows. They are more suited for wastewater flow measurements. See chapter 3 for additional information on flow testing and measurement.

4.2.5.6. Pump Curves. Levels of confidence in modeling results can be significantly impacted by the source of pump curve data. Depending on a facility's age, pump curves may or may not be available and can be difficult to obtain. The best pump curve information is created using recent data taken from certified results of pump tests conducted after pump installation or from pump tests for older facilities. Although older facilities may have certified curves that were accurate at the time of installation, they may not represent the pump's current performance. Operators sometimes trim or replace impellers, which will affect pump performance. In these cases, a pump test may be needed to evaluate current pump performance. In the absence of pump test data, manufacturers' standard curves are the best source of initial pump curve data.

Manufacturers' standard curves are often provided in lieu of curves from actual pump tests for modeling purposes. When no pump curve can be obtained, usually an operating point (design flow and design head) can be used. Most modeling software packages have routines to estimate pump curves based on input of design operating points.

However, the estimated curve generated by the model may not accurately represent actual pump characteristics.

SCADA data can also be used to develop pump curves using recorded flow, suction, and discharge pressures at a pump. When only one pump is operating, it is relatively easy to determine actual points on a pump curve and compare them with the original curve. This can readily identify whether the original curve is correct or if additional testing is required. See chapter 3 for additional information on pump testing procedures.

When adjusting pump curves during calibration, a distinction should be made between the pump head delivered by the pump and the head delivered by the facility, with the second head incorporating all head losses that occur in the pumping facility. Facility-delivered head can be and often is smaller than pump head, particularly if the manufacturer's curve is used to represent the pump. This issue is particularly important for wastewater pumping, where a large portion of the delivered head can be used to overcome the losses in the lift station.

4.2.5.7. Control Valves. Control valves maintain a setting that keeps operating parameters such as flow, upstream pressure, and downstream pressure constant for the duration of the setting. Some utilities receive water from a wholesale water provider through a turnout, which has settings for both the pressure at which the water is provided and the flow setting. Other types of control valves include PRVs and pressure-sustaining valves (PSVs) that allow water transfer from one pressure zone to another and storage facility fill valves that control the rate of flow into a reservoir that is being refilled after having been depleted. Throttling valves may be used to induce head losses within a system for various reasons, such as filling a ground storage reservoir, without dropping the pressure in the distribution system to a level that is too low. In this case, the operator may set the percent open on a throttling valve based on maintaining an acceptable upstream pressure in the system. It is important to gather as much information as possible on control-valve settings during calibration in order to accurately simulate actual performance.

Many control valves have smaller-diameter piping than the adjacent piping, and minor losses may need to be incorporated into the model for the model to accurately simulate actual conditions when the valve is fully open. Some PRV stations have no method for capturing and reporting actual pressures and flows. The actual set point of a PRV may not be accurately recorded and/or may have been adjusted in the field without documentation of the change. Control-valve settings should be verified in the field as part of model calibration whenever possible. It is particularly important to have the model control node elevation match the system control pressure gauge elevation. Alternatively, a model PRV should use the HGL setting instead of pressure setting (if the modeling software allows).

Where secondary or backup PRVs are installed to serve during high-demand periods or during a significant drop in pressure, it may be necessary to install a pressure logger upstream to verify the intended operation.

It should be noted that field adjustment and measurement of PRV set points can be imprecise or inconsistent. The process of setting a PRV or verifying settings often involves turning a nut while listening for movement in the control tubing and/or watching for stem rise, all of which involve some degree of uncertainty and interpretation. Results may vary in a range of up to 5 psi (34 kPa) or more simply depending on which operator performs the adjustment and the consistency of operator training. The precision implied by simulated PRV operation in a model may be misleading, and care should be taken to field verify downstream HGL.

It is important to know what type of control is provided by the control valves included in the model. A PRV will keep a constant pressure downstream of the valve as long as the upstream pressure is higher than the PRV setting and the downstream pressure is lower than the PRV setting. A PSV will prevent upstream pressure from falling below its setting, protecting upstream customers from low pressures. A relatively easy way to calibrate these devices within the model is to compare pressures and head both upstream and downstream of the valve. Common errors in modeling PRV stations include setting the wrong elevation for the valve and/or using an incorrect pressure setting. If flow is measured at or near the PRV, it can be logged and compared with model predictions. PRV manufacturers are now offering equipment that, though rarely used, will measure flow through these devices. This additional information can be very helpful during model calibration.

During model calibration, special attention should be paid to flow control valves (FCVs). Sometimes flow control devices maintain constant flow only for a narrow range of conditions. Once the system is put under stress, such as during fire hydrant flow testing, flow control devices can allow significantly larger flow. However, model FCVs always maintain the desired flow setting. In such situations, the FCV might be better represented in a model by a throttle control valve with a setting that will limit the flow to the desired setting during normal operation but allow larger flows during stress conditions.

4.2.5.8. Unmetered Flows. Unmetered flows are difficult to estimate. Unmetered flows are prevalent at PRV stations. Many utilities provide PRV stations that allow water from higher-pressure zones to be delivered to lower-pressure zones. If there are no meters at a PRV station, flows can be estimated using billing records, although adjustments and assumptions are required to convert billing records from monthly time frames to average daily flows. Inaccuracies in estimated unmetered flows should be considered when evaluating model calibration.

4.2.5.9. Metered Sales Data. The basis of allocation of demand to the model is generally on an average annual day, with spatial distribution determined from an analysis of meter sales data. Many analytical methods can be used to determine the spatial distribution of average day demand using metered sales data. GIS tools are increasingly used to accurately locate each meter in the system, associate sales from that meter to the assigned location, and then use this information to determine the allocated sales. Errors can arise from incorrectly located meters.

4.2.5.10. Nonrevenue Water. Nonrevenue water is the difference between billed authorized consumption and the water delivered to the system. Nonrevenue water generally includes lost water (leakage and unauthorized use), underregistration of meters, and water use for such things as firefighting, flushing of mains and sewers, and street cleaning. Nonrevenue water is usually estimated fairly accurately for the entire system but may be difficult to estimate for smaller areas or individual pressures zones. Assumptions are usually necessary to allocate nonrevenue water to the system. These assumptions may not accurately account for higher loss in older areas or for any unknown, except significant leaks in the system. If possible, nonrevenue water should be determined for discrete areas such as pressure zones, but this is often not feasible. Typical methods of allocating nonrevenue water in the system are to assign it uniformly to all nodes or to allocate it proportionally to all nodes based on demand. Other methods to allocate nonrevenue water may account for pipe age and material. The assignment of nonrevenue water can be considered an adjustment parameter during calibration if such adjustments can be justified. For more information on nonrevenue water, see AWWA Manual M36 (2016).

4.2.5.11. Demand Peaking Factors. Allocated annual average day sales and nonrevenue water are factored, or peaked, in the model to simulate demands that have occurred during the calibration period. The actual peaking of demands by user class and location is typically unknown. However, in some utilities, this information is becoming available with the increased adoption of advanced metering infrastructure (AMI) and service meters that log hourly demand data. Assumptions typically need to be made to peak the allocated demands that may impact the model's calibration accuracies. For large users such as industries, hospitals, and commercial institutions, water usage data can be collected to determine a typical water usage pattern.

Most utilities have very good water production data, along with facility operational reports. This data can be used to create historical demand summaries that can be used to set up models. Production data must be adjusted to take into account leaks and other nonrevenue water. Once summaries have been aggregated by year, month, and day, it is possible to create peaking factors, which are ratios between high-demand periods and the average daily demand. The most common peaking factors are maximum day to average day, which reflects seasonal demands, and peak hour demand to maximum day demand, which reflects diurnal variations. These values are very system specific and depend on the amount of industrial customer usage and the amount and timing of irrigation, among other factors.

4.2.5.12. Pipe Roughness Coefficients. The roughness coefficient (*C*-factor) used in the Hazen–Williams equation is not known throughout the entire system with total certainty because it is not feasible to examine and test every pipe. When the model is first constructed, an initial set of *C*-factors is established for all pipelines. The initial *C*-factors that are used may be arbitrary values based on pipe age, material, and diameter. Some utilities have historical *C*-factor measurements that can be used in regression analyses to determine the estimated *C*-factor for each pipe in the model based on age and material. Additional information on testing to determine *C*-factors is presented in chapter 3.

Regardless of how the initial *C*-factor was established, as part of the calibration process, the *C*-factor can be adjusted to improve the agreement between model predictions and actual measurements. This can be done manually for all pipes, for certain areas of the system, or automatically using software calibration tools.

4.2.5.13. Minor Losses. Minor losses are frequently neglected in distribution system models and instead are integrated with linear pipe losses. However, when a specific area is experiencing a significant head loss, as when a valve is known to be partially closed or certain known fittings are present that cause the system to exhibit a significant head loss, it is important to calibrate this minor loss by assigning an appropriate minor loss coefficient (*K* value) to the pipeline. This can be especially important at pump stations. When comparing suction and discharge pressures at a pump station, it is important to know where the pressure gauge is located. Ideally, the pressure gauge is located adjacent to the pump. There are many fittings and valves inside a pump station that create additional minor losses. It is important that these be taken into account during microcalibration.

4.2.5.14. Connectivity. Whether planning to conduct a steady-state or EPS calibration, it is important to check system connectivity, particularly when models are constructed using data imported from GIS. In many instances, GIS represents pipelines in a way that is not hydraulically correct. For example, nodes that are very close to one another but not connected may appear to be connected when viewing them in the GIS. In the model, pipes must connect to the same node in order to be hydraulically connected. Current modeling software includes tools that allow the modeler to perform connectivity checks, ensuring the model has the correct connectivity. See Chapter 2 for more information on connectivity.

4.2.5.15. Mass Balance. Mass balance is another important error check. The modeler should verify that total node demands for a specific run match the intended scenario.

When all pumps are off, the total demand should be equal to the outflow from storage facilities. Conducting mass balance checks on demands is a good way to perform quality assurance in the model.

4.3. STEADY-STATE CALIBRATION

A steady-state simulation represents a snapshot of the system under one set of conditions. A common approach has been to first perform a steady-state calibration followed by an EPS calibration. However, if an EPS model is developed, the model may be directly calibrated for EPS.

Steady-state calibrations, based on simulation of historical conditions using SCADA and other operational data, are typically conducted during the peak demand hour on a maximum demand day because the demands are highest, resulting in greater energy (head) loss throughout the water system. Another approach to steady-state calibration is to calibrate to a set of induced high flows generated from open and flowing hydrants. Both types of steady-state calibrations are discussed in this section.

Without significant head loss, the HGL can be close to the static grade line, which minimizes potential differences between predicted and measured values. To effectively assess the level of accuracy of model predictions, it is important to stress the system and verify that the model will correctly represent the changes in HGL. A stressed condition with high head losses produces more meaningful comparisons between field measurements and model predictions.

Artificial demands can be placed on the system to induce high head losses and calibrate under lower-demand conditions. Stressing smaller pipes is relatively easy to do using fire hydrants. Flow from hydrants causes high flow rates and measurable head loss. Opening hydrants also provides an opportunity to conduct field flow measurements. Stressing larger pipes requires more effort but can be done by filling tanks and operating large pumps.

During a steady-state calibration process, the most significant comparison values are HGL values (calculated from pressures and node elevations) at selected nodes. The nodes are preselected and measurement is conducted for the specified time period. It is better to select locations in various pressure zones and, preferably, farther away from reservoirs to allow for head loss accumulation. Ideally, multiple observations will be made within the same zone to be able to identify potential boundary condition issues. It is best to make comparisons in terms of hydraulic grade rather than pressure because hydraulic grade accounts for both pressure and elevation. Plus, it is relatively easy to spot anomalous data when hydraulic grade is used.

4.3.1. Peak Hour Calibration

Using flow into the system from production data and changes in storage, it is possible to verify hourly demand if sufficient operational data are available. Historical demand information also provides ratios between maximum day demand and peak hour demand. Based on this information, the model can be set up to simulate a peak hour scenario, which can be used for calibration, assuming that the field data for this condition are available.

Appropriate calibration hours are selected to verify system performance under various operating conditions. As an example, the peak hour is useful when evaluating the system for high head loss and when checking whether appropriate *C*-factors are used in the model. The lower demand period can be used during reservoir refill to look at inflow into reservoirs and to verify total dynamic head and pump station flows.

The model needs to match the state of the system, such as boundary conditions, at the time of calibration. Whatever is occurring in the water system at the time of calibration should be replicated in the model. If pumping stations are off, the model will have pumps set to off, and PRVs should be set at the same setting as in the field. Utilities usually have lists of facilities and their settings, including valves and PRV stations. The starting water levels in reservoirs should match the levels at the time of calibration. This will reduce the level of uncertainties in the model and help focus the calibration efforts.

4.3.2. Fire Flow Calibration

Fire flow calibration involves conducting fire hydrant flow tests. These tests can be important for calibration because they simulate high demand and create significant head loss, which provides a great opportunity to test how closely the model mimics the water system's performance. This type of test not only helps identify incorrect roughness but also verifies functionality and the set points of secondary/backup facilities (e.g., zone check valves, redundant PRVs, fire pumps) and indicates closed valves and/or incorrect connectivity. These tests can also help differentiate between errors caused by hydraulics versus those caused by demands (Walski 1983). Additional information on hydrant flow testing can be found in chapter 3.

Hydrant flow testing provides both static and residual pressures where the testing is conducted. Replication of this test in the model can be done in many ways. First, the flows measured during the flow test are applied to corresponding nodes in the model. Any changes in operation that were observed during the test, such as pumps being on or off, should be applied to the model. Then, the model is run to verify that the residual pressure(s) predicted by the model matches the residual pressure(s) obtained from the hydrant flow test, after adjusting for any elevation difference between the field instrument installation (e.g., hydrant port connection or other tap) and model node (typically ground surface). It is a good practice to convert all measured pressures to head and to make comparisons in terms of HGL rather than comparing pressures. Also, for residual conditions, observed HGL drop should be compared to measured HGL drop. This will ensure that any issues with static HGL are not eliminated from the examination of residual HGL. For example, if under the static conditions the model overpredicts HGL by 5 ft (1.5 m) and observed residual HGL is 6 ft (2 m) larger than the modeled HGL, then the pressure drop difference is 11 ft (3.3 m). While both static and residual HGL differences are less than 10 ft (3.0 m) the HGL drop difference is 11 ft (3.3 m) and might not be acceptable.

If discrepancies are found, the modeler will need to troubleshoot the cause of the differences and make the necessary adjustments to bring observed and calculated values closer. It is important to identify and duplicate boundary conditions (e.g., tank levels, pump operation) in the model to ensure the usefulness of this test method.

4.3.2.1. Identifying Closed Valves. Fire flow calibration can be a useful tool for identifying inadvertently closed or partially closed valves. As an example, at one utility, initial fire flow calibrations did not match well in one area of the system. An additional series of focused fire flow analyses was conducted in the area where an improperly closed valve was suspected. Model analyses were conducted using the focused fire flows. After a series of iterative analyses, the closing of one valve in the model resulted in accurate simulation of all fire flow tests. Operators later verified that there was a closed valve at that location that should have been open. Valve exercise programs, which are routinely performed by some utilities, also provide valuable information for model calibration and troubleshooting.

4.3.2.2. All-Pipes Models. Many utilities have implemented all-pipes models in recent years to leverage other data sets including GIS. These models are used for a range of applications, from addressing new water quality regulations to supporting asset management. While some may argue that all-pipes model calibration requires more measurement points to cover more areas of the system and to ensure that calibration is as thorough as possible, others maintain that it may be easier to calibrate an all-pipes model because there is no need to adjust roughness to account for missing pipes.

Fire flow calibration in all-pipes models can be taken a step further by modeling fire flow out of tested hydrants as emitters. However, a utility should not rely too heavily on this method because most of the head loss is in the hydrant lateral and barrel, which results in calibration of the hydrant rather than the model. If this method is used, both flow from the tested hydrant and residual pressure should be compared with the predicted values from the model.

4.3.3. Presenting Calibration Results

Steady-state model calibration results are typically presented in tables or plotted to show both the predicted and observed values for each measurement point. It can also be helpful to plot the target plus–minus variance in order to show how well the calibration met desired tolerances. For HGLs, the nodes may be listed along with a description of the point. The observed and predicted values, and the differences between these values (both as a number and percentage) can also be presented. Pump station flows and PRV flows may also be presented along with upstream and downstream PRV pressures or HGLs.

The goal in presenting calibration results is to provide assurance to model users that the model represents the real water distribution system as accurately as possible. When discrepancies remain after model calibration, an acceptable explanation for the differences and the model's resulting limitations should be provided. Figures 4-1 and 4-2 present example steady-state calibration results for flows and hydraulic gradients.

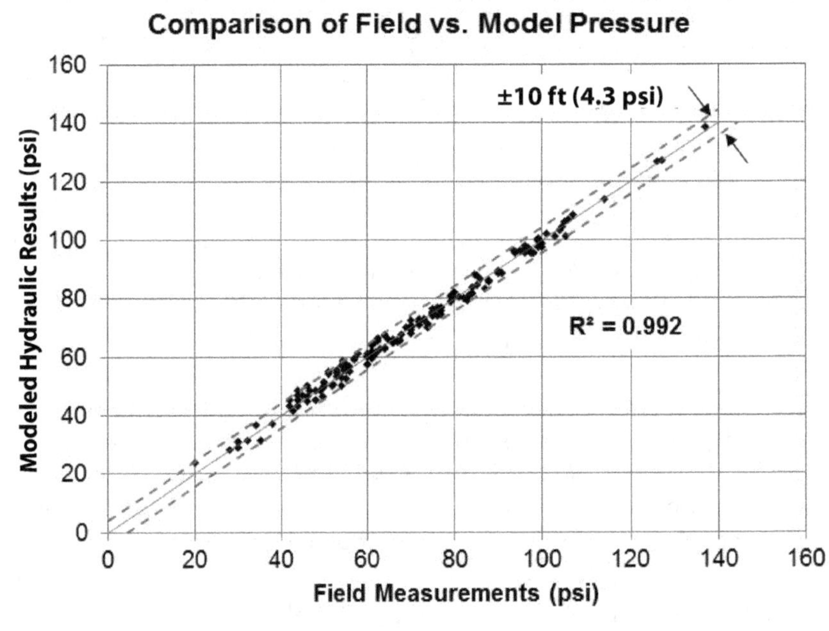

Courtesy Ed Wicklein and Aurelie Nabonnand, Carollo Engineers, Inc.

Figure 4-1 Steady-state flow calibration

Abbreviation: R^2, residuals squared.

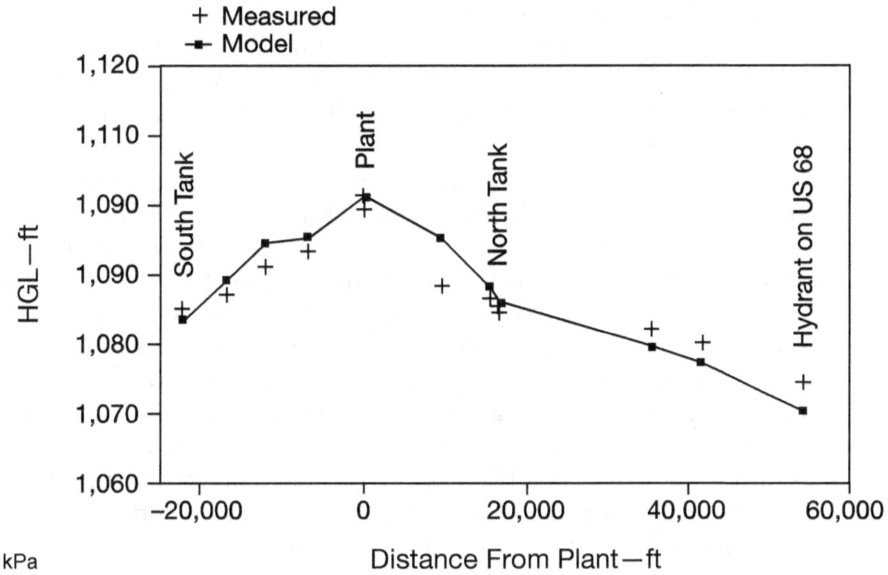

NOTE: HGL ft × 2.989 = kPa
distance ft × 0.305 - m

Figure 4-2 Steady-state hydraulic grade line calibration

Abbreviation: HGL, hydraulic grade line.

4.4. EPS CALIBRATION

EPS calibration methods include comparisons of model results to measurements made in the field over time and to tracer test simulations. Tracer tests simulations as they relate to water quality calibration are covered in more detail in chapter 7. Hydraulic calibration under maximum day conditions is described in the following section. Other demand conditions may be calibrated to ensure that a model simulates system performance under other conditions, such as average day demands for evaluation of energy costs and minimum day demands for evaluation of water quality evaluations.

4.4.1. Maximum Day Calibration

A typical EPS maximum day hydraulic calibration simulates the distribution system for a 24-hour period of high demand. Other periods of time have been used, such as 48-hour or 7-day periods, when calibrating to average day demands. A multiday EPS calibration may be done to simulate changing diurnal usage patterns and operations that may occur during weekdays and weekends or to determine a model's ability to adequately represent system performance over many days. A 24-hour EPS is often used because it simulates the diurnal supply and demand conditions that are typically repeated daily. Longer calibration periods where typical operating conditions might repeat at a period longer than 24 hours may be justified. The benefit of conducting longer calibrations must be balanced with the effort to calculate and input data for demands and operational controls for longer periods.

Similar to peak hour steady-state, high demands are required to stress the system and provide a better comparison of modeled results with recorded data. An EPS calibration normally starts with the same modeled facilities and allocated annual average day demands as a steady-state calibration. Additional data are then required to simulate the system over the designated calibration period.

4.4.1.1. Model Controls. For an EPS calibration, it is necessary to input model controls to tell the model when to activate certain facilities in order to simulate actual conditions. The two most common operational controls required for EPS calibration are pumps and control valves. Initial storage facility water levels and reservoir levels must also be entered accurately.

Model controls for pumps and valves are often initially input using time-of-day, or simulation time, controls that reflect actual times of operation as recorded on the day(s) of calibration. While logical controls that regulate operation of one facility based on specific conditions at some other facility or location in the system are available in the model, for example, pump on–off based on water level in a storage facility, these may make calibration more difficult. It is common for some operators to initiate operation of facilities without a strict set of operational parameters. Use of time-of-day controls for calibration removes an unknown parameter and allows the modeler to focus on flows, pressures, and water levels rather than on whether or not the operational controls are correct. That is, it ensures boundary conditions are accurately modeled during the calibration period. It is important to recognize that calibration using time-of-day controls makes it difficult to extrapolate the model to new situations. Other logical controls should be considered following the initial time-of-day calibration to increase the ability of the model to simulate differing situations.

Pump controls may include on–off status and pump speed. Care should be taken to identify all pumping units that are equipped with variable-frequency drives (VFDs) and to obtain pump speed as a percent of full speed during calibration, whenever possible. Where VFDs were operating during the calibration, it may be necessary to estimate the speed if not recorded. The speed may be estimated based on flow measurement data from the pump station.

Valves that require control include reservoir fill valves, throttling devices that may be modulated, and motor-operated valves that may be moved from the open to closed position during routine system operation. PRVs, PSVs, and altitude valves are not routinely modulated and generally should not have variable set points during a calibration period. For valves that are modulated, it is possible to model the valve as a flow-control valve or as a throttling valve with a specified position (e.g., percent open–closed). Many times the position of a modulating valve is not captured in the historical operational data reports. EPS calibrations can be simplified by simulating valves that modulate as flow-control valves and then inputting recorded or calculated flow rates.

Once the model is calibrated sufficiently with the valve modeled as a flow-control device, the modeler may elect to then calibrate the throttling characteristics of the valve so that modeled flows match recorded flows at given, or recorded, valve positions. This is only possible if both the valve position and the flow through the valve are recorded. Flow rates through reservoir fill valves may not be recorded but can be calculated based on water level in the reservoir and any recorded pumping or outflow from the reservoir.

Storage facility water levels required for EPS calibration can be categorized as those that must be set as boundary conditions at each time interval and those that are allowed to fluctuate based on other parameters within the model. Water levels that must be set at each time interval are those that establish the hydraulic gradients in the system at each time interval. One example is the level of a finished-water storage facility that provides suction to pumps that deliver water from the treatment plant to the distribution system. In this case, a finished-water storage facility, or clearwell, may be modeled as a variable-head reservoir and a pattern input to the model in order to vary the water level in the reservoir to mimic what was recorded during the calibration period. If the water level is not recorded, it may be acceptable to model this situation with a fixed-head reservoir that reflects typical or average levels.

Another example of a water level that requires a known level at each hour is a well supply where groundwater level is simulated by a reservoir. In both of these cases, water levels are necessary to establish the starting supply conditions for model analyses.

Except for facilities that are required to establish starting supply conditions, all other storage facilities are generally allowed to fluctuate. This includes all floating storage facilities and other facilities where water is delivered via throttling valves and then repumped as needed. For these facilities, only the initial starting level in the tank is put into the model. The fluctuation in these storage facilities is generally a calibration parameter where the modeled water level is compared to the recorded water level.

4.4.1.2. Demand Peaking Factors. The second significant requirement for EPS calibration is the addition of data to allow the model to simulate changing demand conditions. Patterns are applied to the allocated demands for each calibration increment. The patterns represent a variable peaking factor that is applied to the allocated demands to vary the demand conditions in the model. Differing demand peaking factors may be developed for each class of demand allocated to the model. Although information on variations in peaking by user class is becoming more readily available as utilities adopt AMI or install newer meters with data logging capabilities, it is important to recognize that this information is still not well known in most cases. However, in some cases, the input of variable peaking factors by class may be required to achieve successful calibration. Examples of this include a golf course known to take the majority of its water at certain times of day or a wholesale customer known to use water according to specific characteristics or whose hourly use may be recorded in operational data reports. An example of typical hourly residential demand factors that might be applied during an EPS calibration is shown on Figure 4-3.

When the system serves large water-using industries, it may be advisable to install data loggers on their water meters to better understand and simulate their use patterns, thereby removing significant unknowns and potential error sources.

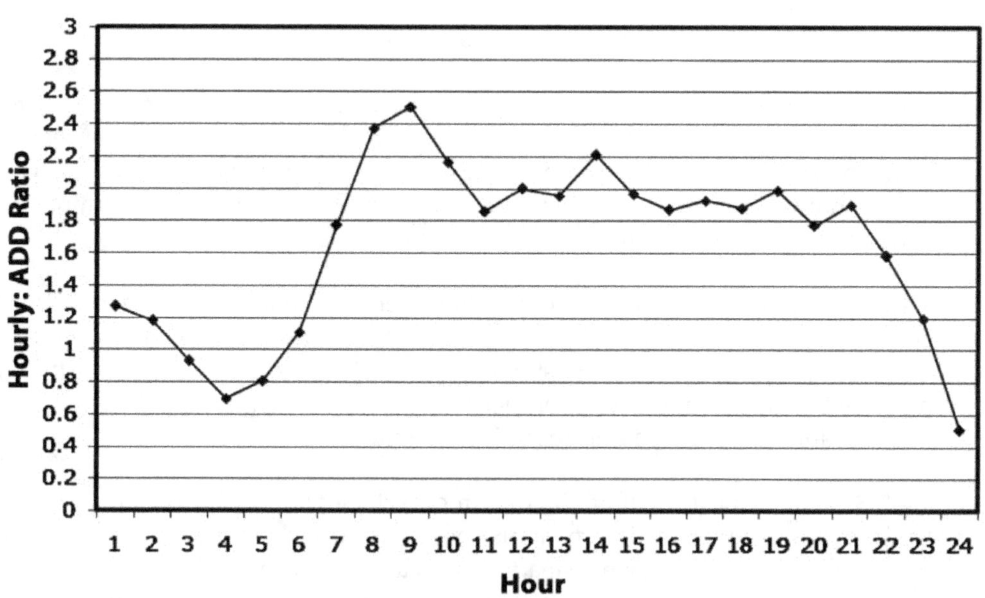

Figure 4-3 Extended-period simulation hourly peaking factors

Abbreviation: AAD, average daily demand.

4.4.1.3. Fire Flows. An EPS calibration can be conducted with a simulated fire flow superimposed on the system. This may be done for the same reasons a steady-state calibration is done – to supplement existing operational data or to conduct microcalibration. Fire flows for EPS calibrations are generally simulated using demand patterns. One way to accomplish this is to assign a demand unit of 1 to a fire flow node. The demand pattern would contain values of zero when the hydrant is not flowing and an appropriate peaking factor when the hydrant is flowing such that the factor demand matches recorded flow at the hydrant.

4.4.2. Adjustments to the Model for EPS

Making adjustments for EPS calibration can be difficult. Many types of model adjustments can be made. The following are some considerations for calibrating extended period models.

4.4.2.1. Pump Curves. There are many reasons why the flow output of a pump station indicated by SCADA is considerably different from model results. Differences in pressure losses on the suction side or discharge side can affect pump output. This is especially true with head losses within station piping. These are often underestimated, and high velocities in the pump discharge piping can lead to much higher head losses. However, if system head losses are correct, the pump curves could be incorrectly characterized. Pump impellers wear over time, potentially reducing pump output. Adjusting the pump curve can help match model results to SCADA data.

4.4.2.2. Valve Positions. It is important to confirm the minor loss characteristics of valves, especially those that are partially closed. A flow control valve works by imparting head loss to match upstream and downstream pressures for the flow set point. Throttle control valve positions are not normally set with precision. When the valve is throttled significantly, a small difference in position can make a large difference in the head loss through that valve. In addition, PRV settings can sometimes drift from their set points over time, especially when not routinely checked and reset.

4.4.2.3. Control Logic. It is possible that the control logic programmed into the model does not match what was programmed into the SCADA system. The response time of some instruments may be long enough that a pump or valve changes its actual status at a time that is different from what is indicated by the model. More common is a change in status that occurs between model time-steps.

4.4.2.4. Diurnal Demand Patterns. Adjustments to demand patterns for specific customers or localized areas may be required to achieve acceptable calibration. As previously mentioned, a large irrigation customer, wholesale customer, or large industrial customer may have a significant impact on actual system performance. Specific diurnal patterns may need to be estimated and input into the model.

4.4.2.5. Pipe Condition. If other adjustments do not seem to be effective or if local anomalies prove difficult to calibrate, adjustments to pipe *C*-factors may be necessary. Adjustments that are made during steady-state calibration may have overlooked some areas not covered by hydrant testing, which could affect how different parts of the system are served. Specifically, the *C*-factors for transmission mains have a larger impact on overall system performance and facility trends than the *C*-factors for smaller-diameter mains.

4.4.2.6. Closed Valves. Improperly closed valves may also be a cause of poor calibration results. It may be necessary to assume closed valves at certain locations to achieve acceptable calibration results. If closed valves are assumed in the model, they should be field verified. Additional field testing may be warranted to identify and verify the locations of assumed and potential improperly closed valves.

4.4.2.7. SCADA Errors. The possibility of errors in the SCADA system should not be discounted as a reason why model results do not agree. Older instruments such as an analog 4–20 mA communication protocol can show drift in their readings, so the model results may actually be more accurate than the SCADA readings. Some SCADA values are calculated within the SCADA system based on the summation of values from several instruments. These types of instruments are more prone to calculation errors than other instruments.

4.4.3. Presenting EPS Calibration Results

EPS model calibration results are typically presented in graphs in which model-predicted and field-observed values are shown for each measurement point. Typical values that are compared include flows, pressures, and water levels. The guidelines provided in this chapter can be used to assess the quality of calibration results. Figure 4-4 is an example of an EPS calibration graph that compares the modeled water level to the recorded water level.

Care must be taken when evaluating statistical calibration accuracies. In many cases, a pump station only pumps for several hours during the day. When the pumps are not operating, instances where pumps are controlled by time-based controls in the model inherently match the operating data and may be excluded in calibration statistics. Flows at hydropneumatic pump stations should also not be included in calibration accuracies. Hydropneumatic pump stations are regularly used to increase pressures to smaller areas of the distribution system.

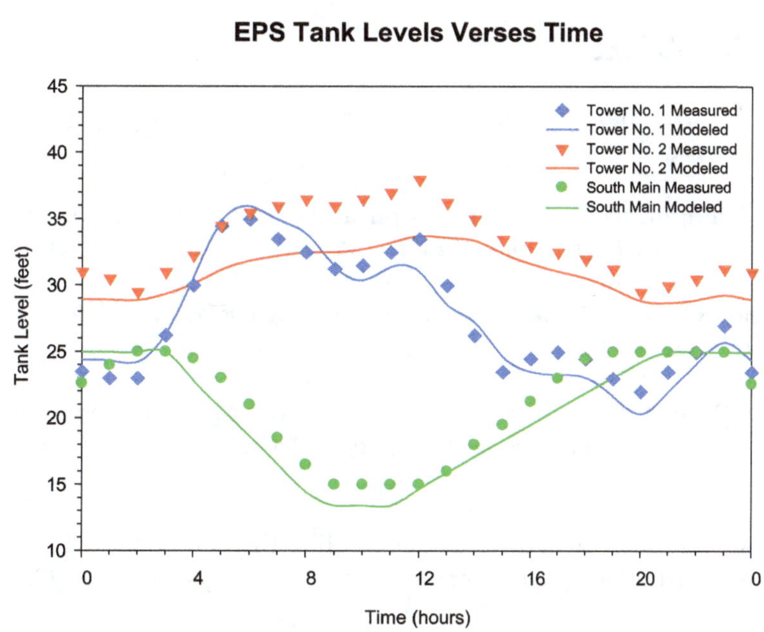

Courtesy James Cooper, Arcadis

Figure 4-4 Extended-period simulation water level calibration

Abbreviation: EPS, extended-period simulation.

Calibration accuracy can be calculated in many ways. Results of all statistical analyses should be documented sufficiently to identify the statistical methods used, the assumptions made, and a discussion of deviations. Forced data, such as using flow control valves or wells modeled as negative demands, should not be included. For example, the calibration could be determined based on the average variance, or the variance between the modeled result and field data could be computed at each time-step.

For EPS calibration, it is useful to plot tank levels over time (see Figure 4-4) to compare the frequency and rate of tank fill–drain cycles with observed SCADA information. The simulation time should be long enough that fill–draw cycles reach a steady state. This strategy can also help to identify or diagnose a demand imbalance that might cause simulated tanks to fully drain when no such field condition exists.

4.5. REFERENCES

American Water Works Association (AWWA), Engineering and Computer Applications Committee, 1999. Calibration Guidelines for Water Distribution System Modeling. In *Proceedings of AWWA Information Management and Technology Conference*. Denver, CO: AWWA.

AWWA, 2016. Manual M36, *Water Audits and Loss Control Programs*. Denver, CO: AWWA.

Walski, T.M. 1983. Technique for Calibrating Network Models. *Journal of Water Resources Planning and Management*, 109:4:360–372.

Water Authorities Association and WRc, 1989. *Network Analysis—A Code of Practice.* Swindon, UK: WRc.

4.6. ADDITIONAL RESOURCES

AWWA, 2012. Manual M6, *Water Meters—Selection, Installation, Testing, and Maintenance.* Denver, CO: AWWA.

AWWA Engineering Modeling Applications Committee, 2013. Defining Model Calibration. *Journal of the American Water Works Association*, 105:7:60.

Cesario, A.L., 1995. *Modeling, Analysis and Design of Water Distribution Systems*. Denver, CO: AWWA.

Chase, D., 1999. Calibration Guidelines for Water Distribution System Modeling. In *1999 Proceedings of the AWWA Information Management Technology Conference*. Denver, CO: AWWA.

Cruickshank, J.R., & Long, S.J., 1992. Calibrating Computer Models of Distribution Systems. In *AWWA Computer Conference Proceedings.* Denver, CO: AWWA.

Edwards, J.A., Cole, S., & Brandt, M. 2006. Quantitative Results of EPS Model Calibrations With a Comparison to Industry Guidelines. *Journal of the American Water Works Association*, 98:11:72–83.

Grayman, W.M., Rosssman, L.A., Deininger, R.A., Smith, C.D., Arnod, C.N., & Smith, J.F., 2004. Mixing and Aging of Water in Distribution System Storage Facilities. *Journal of the American Water Works Association*, 96:9:70–80.

Hirrel, T.D., 2008. How Not to Calibrate a Hydraulic Network Model. *Journal of the American Water Works Association*, 100:8:70–81.

Ormsbee, L.E., & Lingireddy, S., 1997. Calibrating Hydraulic Network Models. *Journal of the American Water Works Association*, 89:2:42–50.

Speight, V., & Khanal, N., 2009. Model Calibration and Current Usage in Practice. *Urban Water Journal*, 6:1:23–28.

Speight, V., Khanal, N., Savic, D., Kapelan, Z., Jonkergouw, P., & Agbodo, M., 2010. *Guidelines for Developing, Calibrating, and Using Hydraulic Models.* Denver, CO: Water Research Foundation.

US Environmental Protection Agency (USEPA), 2003. *The Stage 2 DBPR Initial Distribution System Evaluation Guidance Manual*, EPA-815-D-03-002 (Draft). Washington, DC: USEPA.

USEPA, 2005. *Water Distribution System Analysis: Field Studies, Modeling and Management, A Reference Guide for Utilities*, EPA-600-R-06-028. Cincinnati, OH: Office of Research and Development.

Walski, T.M., 1990. Sherlock Holmes Meets Hardy-Cross or Model Calibration in Austin, Texas. *Journal of the American Water Works Association*, 82:3:34–38.

———, 1995. Standards for Model Calibration. In *AWWA Computer Conference Proceedings.* Denver, CO: AWWA.

———, 2000. Model Calibration Data: The Good, the Bad, and the Useless. *Journal of the American Water Works Association*, 92:1:94–99.

Walski, T.M., & O'Farrell, S.J., 1994. Head Loss Testing in Transmission Mains. *Journal of the American Water Works Association*, 86:7:62–65.

Walski, T., Chase, D.V., Savic, D.A., Grayman, W.M., Beckwith, S., & Koelle, E., 2003. *Advanced Water Distribution Modeling and Management.* Waterbury, CT: Haestad Press.

This page intentionally blank.

Chapter 5

Steady-State Simulation

5.1. INTRODUCTION

Model simulations are divided into two main categories: steady-state and extended-period simulations. A steady-state model simulation computes the state of the system (e.g., flows, pressures, pump operating attributes, valve positions) at a given point in time. Steady-state analysis is sometimes compared with taking a snapshot of the distribution system. Typically, these snapshots are taken under worst-case or system-defining conditions such as peak demand times, fire protection usage, and system component failures. Steady-state simulations are often used to identify improvements for basic infrastructure-related problems.

Extended-period simulation is a technique for modeling a distribution system using a series of steady-state analyses; the simulation is performed over a specified time period at selected intervals. Extended-period simulations are useful in seeing the effects of water usage over time in order to observe system behavior such as tank level fluctuations and pump and valve operations. Extended-period simulations are often used for master planning, operational studies, and water quality simulations. More discussion on this topic can be found in chapter 6.

This chapter details the following general topics:

- *Selecting limiting conditions for design scenarios:* The modeler determines the design conditions under which the model predicts distribution system performance.
- *Comparing with design criteria:* A model compares distribution system performance with a set of performance criteria to determine the system's adequacy.
- *Developing system improvements:* The processes for several categories of model simulations are described to explain the process of steady-state modeling.

► **TYPES OF PROBLEMS ANALYZED WITH STEADY-STATE SIMULATIONS**

- Sizing mains, pumps, and reservoirs
- Performing fire flow analysis
- Sizing fire protection systems
- Assessing the impact of system changes such as new large customers or housing developments during maximum demand conditions
- Planning water main replacement or rehabilitation projects
- Planning the layout of a water distribution system
- Establishing set points for control valves and variable-speed pumps
- Locating hydraulic restrictions such as undersized pipes or closed valves
- Determining pressure zone boundaries

5.2. SYSTEM PERFORMANCE ANALYSES

A key use for steady-state simulations is to analyze the system under the most severe demand conditions. All things being equal, it is assumed that if a distribution system operates satisfactorily under the most severe demand conditions, it will operate satisfactorily under all conditions. The most limiting demand conditions are points at which system demands and pressures are at their highest and lowest points.

The most common steady-state scenarios are average day, maximum day, peak hour, minimum day, and maximum day plus fire flow. These demand conditions are coupled with minimum and maximum tank levels, pumps being turned on and off, and/or control valves being opened and closed to ensure the system can handle worst-case conditions. The choice of which condition to model depends on which question the modeler is trying to answer. A common approach is to initially use steady-state simulations to evaluate the system and develop the conceptual design of improvements, and then use extended-period simulations to verify the final design. Table 5-1 summarizes some typical analysis types and the corresponding conditions used to simulate the network.

Table 5-1 Typical model scenarios

Purpose of Analysis	Recommended Steady-State Demand Scenario
Studies of normal operation	Average day, peak hour
Production and pumping requirements	Maximum day, peak hour
Sizing of pipelines—small systems	Maximum day plus fire flow*
Sizing of pipelines—large systems	Peak hour, maximum day plus fire flow*
System reliability during emergency or planned shutdown	Condition when the emergency or shutdown is likely to occur
Model calibration	Condition during time when measurements were collected

Note that sizing of storage facilities and tank filling capability analyses are best studied using extended-period analyses as discussed in chapter 6.

*Both maximum day plus fire flow and peak hour conditions are generally studied for sizing of pipelines regardless of system size.

The following sections discuss the major demand scenarios that are normally studied under steady-state conditions; the more detailed discussion will assist the modeler in analyzing the system under varying conditions. It is critical to have a calibrated water model prior to running any of the following scenarios. A noncalibrated model could provide erroneous results.

5.2.1. Maximum Day Demand Conditions Analysis

Maximum day demand is the largest quantity of water consumed on any one day of the year, assuming normal system operations. Maximum day model runs assess the ability of the system to meet demands with adequate residual pressures. In a maximum day model run, system pump capacity (or inflow conditions for gravity feed systems) should be able to meet or exceed the demand of the system while tanks are filling. Pumps that are not able to meet or exceed the maximum day demands should be of concern. Maximum day is a critical condition because it is the largest demand to be supplied solely from production, without using system storage.

To execute a maximum day demand simulation, the following model settings are suggested unless a specific operational condition is being examined:

1. Set the model demand conditions to simulate a maximum day condition. Maximum day demand is modeled using peaking factors from average day demands or input directly from the demand data. Peaking factors are obtained from field measurements or supervisory control and data acquisition (SCADA) records.
2. Set the initial or operational conditions for the pumping facility that serves the pressure zone of interest with the proper number of operational pumps while allowing for some spare pumping capacity. If not assessing existing pumps, the modeler can introduce an inflow rate at the source point(s) instead.
3. Set tank elevations at a specifically determined point such as near their normal low operating level or near their overflow elevation.

The results of the existing system maximum day model analysis can be used to identify deficiencies in the existing system and plan short-term improvements to address these deficiencies. The goal for future year maximum day model runs is to identify water lines and facility improvements that will enable the system to provide an adequate water supply to customers under future demand conditions as well as to address potential future deficiencies and corrective measures.

5.2.2. Peak Hour Demand Conditions Analysis

Peak hour demand represents the single hour of the maximum day with the highest system demand. Systems are typically designed to meet this demand using both the production pumping capacity and system storage capacity. Thus, the peak hour model runs are used to assess the ability of all distribution system components to maintain sufficient residual pressures under worst-case conditions and to maintain reasonable head loss within piping systems. Peak hour is most often the limiting design condition for large water systems as it usually exceeds maximum day plus fire flow in large systems that have a maximum day greater than 7 to 10 mgd (30 to 40 ML/d).

To execute a peak hour demand simulation, the following model settings are suggested, unless a specific operational condition is being examined:

1. Set model demand conditions to simulate the peak hour demand condition. Peak hour demand is modeled using peaking factors from maximum day demands or

input directly from the demand data. Peaking factors are obtained from field measurements or SCADA records.

2. Set the initial or operational conditions of the pumping facility that serves the pressure zone of interest to the desired setting. One option is to turn on all available pumps, and then start turning them off one by one until the storage is slightly draining.
3. Set tank elevations at a point near the normal low level. If this information is unknown, it is reasonable to assume the tank elevation to be half-full as a worst-case scenario.

The peak hour demand simulation allows comparison of pressures under worst-case conditions for various alternatives and helps justify decisions for proposed improvements. If the model predicts pressures that are not within design criteria, additional improvements can be analyzed. Consider using peak hour plus fire flow to size pumps and water mains for systems with little to no system storage. Additional discussion on pressure and piping design criteria is provided later in this chapter.

Tank depletion rates during the peak hour should be checked against tank capacities. A rule of thumb is that a tank should not exceed a drain rate, in million gallons per day, equal to three times its volume, in million gallons, which assumes using half the volume in four hours. Further discussion on sizing of water storage tanks is provided later in this chapter and in chapter 6.

5.2.3. Average Day Conditions Analysis

Average day demand is defined as the total water used in a year divided by the number of days in the year. This assumes the system is not affected by seasonal or intermittent demand changes. In these instances, the modeler must determine the appropriate average day demand that applies. Average day simulations model the normal operation of a system under typical conditions. Calculated flows are compared to annual average production and billing records. Steady-state simulations of average day demand are often used as the basis for extended-period simulations that study power costs or water quality.

Pump conditions can be verified and new pump selections can be optimized using average day model runs. The pumps should be operating at or near their selected design points at high efficiency while supplying average day demands.

5.2.4. Fire Flow Analysis

Fire flow simulations are the most common application of steady-state modeling and are often referenced in regulations as a controlling scenario. The results from a fire flow simulation indicate whether the system is capable of providing adequate fire flow protection under the most stressed system conditions.

► KEY INFORMATION PROVIDED FROM A FIRE FLOW SIMULATION

- Available fire flow
- Residual pressure at the fire flow location
- Minimum residual pressure within the zone where the fire flow takes place
- Minimum overall system pressure
- Maximum pipeline velocities

The sophistication of current modeling programs allows the modeler to set the desired fire flow rate, as well as an upper limit, to determine available fire flows given a minimum allowable system pressure. Some programs can determine the available fire flow at all applicable points in the system with one model execution. Depending on the situation, several limiting parameters such as residual pressure or pipe velocity should be set to determine fire flow availability.

Minimum residual pressure should be set such that the fire flow will be limited based on the point at which pressures drop below a predefined level within the water system. The limiting residual pressure can be set for the whole system, the same pressure zone as the fire flow, a different pressure zone, or the point of the fire flow depending on the situation. As a general rule, the minimum residual pressure is set to 20 psi (138 kPa) in the United States within the pressure zone. However, residual pressures should be set higher where required. Many commercial structures have fire suppression systems, and the analysis needs to include pressure requirements of these systems.

When running such scenarios, the modeler needs to be aware of points in the system that may already be below 20 psi (138 kPa), typically near water storage tanks or on the suction side of booster pumping stations. If these junction nodes are not removed from consideration as limiting conditions, results will indicate that no fire flow is available. The modeler also needs to make sure that high ground elevation points are appropriately identified in the model as well and included as limiting locations. If these points are missing, the results could be incorrect.

Because limiting fire flow due to pipe velocities is an option in most current models, consideration can be given to its use to avoid excessive pipe velocities. Maximum pipe velocities can also be a major concern in a water distribution system. High pipe velocities can lead to high head loss conditions and possibly transient flow (water hammer) concerns, potentially causing damage to the water system. For more information on how pipe velocity changes can lead to transient flow, see chapter 8.

To execute a fire flow simulation, the following model settings are suggested, unless a specific operational condition is being examined:

1. Set model demand conditions to simulate either maximum day or peak hour demand condition. Sometimes regulatory requirements dictate what demand conditions are required.
2. Set the initial or operational conditions of the pumping facility that serves the pressure zone of interest to the desired setting.
    - One option is to keep the same settings as the base condition (maximum day or maximum hour), which would provide the initial fire flow available within the system.
    - A second option is to select settings based on the base demand (such as maximum day) plus the highest fire flow requirement. Determine which pumps will turn on during the fire flow condition and add those to the settings of the base condition.
3. Set tank elevations at a point near the normal low level. If this information is unknown, it is reasonable to assume the tank elevation to be half-full as a worst-case scenario.
4. Set the desired fire flow rate for individual nodes in the model as well as the limiting parameters such as pressure and velocity. For skeletonized models, localized piping may need to be added to allow for more accurate results. Fire flow requirements within distribution systems typically range from 500 to 3,500 gpm (approximately 100 to 800 m^3/h). Demands greater than 3,500 gpm (approximately 800

m^3/h) are typically obtained using alternative methods. See AWWA Manual M31 (2008) for further discussion concerning required fire flow rates and durations.

These results help determine if adequate fire flow is available or if fire flow is inadequate due to inadequate pipe sizes or high pipe velocity, or if fire flow is limited due to loss of pressure at a localized high point. System improvements can then be analyzed to ensure adequate fire flow is provided. To determine if tank storage is adequate for a specific fire flow condition, it is recommended that an extended-period simulation be performed over the duration of the fire flow.

5.2.5. Minimum Day Demand Conditions Analysis

Minimum day demand is the day during which customers consume the smallest amount of water. Generally, minimum day runs assess the maximum pressures the system will have to endure during the day; some conditions may exist near pump station discharges with high pressures as well under different demand conditions. Excessively high pressures can damage system components. Coupled with transient flow conditions that could occur as pumps turn on and off, high pressure can be a major concern. Therefore, this analysis can help the modeler to determine key areas of vulnerability and provide solutions to mitigate high-pressure concerns.

To execute a minimum day demand simulation, the following model settings are suggested, unless a specific operational condition is being examined:

1. Set model demand conditions to simulate the minimum day demand condition. Minimum day demand is modeled using peaking factors from average day demands or input directly from the demand data. Peaking factors are obtained from field measurements or SCADA records.
2. Set the initial or operational conditions for the pumping facility that serves the pressure zone of interest with the proper number of operational pumps to meet demand. If not assessing existing pumps, the modeler can introduce an inflow rate at the source point instead.
3. Set tank elevations at a point near their overflow elevation or the point where the pumps are turned off and the tanks are considered full.

The results of the minimum day demand simulation indicate where pressures in the system are high or above desired levels under a worst-case condition for high pressures. These results may be used to indicate where pressure-reducing valves (PRVs) may be needed.

5.3. SYSTEM DESIGN CRITERIA

Design criteria define system capabilities by specifying the performance requirements of the system components. Thus, whether the objective of the analysis is to design a new system or to improve an existing system, the design criteria define the potential solutions and are the standard against which system performance, both observed and predicted, is compared to determine sufficiency of service.

5.3.1. Pressure Design Criteria

When assessing the adequacy of a system, the first parameter to check is usually the predicted pressure. Generally there are three design pressures that are defined for each community: maximum pressure, minimum pressure during peak hour, and minimum pressure during fire flow. The range of pressure fluctuations at a single point experienced over a 24-hour period may also be analyzed; pressure is typically kept to less than 20 or 30 psi (140 or 207kPa).

► **TYPICAL PRESSURE DESIGN CRITERIA**

- 90–110 psi (620–758 kPa) during minimum demands
- 35–50 psi (240–345 kPa) during maximum hour
- 60–90 psi (414–620 kPa) during average demands
- 20 psi (138 kPa) minimum during maximum day + fire flow

5.3.1.1. Maximum Pressure. As noted, the maximum pressure is typically simulated during the minimum demand scenario when using a steady-state model and refers to the maximum pressure that customers experience. It is important to conform to any limits imposed by local building codes or regulatory agencies as well as the design conditions of the water mains. The maximum pressure limits are often in the range of 90 to 110 psi (620 to 758 kPa). However, in mountainous regions, pressures ranging from 100 to 130 psi (690 to 896 kPa) are not uncommon due to extreme topography. Additionally, in some communities, the maximum allowable pressure is limited to 80 psi (550 kPa) due to internal plumbing and water heater constraints. In these instances, the distribution system may be designed for the higher pressure, with PRVs installed on the service lines to customers.

5.3.1.2. Minimum Pressure–Peak Hour. Minimum pressures are simulated during increased demands such as peak hour or maximum day plus fire flow conditions. The minimum pressure during peak hour refers to the minimum pressure that customers experience during normal system operation. This value is typically in the range of 35 to 50 psi (240 to 345 kPa) and ensures that there is adequate pressure available for two-story buildings. In the past, minimum pressures as low as 20 to 25 psi (138 to 172 kPa) were acceptable in the design of water distribution systems; however, if this pressure is used, the utility can expect customer complaints of lower pressure. A minimum pressure of 20 psi (138 kPa) is required to ensure that the system remains sufficiently pressurized to prevent water main damage caused by collapse. Minimum pressure is also necessary to prevent backsiphonage of nonpotable water from customer properties or infiltration of groundwater around pipes when very low pressures exist in the water lines.

When selecting minimum pressure constraints, the modeler must be aware of the elevations of facilities being served in relation to node elevations in the model. Minimum pressure at the model node may not provide adequate pressure at the facility being served. Multiple-story buildings may need to be evaluated on a case-by-case basis since some are equipped with booster pumps that accommodate the higher floors. Minimum pressure also affects the design of lawn irrigation systems and fire suppression systems. It is important to note that where residential fire sprinkler systems are required by legislation, the minimum acceptable pressure is 50 psi (345 kPa) for them to operate properly. Backflow preventers are required in many commercial, office, industrial, and retail buildings. Use of these devices results in a 5 to 15 psi (34 to 103 kPa) head loss, so minimum system pressures should be set accordingly. It is also important to check regulatory requirements prior to determining the minimum acceptable pressure during peak hour demand conditions.

5.3.1.3. Minimum Pressure–Maximum Day Plus Fire Flow. The minimum pressure is the minimum desired pressure in the distribution main that is closest to the flowing hydrant. The recommended minimum pressure during fire flows is typically 20 psi (138 kPa). This value has been recommended by the National Fire Protection Association in the United States and by the Fire Underwriters Survey in Canada. It is important to note that some communities require higher fire flow residual pressures, and local regulations should be followed.

The value of 20 psi (138 kPa) is used because it allows for adequate supply of water to pumper vehicles while friction losses in the hydrant branch, hydrant, and suction hoses to the pumper vehicle are overcome. This standard assumes typical hydrant branch configurations. If the hydrant branches are very short or very long, the standard is adjusted to reflect the decreased or increased losses because of friction in the hydrant branch. Some modeling programs allow for data input related to hydrant branches to account for such situations.

5.3.1.4. Other Pressure Constraints. Depending on the extent of distribution system deficiencies, the distribution system model may predict negative pressures. Predicted negative pressures result in creation of a vacuum in the system and potentially backflow contamination or cavitation. Though negative pressures can only drop to the value of vapor pressure (–14.7 psi [–101.3 kPa]) at ambient temperature, the model could predict greater negative pressures (which are not physically possible) because the mathematical model is forced to supply the full amount of all nodal demands. The significance of these excessively low predicted negative pressures is that system demands cannot be met and there are serious deficiencies either in the system or in the model. In some cases, negative pressures could occur in limited areas of the system, for example in areas of high elevation. However, in general, a deficient distribution system simply fails to meet demands. Most modeling programs allow the user to input pressure-dependent demands in order to better represent conditions under low pressures. However, evaluation of these conditions is more achievable for networks of discrete regions rather than system-wide all-pipes models due to their intense computational needs.

5.3.2. Piping System Design Criteria

It should be noted that piping design criteria are rules of thumb and are provided as a guide for identifying areas where improvements are most needed. Many modelers and utilities consider them to be secondary criteria that can be used to identify causes of low-pressure areas that do not necessarily require improvements.

It is important to understand that when simulated pressures are adequate during low-flow conditions but drop below standards during high-demand periods or fires, some pipes may be undersized. To find bottlenecks in the system, pipes with the highest velocities and head loss rate conditions should be identified. It should also be recognized that higher velocities within pipe could make the pipe more vulnerable to transient flow problems (see chapter 8). Therefore, it is important for the modeler to consider these conditions when designing piping systems. Although none of these conditions are without exception, they can indicate inefficient system design or operations due to aged pipes, partially or fully closed valves, or improperly sized tanks, all of which require additional pumping and result in wasted energy. The operating costs can be substantial over the life of the pipe. It is important to note that as velocity within a pipe doubles, the head loss rate increases by approximately a factor of 4 to 6.

> ► **TYPICAL PIPING DESIGN CRITERIA**
>
> - Velocities greater than 4–6 ft/s (1–2 m/s)
> - Small pipe diameters (less than 16 in. [406 mm]) having head losses greater than 5–7 ft/1,000 ft (5–7 m/km)
> - Large pipe diameters (16 in. [406 mm] or greater) having head losses greater than 2–3 ft/1,000 ft (2–3 m/km)

5.3.2.1. Pipeline Sizing. The design of new pipelines should address two basic features: sizing and routing. Pipe size is determined by the maximum flow rate carried, typically either the maximum day plus fire flow condition or the peak hour condition for distribution mains. Typically, either the peak hour flow rate or the maximum storage replenishment rate is the maximum flow for large transmission mains. The pipe size is selected to ensure that velocity and head loss rate, as previously described, are not exceeded. The steady-state model is a useful tool when determining the required flow rate in looped systems and iteratively selecting the most efficient design.

5.3.2.2. Pipeline Routing. The second pipeline design feature that is established is routing. In general, looping of mains should be incorporated wherever possible. A looped system provides supply from two or more directions for large demands such as fire flows, which keeps velocities lower and provides backup and redundant supply paths in the event that a section of pipe is removed from service. The location of the new pipe is determined by the availability of rights-of-way, construction easements, and common sense. When locating new pipelines, consideration should be given to the projected growth characteristics of the service area and the locations where major demands are expected in the future. Routing studies consider the impact of development staging. Development restrictions may be required to limit the number of properties that need large fire flows until a certain portion of the network and looping are completed. Steady-state models are used to validate selected routes and proposed demand staging.

5.3.3. Pumping Systems Design Criteria

Distribution systems have both high-service pumps and booster pumps. High-service pumps obtain water from a reservoir that is open to the atmosphere (usually the system water source or finished water). Booster pumps obtain water from within the pressurized distribution system. In systems with elevated or high-level ground storage, constant-speed pumps are typically used for high service and booster pumping to meet maximum day demands. Storage is used to equalize the pumping rate over the range of water demands and provide additional supply during peak hour and emergency conditions. For systems without such storage tanks or with insufficient equalization storage, it is generally desirable to use VSPs to efficiently change the pumping rates to meet system demands.

Regardless of type, pumps must be sized to meet the full range of system demands, and all of the limiting demand conditions must be considered. Individual pumps and combinations of pumps are sized to meet the range of demands from minimum day to average day to maximum day plus fire flow. However, the pump installation should, at a minimum, be designed to deliver the maximum day demand. The duration of each condition should be considered when selecting the appropriate pump size. The best choice may be a series of parallel pumps of varying sizes that can be operated to most efficiently match the pumping requirements. The steady-state model is used to define the flow rate and head requirements for each condition.

A typical pump rating curve and a system head curve used to select pump size are shown in Figure 5-1. The pump operates where the two curves intersect, and the pump should be selected to operate at or near its highest efficiency during average day conditions as indicated by the pump's efficiency curve.

The system head curve represents the required head for a range of flow rates – for example, minimum day, peak hour, maximum day plus fire flow. It is a relatively easy curve to develop for a single pipe. The system head curve is calculated by adding static head, head loss through the pipe, and pressure required at the end of the pipe. However, the situation is more complex in a pipeline grid where water is conveyed to many points and via various routes. The model is used to simulate the varying demand conditions, and the required system head curve is developed based on the pressure standards. This curve is plotted and the pump is selected based on the percentage of time each condition is expected to occur. Figure 5-2 illustrates this selection process. Some models provide the modeler with the system head curve as an output under specific system conditions.

According to the US Department of Energy (2006), electricity represents approximately 75 percent of the cost of municipal water processing and distribution. Of this, pumping typically accounts for 75 to 80 percent of the power consumed by water utilities. Consequently, the most efficient pumps for the range of demands must be selected. Figure 5-3 shows the typical efficiency range for a pump rating curve. Specific information on pump efficiency is available from the pump manufacturer. The modeler and designer select pumps that show the highest efficiencies over the expected ranges of demands. For multiple pump applications, establishment of the optimum sequential ordering of pumps has a substantial impact on power consumption and cost to the utility. Pump optimization strategies are best evaluated using extended-period simulations. Additional discussion on this topic is provided in chapter 6.

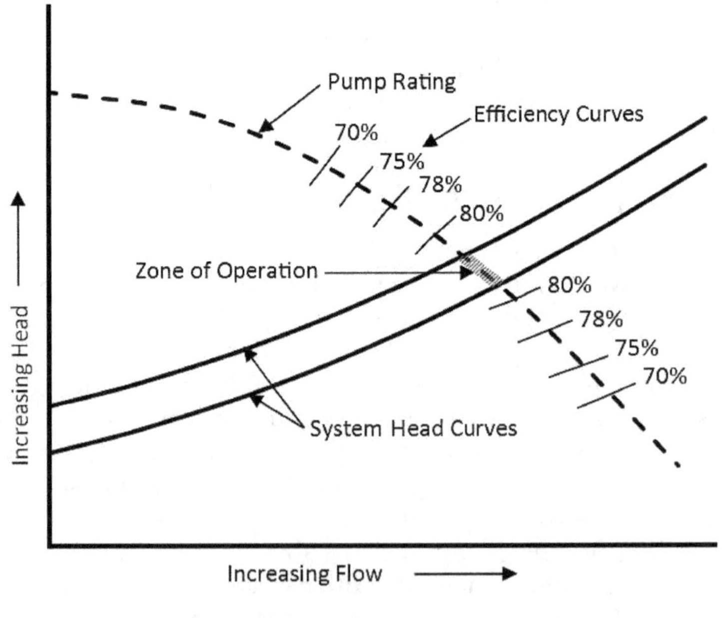

Figure 5-1 Pump rating curve versus system head curve

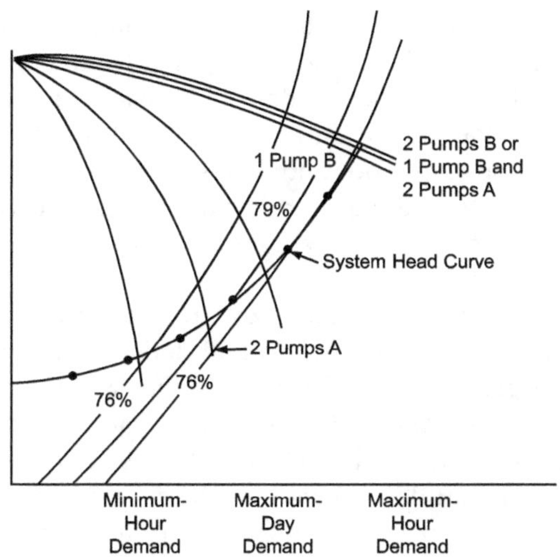

Figure 5-2 Multiple pump rating curves

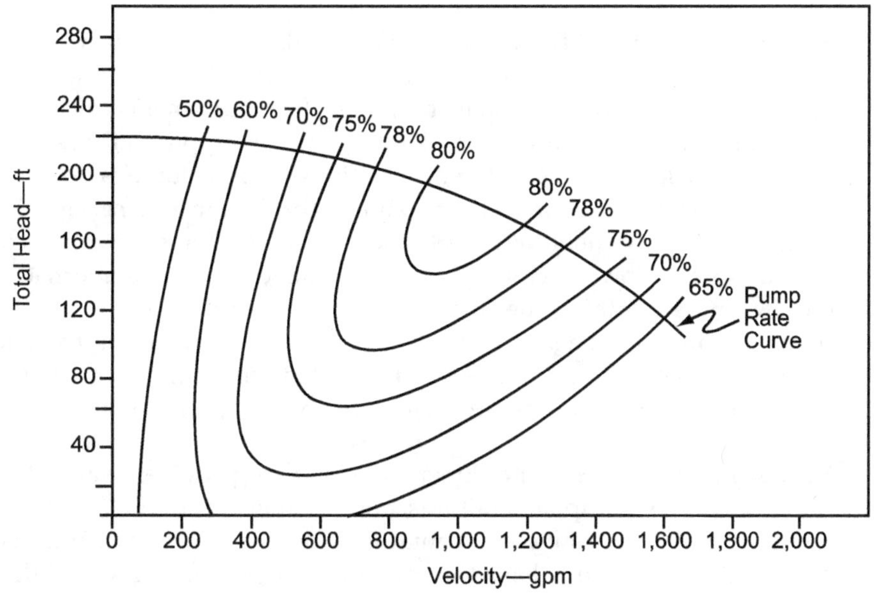

Figure 5-3 Example pump efficiency curve

5.3.4. Storage Facility Design Criteria

Storage facility design criteria must address many interrelated factors including system storage requirements, elevated or high-level ground storage versus ground storage, and the number and location of storage facilities. Once system storage requirements are determined, the steady-state model can be used to iteratively select the number and location of storage facilities that will best support the distribution system, especially during fire flow. Potential elevated tank sites can be identified by first looking at areas in the system with higher ground elevations. Higher ground elevations allow for shorter elevated tanks, resulting in more cost-effective storage. Once areas of high ground elevations are identified, specific sites can be identified based on proximity to large-diameter water mains, proximity to existing source and storage, availability of land, and land usage and zoning. Once potential sites have been identified, steady-state modeling during maximum day plus fire flow can be performed to determine each tank site's benefit to the system pressure and fire flow availability. Tank operation and proper pump controls are best analyzed using an extended-period model, however.

The volume of storage required is classified into three primary components: equalization storage, fire storage, and emergency storage.

5.3.4.1. Equalization Storage. Equalization storage is the amount of water required to meet demands in excess of production and delivery capabilities. This storage is generally less expensive to provide than the production facilities, pumping, and piping that are required to meet all instantaneous demands. The amount of equalization storage that a community maintains should be determined based on a comparison of production capabilities versus the demands expected on the system. In most communities, this is based on the maximum day condition and the ability to ensure that the equalization storage is sufficient to meet demands that exceed the production capabilities. Equalization storage volume is typically calculated by plotting the hourly diurnal curve for maximum day demands, as shown in Figure 5-4. The area under the diurnal curve but above the average of the hourly demands for the maximum day is the equalization volume required.

5.3.4.2. Fire Storage. The fire protection needs of a community typically are determined by the Insurance Service Office in the United States, the Insurance Advisory Organization in Canada, and the International Fire Code. The requirements calculated by these organizations are based on building types, land use, water supply facilities, and the response capabilities of the local fire department. The fire storage volume is determined by multiplying the required flow duration by the maximum fire flow in each service area of the distribution system. AWWA Manual M31 (2008) includes tables that show the recommended duration for various fire flows. Factors such as redundant piping and alternate supplies are considered when determining the final requirements for the community. Large fire flows may require more than one storage facility and a secondary piping route to the site. Typically, fire storage is obtained from reservoirs located within the same pressure zone as the fire, although some smaller pressure zones must rely on pumping to supply needed fire flows.

5.3.4.3. Emergency Storage. Emergency storage provides water during events such as pipeline failures, equipment failures, power outages, pumping system failures, water treatment plant failures, raw water contamination events, and natural disasters. The amount of emergency storage is based on an assessment of the risk of failures and the desired degree of system dependability. As an example, a system operator may desire a full day of emergency storage in order to provide enough storage in the event of a system-wide power failure. An assessment must be made as to the type and nature of the emergency condition, including the frequency, intensity, and duration. In general, a vulnerability analysis, as described in AWWA Manual M19 (2001), should be used to determine emergency storage requirements. Some state or provincial regulations indicate the minimum emergency storage required for the community based on the average or maximum daily demands. In addition, water quality should be a part of the equation when determining the amount of emergency storage to provide. The utility may relax its minimum pressure requirements under emergency conditions as long as negative pressures are avoided.

5.3.4.4. Effective Storage. Effective storage is the volume of water available in a tank above a level that will provide pressures above the minimum allowable pressure under peak hour or maximum day demand conditions during normal operating conditions, generally, 35 to 50 psi (241 to 345 kPa). The level at which this occurs is referred to as the *minimum effective elevation*. The minimum allowable pressure and demand conditions for identifying the minimum effective elevation may be based on regulatory requirements or determined by the utility. Any storage within a tank that is below the minimum effective elevation is considered ineffective storage; however, it may be available for emergency purposes because it may still provide pressures above the absolute minimum pressure of 20 psi (138 kPa). Providing all storage above the minimum effective elevation may not always be practical but should be considered when placing a tank.

5.3.4.5. Storage Allocation. When evaluating system storage and developing operational plans for its use, the allocation of storage for equalization, fire, and emergency must be addressed. It is useful to consider storage schematically. Figure 5-5 shows how the three storage components are positioned in the reservoir. Storage for equalization occupies the top portion of the reservoir. During the day, this volume increases and decreases as water is pulled from the reservoir when demand exceeds production rates. Fire storage is positioned in the reservoir under the equalization storage. Emergency storage occupies the bottom portion of the reservoir. Levels are designated to the different storage components to ensure that the volume has the correct hydraulic grade line (HGL) for its intended purpose when dealing with elevated or high-level ground storage reservoirs. A fire may occur when the equalization storage has been completely consumed for the day, and the remaining HGL must be sufficient to deliver the required flows at the proper pressures. Some utilities meet water quantity and pressure needs for fire flows and emergencies within a region by relying on several storage tanks, in addition to PRVs and/or pumps.

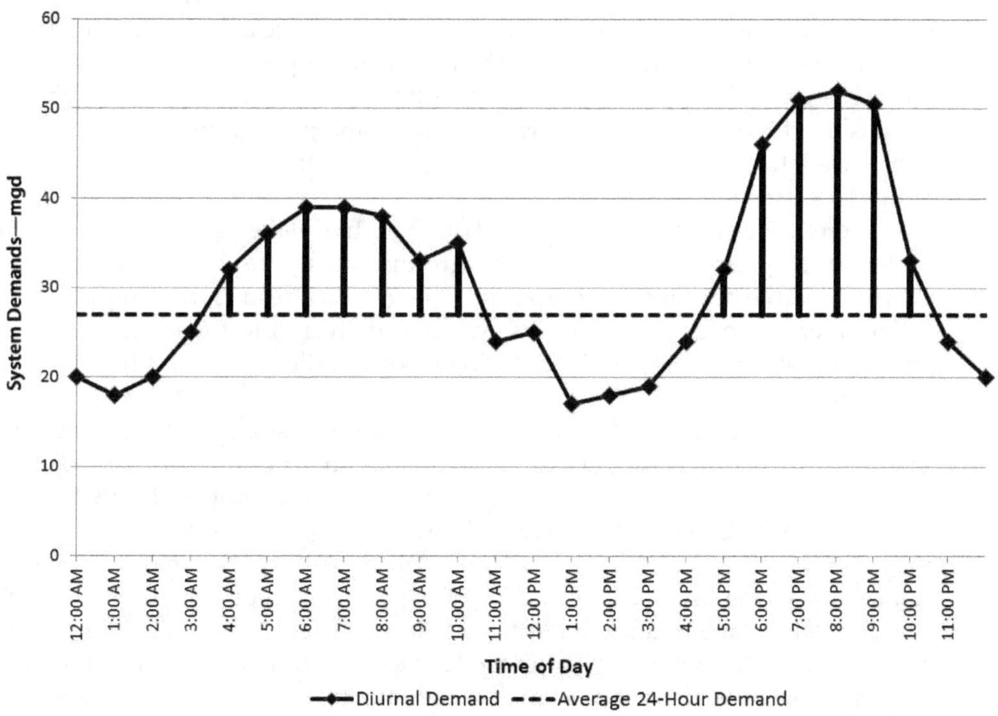

Figure 5-4 Equalization storage requirements for maximum day conditions

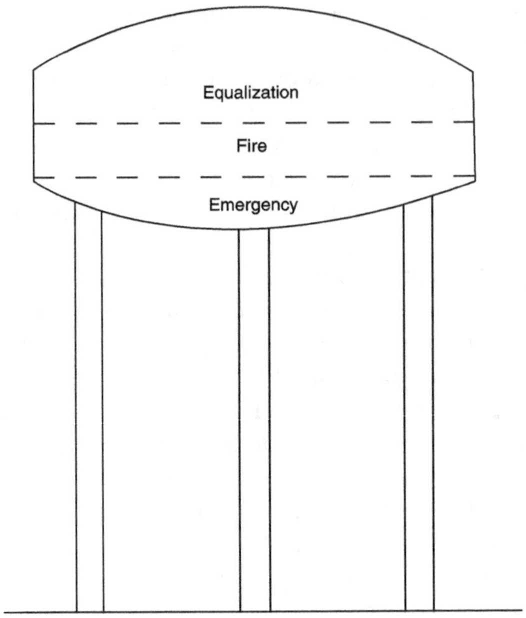

Figure 5-5 Storage allocation

5.3.4.6. Types of Storage. The determination of the type of storage to construct includes high-level ground storage, elevated storage, and ground storage with pumping (Figure 5-6). If there is high ground in or near the service area, high-level ground storage is connected directly to the system. This type of tank is generally found in mountainous or hilly areas where high ground is not difficult to find. However, in areas where the topography is much flatter, elevated storage, ground storage with pumps, or a combination of both is used. If a ground storage tank's height is greater than its diameter, it is commonly known as a *standpipe*.

There are advantages to elevated storage, high-level ground storage, and ground storage tanks. Both elevated and high-ground storage tanks can maintain the system HGL without the need for continuous pumping, which results in a simpler operation. Both are considered more reliable because water is delivered to the system from the tank by gravity and water can be provided during a power failure. Also, both provide more stable system pressures.

The major advantage for both high-level ground storage and ground storage is the reduced construction costs per unit of volume. Both types of ground storage can be built to hold large volumes of water, whereas the volume size limit of elevated storage is about 4 mil gal (15 ML). Ground storage facilities with pumps can be adapted to varying system hydraulics by changing pumps and controls. In contrast, elevated or high-level ground storage tanks must operate at the HGL defined by their geometry. Ground-level tanks can have a disadvantage because a large percent of the tank's contents falls below the minimum level for effective storage. This can lead to low turnover and stagnation unless methods to enhance turnover are provided. Water quality issues in tanks are described in more detail in chapter 8. Frequently, as water systems expand, the required HGL and water volumes change. This sometimes results in situations where either there is little turnover in the elevated reservoir because the system HGL is higher than the tank or, conversely, the tanks deplete too quickly during high-demand periods. These impacts must be considered in storage analyses of future conditions.

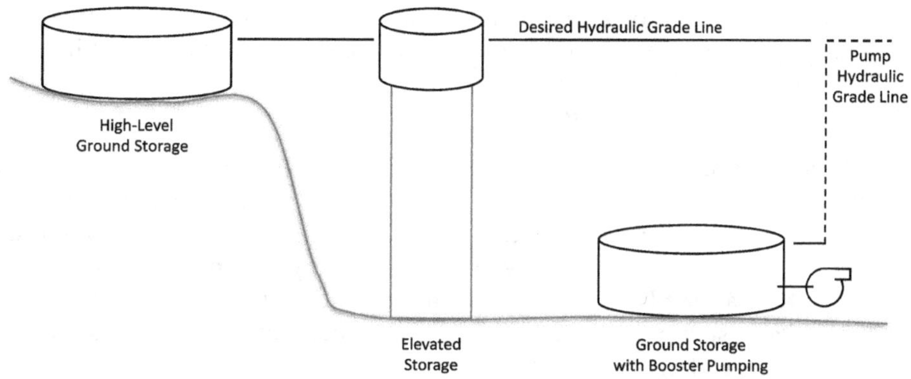

Figure 5-6 Types of storage and elevation

5.3.4.7. Number and Location of Storage Facilities. The factors to consider when determining the number and location of storage facilities include reservoir size, location of other system components, hydraulic elevation, and fire flows. System reliability, available land, and stakeholder input can strongly influence the selection of storage locations.

The required number of reservoirs is determined using a trial-and-error analysis. The analysis determines the cost-effective number of storage reservoirs, taking into consideration the impacts of cost, piping, pumping, storage, and water quality. A key factor to establish is the limiting size of each reservoir. The reservoir type (elevated versus high-level ground versus ground storage) determines the limiting factor. In general, the reservoir should be as large as is practical but small enough to be filled within a reasonable amount of time. The reservoir size should be compatible with the size of other reservoirs in the system. A system with tanks of significantly different sizes could be difficult to operate, resulting in water quality issues. Water quality issues should be considered when designing and operating tanks. Operations that promote mixing and turnover within the tank, for example, reduce water stagnation and maintain water quality. The use of individual reservoirs for each region to be supplied is compared with the cost of using regional reservoirs to supply multiple areas. The cost comparison includes reservoir construction and operating costs, as well as the associated pumping and piping required for each alternative.

5.4. DEVELOPING SYSTEM IMPROVEMENTS

A distribution system model can be used in many ways, but the primary uses are system capacity assessment, master planning design, operational studies, energy and optimization studies, and water quality studies. The primary use of most steady-state models is for master planning or for developing and evaluating potential system improvements. Other uses of steady-state models include assessment of proposed local system changes such as subdivisions or new large customers, determination of control-valve settings, and emergency planning.

In system modeling, a field-calibrated model is used to evaluate a distribution system for both existing and projected design scenarios. The design criteria that have been established before any scenarios are evaluated are used to ensure that the proposed system meets the customers' needs. Before running each analysis, the modeler prepares an outline or index of scenarios that describe each situation to be simulated (e.g., 2025 peak hour, 2050 peak hour). It is suggested that all runs be executed within the same model and that these different runs be broken out into different scenarios using the modeling software's scenario manager. Each scenario is either named or numbered, and results of each analysis are documented. Notes are taken on each scenario and its results so that the evaluations are readily understood by anyone who refers to them in the future.

5.4.1. Master Planning

Master planning is the development of long-range plans for the growth or reduction of a water distribution system. Typically, master plans are developed for 5-, 10-, and 25-year projections. They include assessment of the most cost-effective solutions for locating new production or storage facilities and for expanding the distribution system to support growth. Master plans should also analyze how to address existing system deficiencies, optimize system performance, and interconnect separate systems for emergencies.

Master planning not only looks at the construction and expansion of water infrastructure. Planning also assesses the impact of changes in average and maximum customer demands caused by water conservation, population declines, or shifting demands to other times of the day through customer education programs. By increasing or decreasing demands in the model, the utility can assess the changing effects of demands during specific time periods.

5.4.2. Subdivision Planning

Subdivision planning is a localized analysis that considers the staging and growth patterns of new areas in the distribution system. The master planning model identifies the transmission main that supplies the volume of water required to the area from the production source. The subdivision model outlines the details of how the area should grow in a sequential manner.

The modeler considers the effects of staging from the perspective of system reliability by assessing the number of lots on a single feed. Fire flows are confirmed at each stage and, if insufficient flows are available, contingency plans are developed to either limit high-density development or provide additional looping to safeguard property.

In some communities, adjacent developments can affect a neighborhood's ability to grow in a cost-effective manner. Utilities use subdivision modeling to identify these problems and address them before they occur.

5.4.3. Outage Planning

If a utility maintains a calibrated model of its transmission and distribution system, the model can be used to assess the impact of an upcoming outage on the system. The modeler should assess three main issues: reservoir volumes, minimum system pressures, and firefighting capabilities. Mitigation strategies (e.g., temporary tank, PRV) should also be modeled to determine necessary practices to maintain a minimum level of service.

Reservoir volumes are reduced either because of a reservoir shutdown for cleaning or maintenance or because of shutdown of a transmission main that fills the reservoir or supplies water to a neighborhood. In these cases, the duration of the shutdown is assessed against production capabilities, and the network is analyzed to ensure the needed water is moved to the affected locations to maintain customer supply at adequate pressures. Shutdowns of large pipelines or other facilities can have a similar effect.

The minimum pressures in the water distribution network are analyzed during the planned shutdown. If there is a major change and the shutdown is expected to cover an extended period, residents in the affected area should be notified in advance. It is important to identify critical customers such as hospitals, nursing homes, daycare centers, dialysis centers, and large industrial or commercial customers that should not lose water service. The utility call center should also be notified so that qualified personnel can respond effectively to any customer complaints caused by the reduced level of service.

Finally, the firefighting capabilities in the immediate area of the shutdown are assessed. Contingency plans are developed with the local fire department to identify how to respond to a fire in the affected area during the shutdown.

5.4.4. Local Impact Analysis

If a utility has a model that is well calibrated for current conditions, the model can be used to assess the potential impact of localized changes in the system. New customers may want to know what level of service they can expect if they locate to a certain area. The model can identify the range of pressures that can be expected as well as the available fire flows. If a new customer could create a potential deficiency in the system, the utility can use the model to assess what improvements are needed to provide adequate service to that customer while maintaining equal or better service to existing customers.

5.4.5. Control-Valve Set Points

Modeling analyses can identify smaller areas within the system that may require adjustment of pressures or flows. For example, customers in an area of lower elevation may experience higher pressures (typically greater than 100 to 110 psi [700 to 758 kPa) than customers at higher elevations. Generally, such small areas could be served by one or more PRVs that lower pressures to acceptable levels. The model is used under a range of demand conditions to determine the pressure settings for the valves as well as the size. At a particular location, two or more PRVs may be used to cover the expected range of flows to the area. Care must be exercised when designing parallel PRVs to avoid convergence issues with the model and an excessive set point. Doing so provides the utility with more control and helps avoid cavitation in the valves. In other cases, areas with higher elevations could experience excessive drops in pressure as water moves through the area to serve adjacent lower-elevation neighborhoods. In these cases, a pressure-sustaining valve could be used to restrict the amount of flow that passes through the higher-elevation neighborhood to prevent excessive reductions in pressure.

5.4.6. Continuing Use of the Model

A utility that invests the time, effort, and money to perform a computer-assisted analysis of its water distribution system should consider maintaining the model for continued use. A model is a corporate asset and should be treated as such. This requires the support of a modeling team that ensures that the model remains current and sufficiently calibrated for its intended use. The utility should develop a protocol or work flow that defines the process and frequency for model updates from GIS files, as-built drawings, and other data sources. Chapter 9 describes ongoing model maintenance.

5.5. REFERENCES

American Water Works Association (AWWA), 2001. Manual M19—*Emergency Planning for Water Utility Management.* Denver, CO: AWWA.

AWWA, 2008. Manual M31—*Distribution System Requirements for Fire Protection.* Denver, CO: AWWA.

Fire Underwriters Survey, 1981. *Water Supply for Public Fire Protection—A Guide to Recommended Practice.* Toronto, Ont.: Insurance Bureau of Canada.

National Fire Protection Association (NFPA), 1985. NFPA Standard 291-1985, *Recommended Practice for Fire Flow Testing and Marking of Hydrants*. Quincy, MA: NFPA.

US Department of Energy, 2006. Energy Demands on Water Resources—Report to Congress on the Interdependencies of Energy and Water. Washington, DC: DOE.

5.6. ADDITIONAL RESOURCES

Boulos, P., Lansey, K., & Karney, B., 2006. *Comprehensive Water Distribution Systems Analysis Handbook for Engineers and Planners.* Broomfield, CO: MWH Soft.

Cesario, A.L., 1995. *Modeling, Analysis and Design of Water Distribution Systems.* Denver, CO: AWWA.

Chase, D. 1999. Calibration Guidelines for Water Distribution System Modeling. In *Proceedings of the AWWA Information Management Technology Conference.* Denver, CO: AWWA.

Cruickshank, J.R., & Long, S.R. 1992. Calibrating Computer Models of Distribution Systems. In *AWWA Computer Conference Proceedings.* Denver, CO: AWWA.

Doe, S.R.K., & Duncan, C.T., 2008. Naming Conventions and Documentation for Distribution System Modeling. *Journal of the American Water Works Association,* 100(9):132–138.

Electric Power Research Institute, 1999. *A Total Energy and Water Quality Management System.* Palo Alto, CA: EPRI.

Walski, T.M., 1995. Standards for Model Calibration. In *AWWA Computer Conference Proceedings.* Denver, CO: AWWA.

Walski, T.M., & O'Farrell, S.J. 1994. Head Loss Testing in Transmission Mains. *Journal of the American Water Works Association,* 86:7:62–65.

Walski, T., Chase, D.V., Savic, D.A., Grayman, W.M., Beckwith, S., & Koelle, E. 2003. *Advanced Water Distribution Modeling and Management.* Waterbury, CT: Haestad Press.

Chapter 6

Extended-Period Simulation

6.1. INTRODUCTION

Extended-period simulation (EPS) is a technique for modeling a distribution system where a series of steady-state simulations at regular specified intervals are performed over a specified time period. While a steady-state analysis is useful for sizing pipelines and supply facilities, an EPS analysis provides a great deal more information about system operating characteristics and how the water system responds to changing demands and emergency situations. This information helps properly size booster pumps, determine fluctuating reservoir levels and size of the reservoirs, refine water supply and distribution system improvements, determine the best locations and sizes of future storage and pumping facilities, and set system control valves. An EPS model is also needed to perform all water quality and energy analysis simulations. EPS can also be developed from an existing steady-state model. It is recommended that a calibrated steady-state model be used as the starting point for an EPS model; this will reduce the number of parameters needed during EPS calibration.

While the length of an EPS scenario depends on the use of the EPS scenario, a typical extended-period analysis used strictly for hydraulics (no water quality) is often simulated for a 24-hour period or for multiple 24-hour periods during average and maximum demand days. Water quality runs are typically much longer because the shorter average or maximum day EPS runs are repeated over several days, weeks, or even months until the system water quality parameters being examined stabilize and reach a steady, repeatable pattern.

> ▶ **KEY USES OF EXTENDED-PERIOD SIMULATIONS**
>
> - Evaluate storage capacity and operation
> - Evaluate pumping capacity and controls
> - Evaluate operational controls and set points
> - Evaluate infrastructure interaction
> - Analyze water quality
> - Examine what-if scenarios such as shutdown and emergency scenarios
> - Optimize energy use
> - Operator training

EPS scenarios are analyzed at user-selected time intervals, typically in 1-hour intervals so that a 24-hour analysis is analyzed at a minimum of 24 periods. The minimum number of periods used depends on the simulation length and the interval period selected. Smaller time intervals – for example, minutes – can be used depending on specific system constraints and characteristics. Some applications of EPS are described in the following sections.

6.1.1. Reservoirs and Tanks

When one is simulating storage tanks and reservoirs, the volumes used as floating storage, minimum emergency reserve, and speed of refilling are easily calculated. An EPS uses a steady-state analysis subroutine to determine pressures and flows in the distribution system and a subroutine to perform a time integration of reservoir node flow in or out of the system to track reservoir volume changes as the EPS run progresses. This allows the model to dynamically track the water level changes during the simulation period as the level reacts to system demands and pump flows.

6.1.2. Pressure Zones

When one is modeling individual pressure zones in a steady-state model or independently, it is nearly impossible to accurately predict how the boundary conditions of one zone impact an adjacent zone under varying supply and demand conditions as conditions change throughout the day. Experience shows that entire water systems with multiple pressure zones are best modeled together in order to accurately simulate real water system operation when pressure zones are interconnected. For these complex, interconnected water system pressure zones, an EPS model simulates how an entire water system interacts dynamically. This is done by varying reservoir water levels, turning booster pumps on and off, and operating pressure and flow control valve between pressure zones.

6.1.3. Water Quality

A prerequisite for modeling water quality in distribution systems is to have a model that has been fully calibrated for extended-period hydraulic conditions, because water quality degrades over time and cannot be analyzed using only steady-state, snapshot-type model runs. Water quality parameters that can be simulated include age of water in the system, percent of water derived from a source, and growth or decay of one or more chemical constituents.

> **► KEY DATA FOR EXTENDED-PERIOD SIMULATIONS**
>
> - Diurnal demand curves
> - Operational controls
>   - Pumps
>   - Valves and pressure-reducing valves
>   - Flushing schemes
> - Tank shape (depth–volume) curves

6.2. EXTENDED-PERIOD SIMULATION BASICS

All of the data used in steady-state simulations are also used in an EPS. However, additional data are needed to describe how demands, pumps, valves, and reservoir levels change over time. Changes in demand are identified in the model using diurnal curves, which tell the model how demand varies over the course of the simulation. Operational controls are used to control the operation of pumps, valves, and pipes in the model to simulate actual system operational changes that occur during the simulation period.

This section describes the basic information used for an EPS with a model. Additional information used for a water quality model is provided in chapter 7.

6.2.1. Objective

Steady-state water system modeling can describe system performance under a variety of demand conditions such as normal conditions of average day or maximum day demands or non-normal system conditions such as maximum day demand plus fire load, peak hour demand, or low-flow conditions with minimum system demands (e.g., to assess water transfer to storage). The steady-state evaluation is essentially a snapshot of the system under a fixed and defined demand situation. However, steady-state modeling cannot be used to assess what happens before or after these conditions occur and it cannot be used to assess how the system responds to changes in system demands and operating conditions over time. An EPS is an extension of a steady-state simulation in that EPS is a sequential series of steady-state evaluations that are conducted over time to account for changes in system conditions. The EPS model also calculates flow in and out of storage tanks during the simulation and adjusts the tank level based on the shape and volume characteristics to reflect the expected tank levels. This allows the model to dynamically simulate system operation over time, which is the primary basis for an EPS.

An important use of modeling is to identify and remedy major system performance deficiencies. In addition to the normal steady-state indicators such as pipe velocity and junction pressure, EPS performance indicators include storage use, pump capacity, flow reversal, energy usage, and water quality.

6.2.2. Time Intervals and Simulation Duration

The EPS time interval is set based on two factors: the simulation period, which is the interval or duration necessary to achieve the objectives of the simulation, and the time-step interval, which is the minimum time between calculations and must be appropriate for equipment controls to function properly. EPS modelers have to first decide what length of time will be used for the simulation period. The most common simulation length is 24 hours, although lengths of two to three days to a week are becoming more common based on the ultimate goal and use of the model. Selection of the simulation period for the EPS model is often governed by the length of time needed to calibrate the model but ultimately depends on the variation in demand and system controls. The ultimate goal is to create a simulation length where system operation and demand variation are captured and repeatable. For example, if system controls on a tank will turn on pumps once a week, a 24-hour simulation period may not fully replicate operation for the entire week. The modeler must use good engineering judgment to determine the most appropriate simulation period length to use based on the ultimate primary use of the EPS scenario. It should be noted that available budget may also influence the final decision, as longer simulation time frames are needed to develop and calibrate the model and thus will cost more to produce.

The time-step interval used for the simulation period must also be selected. EPS assumes that system changes that happen during a time-step are linear. Any nonlinear changes must happen at the time-step. The most typical time-step used in EPS model runs is a 1-hour interval. This means a 24-hour simulation period or run would be separated into 24 separate periods. Each time-step interval used in the model typically would require a separate point in the diurnal curve and is the minimum number of instantaneous points that the model would analyze over the simulation period. In addition, each time-step selected generally represents a point that would be used during calibration when comparing model results with field results. With the expansion of available supervisory control and data acquisition (SCADA) data, which are currently available in increments as small as less than 1 minute, modelers are increasingly using smaller than 1-hour time-steps in EPS models, as it is easier to get the data needed. Again, good engineering judgment must be used when selecting the appropriate time-step interval so that the model represents system operation accurately enough for the conditions modeled. It is important to note that the model initiates a new time-step if there is a control for any interval operating as specified at that time. Water quality models generally use much shorter time-steps than strictly hydraulic models.

If diurnal demand patterns based on a smaller time interval are used, the modeler should be aware that small errors in the recorded tank levels can cause oddities in calculated flow into or out of the tank as the time interval decreases from the typical 1-hour value. These oddities occur when the tank volume becomes a greater and greater portion of the demand–supply–storage mass balance as the time interval shrinks. This, in turn, magnifies small errors in the recorded tank levels and can result in diurnal patterns that are very jagged or potentially negative rather than smooth in appearance. As the time-step decreases, the error in volume due to measurement limitations of the tank level can introduce more error in the mass balance. Users should be aware that the accuracy of diurnal curves can be impacted by this factor and be cognizant of this when choosing a time interval that is smaller than 1 hour.

6.2.3. Simulating Different Demand Conditions

Utilities establish an acceptable level of service to satisfy customer demands based on local ordinances and applicable regulations. More information on common design criteria including typical minimum and maximum pressure, velocity, and head loss constraints is provided in chapter 5. Most utilities have a goal of providing continuous delivery of high-quality drinking water to all customers within a reasonable range of pressures. This goal assumes that adequate pumping, storage, and redundant critical system components are available to provide an acceptable level of service during normal or maximum demand conditions and even during emergency operations (e.g., firefighting, equipment maintenance, or natural disasters). Regular water outages or other lengthy interruptions in water service are not tolerated in most US communities. However, the acceptable level of service may differ between normal and emergency conditions. Extended-period analyses are often used to verify that the proposed operation satisfies the utility's design criteria and can also help identify operating conditions that can lead to substandard levels of service.

6.2.3.1. Maximum Day. Often, a water system is designed to sustain continuous maximum day demands supplied from the water source over a 24-hour period rather than relying on storage to meet the maximum day demand. Water storage facilities are capable of supplying instantaneous peak demands in excess of maximum day supply and of storing water during low-demand periods. This philosophy ensures that the system meets the system demand under any conditions, minimizes fluctuations in supply, and optimizes storage requirements. Conditions are deemed acceptable if the storage tank volumes at the end of the 24-hour period return to their initial levels.

6.2.3.2. Maximum Week. If the supply is constrained or if the possibility of a maximum day event occurring over a series of consecutive days is not great, the system can be designed for a longer period of recovery. Again, the supply over that longer period should equal the total system demand over that period, and the storage should be capable of supplying instantaneous demands over the available supply and of storing supply in excess of low-usage periods. This philosophy allows for a smaller quantity of supply relative to maximum day demand but does require additional storage. As discussed in chapter 7, this philosophy may have adverse impacts on water quality.

6.2.3.3. Time of Day. Other operating philosophies allow for variations in conditions according to the time of day. Some utilities only operate their treatment facilities for one or two shifts during a day. Some power suppliers make more attractive rates available to water utilities if the power demand occurs during the power suppliers' off-peak hours. The water utility can then respond by operating more power-using facilities such as treatment plants and pumps during off-peak hours. These time-of-day philosophies can result in cost savings for labor and power; however, they may require additional, possibly significant, capital cost outlays to provide water during these specified times of day through additional storage volumes or additional pumping capacity.

6.2.3.4. Minimum and Average Day. An EPS of minimum and average demand conditions is often used to plan daily operations or to assess water quality. As discussed in chapter 7, water quality analyses generally require simulations of very long duration that exhibit repeating cycles of tank filling and draining as supply sources are turned on and off on a regular schedule.

6.2.3.5. Seasonal. Many water systems have water sources that are used differently at different times of the year. These operating conditions could require as many separate EPS configurations as there are different operations. With multiple operating configurations, the most extreme configurations in terms of the performance criteria being evaluated should be selected. Often the variations are dictated by differences in water quality or availability of the source flows.

6.2.4. Diurnal Demand Variations

Diurnal curves need to be developed for EPS runs to show how water use for an individual or group of given customers/users varies over the simulation period. Diurnal curves are developed using a mass balance over a region, where all flows in and out of the region as well as variations in tank water storage are added together to determine the demand for that period. Typically, a mass balance is calculated on an hourly basis for each hour of the EPS period used in the model (see section 6.2.2 regarding time interval selection).

Developing diurnal curves that define variations in demand in the system over time can be a time-consuming component of extended-period modeling. It is also one area of modeling that requires the most use of the modeler's engineering judgment regarding what is reasonable and what is not. The modeler must decide whether to use a single diurnal curve for all demands in a zone or to associate different curves with different categories of customers (e.g., residential versus commercial) and then make sure the combination of different customers and diurnal curves matches the system-wide patterns. Separate diurnal curves are often calculated and assigned to large users, especially if they have a unique water use pattern. Development of diurnal curves is discussed in chapters 2 and 5 as well as later in this chapter. Figure 6-1 presents typical diurnal demand patterns seen in a distribution system with multiple user types. See the inset for simple checks that can be made to avoid the most common errors associated with diurnal curves.

► KEY INFORMATION TO CHECK FOR DIURNAL PATTERNS

- Pattern is generally fairly smooth
  - Large, sharp increases or decreases in demand may indicate missing data or an error
- Make sure to complete a mass balance on the correct time period; flows and tank levels should be on the same hour
- Sum of total diurnals
  - Average demand multiplication factor equals 1
  - Sum of total points equals the number of simulation intervals; that is, 24 points sums to 24, 72 interval diurnals sum to 72

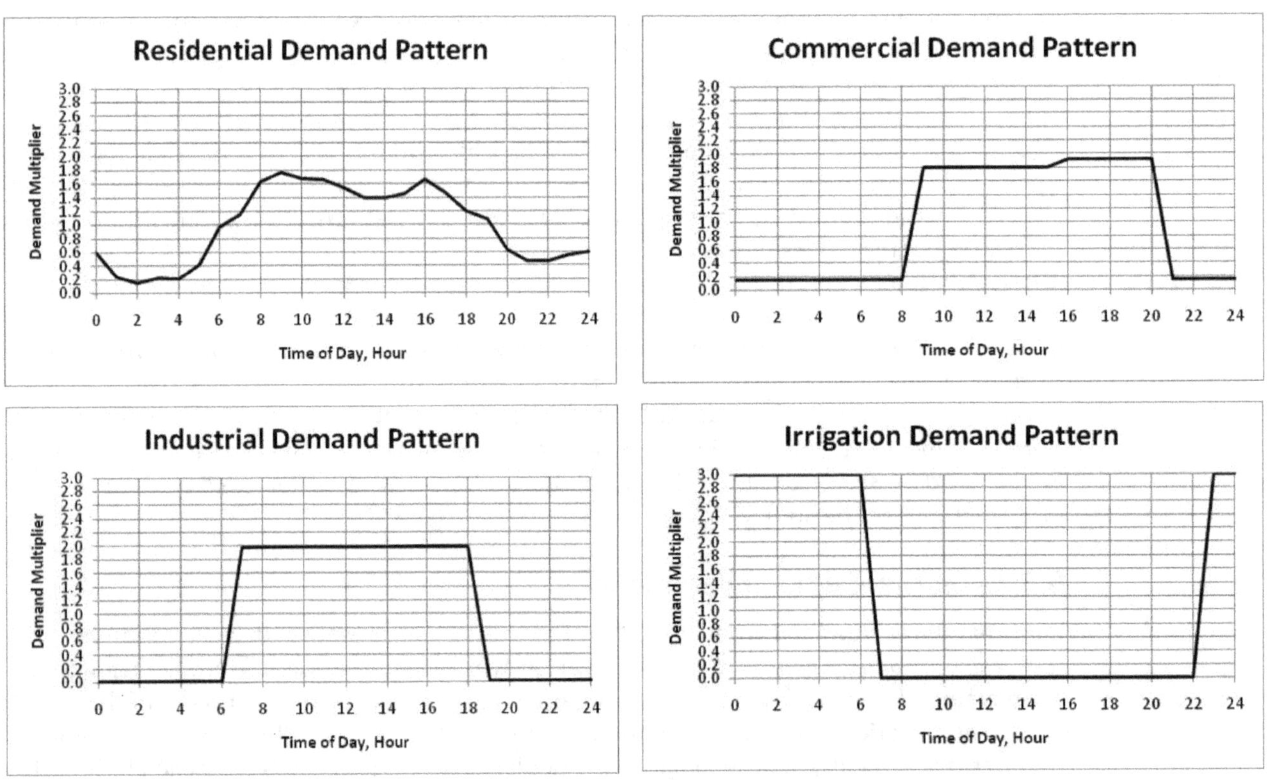

Figure 6-1 Typical diurnal demand patterns for different use categories

6.2.5. Verifying System Operation

Before any steady-state or EPS model is used extensively, calibration and validation efforts should be completed to verify that model results reasonably reflect the actual system performance. Information on how booster pumps are controlled in an extended-period computer run is important when turning pumps on and off in a sequence that is similar to the sequence in the real system. Other system equipment such as isolation and control valves must have information related to how they are controlled. Reservoir water levels and pumped flows during an extended-period analysis should agree with historical field observations of water level fluctuations and flow rates. Regulating valves and general distribution system pressures contained in computer output should also be similar to conditions experienced in the field.

6.3. EXTENDED-PERIOD SIMULATION SETUP

6.3.1. Initial Conditions

The initial hydraulic conditions of an EPS are the same as the boundary conditions used in a steady-state analysis in terms of tank levels and pump run status. Care must be taken to match the time of day at which the model simulation begins with the time period used in the diurnal curves. Typically, an EPS begins at midnight when system conditions can be markedly different from peak demand conditions when steady-state simulations are often run. The initial conditions used for the EPS will be based on operational constraints and on the scenario that is in place or to be tested by the modeler. These typically include initial tank levels and the pumps and major valves that are in operation at the start of the simulation. Additional efforts are required to determine the initial conditions necessary for water quality modeling, as discussed in chapter 7.

► KEY INITIAL CONDITIONS FOR EXTENDED-PERIOD SIMULATION ANALYSES

- Tank levels: Set low enough to prevent overflow but high enough to maintain adequate pressures
- Pumps: Compare pump on–off status with controls to avoid conflicts
- Control valves: Confirm flow or pressure settings
- Isolation valves: Confirm open–closed status (may be represented as pipes)

► KEY TANK INFORMATION FOR EXTENDED-PERIOD SIMULATION ANALYSES

- Operating range: Maximum, minimum, and initial levels
- Shape or depth/volume relationship
- Inlet–outlet piping configuration

6.3.2. Tanks and Reservoirs

While tanks and reservoirs are used fairly interchangeably in the water industry to describe water storage devices, most model software packages treat them differently. In software that uses the EPANET simulation engine, tanks are used to describe water storage devices with limited capacity, while reservoirs are used as infinite water sources or water sinks with a fixed head or a head that varies independently from the system based on a known pattern. This section focuses on the modeling requirements needed to set up storage devices within the model for extended period analyses. More information on selecting tank sizes and locations is provided in chapter 5.

One of the most important pieces of information required for an EPS is the amount of service storage available and the characteristics of that storage. Required information on storage facilities for EPS modeling includes the following:

- Tank elevation: Represents the elevation of the physical bottom of the tank;
- Overflow elevation: The maximum water surface elevation before water will overflow;
- Minimum water level: The lowest water level in the tank where water can exit the tank. Note: Many models generally ignore the tank volume below any minimum height assigned to the tank for usage and for percent full calculations, so caution should be used when assigning a minimum height to the model.
- Tank physical area attributes: Uniformly circular, rectangular, irregular at bottom or top;
- Storage volume/unit of height: The amount of water contained per unit of height;
- Circular tanks or constant area prismatic tanks: Assign the tank diameter or equivalent diameter for prismatic tanks to the model. For tanks that have a variable area with depth, a depth-to-volume relationship curve based on reservoir capacity (often in cubic feet) is required for the model. An example of this relationship is shown in Figure 6-2 for a spherical tank.

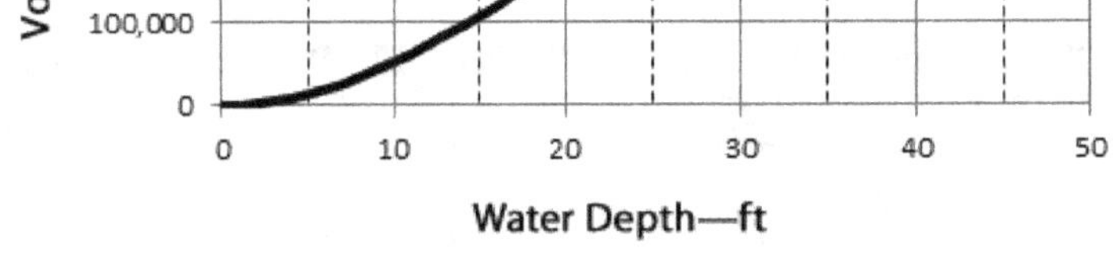

Figure 6-2 Depth–volume relationship for a spherical tank

- Effective height: The difference in elevation between the maximum and minimum water surface elevation;
- Piping configuration of inlets and outlets: Some tanks have a single pipe that serves as both an inlet and an outlet. Other tanks have separate and sometimes multiple inlet and/or outlet pipes. If a tank has an inlet pipe that discharges into the tank above the maximum water level, a node can be added with an elevation above the overflow elevation of the tank. The node is connected to the tank by means of a check valve to allow flow into the tank only. Some modeling programs accommodate differences in tank inlet and outlets and even simulate flow into the tank above the high water level.

As noted, model reservoirs serve as an infinite water source or sink with a fixed head or a head that varies independently of the system. They are used to represent a water source or a known location of fixed hydraulic head in the network for groundwater and turnout supplies, as well as water treatment plants in the network. It is critical to correctly assign head values to system source reservoirs to ensure proper model operation. Source reservoir head values represent the water level that water would rise to if allowed and are calculated as the elevation plus the system pressure converted to feet of water.

6.3.3. Pump Information

Pump station setup requirements for EPS runs are more detailed than in steady-state model simulations. Accurate individual pump curves are essential for an accurate extended-period model because the fluctuations in suction and discharge pressure at a pump station that will occur during the EPS will impact where the pumps will operate on their pump curves. The actual station operating parameters, or control set points, that are used to turn pumps on and off based on the value of a system control point (e.g., tank water level or system pressure) are not required for steady-state runs but are essential for EPS runs. Model pump controls require the user to input control set points and to identify the location of the controlling point in the system (reservoir or system junction point). Key pump station information for the model that is pertinent for EPS runs includes the following:

- Certified pump head capacity curves for each pumping unit within the station pump; controlling point location:
  - Multitype pump curves with six or more points are often the most accurate, but three-point or one-point curves can be used by most packages when full pump curve data is unavailable;
  - Certified, field-tested pump curves are preferred;
  - Pump efficiency curves can also be used, especially when conducting energy analyses;
- Pump control information:
  - Control set points to turn a pump on or off, adjust a pump's speed or a valve's setting, or to open or close a pipe or valve;
  - Controlling elements – generally tank levels or node pressures but sometimes flow through a pipe or valve – must be identified;
  - Pumps may also be controlled on a time basis;
- Initial status (on–off) for each pump:
  - Most modeling software assumes a pump is running initially unless an initial status or pump control identifies the pump as off at the start of a simulation;

- Detailed piping configuration of suction and discharge piping, including estimates of minor loss coefficients for fittings. It is often useful to identify key junctions to be used to compare model pump station suction and discharge pressure and discharge flows when laying out a pump facility in the model. The elevations in the model must match the actual elevations of the sensing instruments.

Care should be taken to not set pump on–off pressure settings too close in order to avoid several short-duration time-interval calculations. Also, modelers should check that pump starts and stops do not adversely affect other modeled components such as other nearby system booster pumps and control valves and that system pressure data accurately reflect real system reactions.

6.3.3.1. Booster Stations With Variable-Frequency Drives. Booster stations in the model need to have controls created in the model to allow the software to turn pumps on or off as needed during the EPS. Most models represent each pump individually and establish individual control settings to simulate actual pump settings in the real system. For example, as system pressure or reservoir level drops, additional pumps "turn on" in the model.

When one is modeling pumps with variable-frequency drives, similar pumps operating in parallel are typically modeled as one pump, with the number of parallel pumps indicated and controlled via a discharge pressure or flow or a remote pressure setting. Many software packages allow pumps to be controlled by a pressure or flow set point and modify pump speed to match the parameter controlled. If the modeling package does not have variable-speed capabilities, the modeler can often reasonably simulate variable-speed pump operation by placing a flow control or pressure control valve downstream of the pumps to control the pump flow by discharge pressure or by flow.

6.3.3.2. Wells. Creating wells within the model requires the user to include additional data in the model that booster pumps do not require. As mentioned in chapter 2, the most critical additional piece of information is that a water source or reservoir is needed to represent groundwater supply that the well pump draws water from. The head of the groundwater reservoir should be taken as the pumping groundwater level when the pump is operating. The water level drawdown represents the difference between the static water level in the well and the pumping water level in the well when the well pump is operating. If necessary, the variation of well drawdown with flow rate can be simulated using a control valve or general-purpose valve to account for the additional head loss as the flow increases between the source reservoir and the pump.

► KEY PUMP INFORMATION FOR EXTENDED-PERIOD SIMULATION MODELING

- A good pump curve ideally based on a certified pump curve
- Pump initial status: Is the pump running at the start of the simulation?
- Pump control details: How the pump is turned on and off
  - On and off set point values
  - Identify what parameter and element are evaluated for control (tank level, system pressure)
  - Time controls can also be used

The well groundwater pumping level and pump curve are generally input into the EPS model for well pumps. As with booster pumps, controls for well pumps also require the location of the controlling point in the system (tank, pipe, or system junction point) and that the initial condition be input to allow the well to function in the model as it would in the real world. Alternately, the pump curve can be modified to represent the discharge pressures at the well if the groundwater levels and drawdown information are unknown. More information on modeling well supplies can be found in chapter 5.

6.3.3.3. Other Supply Sources. Turnout connections to regional water supply systems, major transmission pipelines, and interconnections to other water distribution systems are associated with many water systems and can be modeled in multiple ways. Modelers can use a negative demand at each node that corresponds to these sources. These demands can be constant or vary following a predefined pattern. These connections can also be modeled as constant-head or pattern-defined varying-head reservoirs. The primary downside to using negative demands is that they can create unrealistic head conditions in the model. If used, negative demands should always be checked to make sure the system head is correctly represented by the negative demand.

An alternative method for modeling a source supply is to simply add a source reservoir and a control valve to the model to represent water supply from the supply source. The head of the reservoir only needs to be high enough to allow the highest expected flow into the system and may only need to be 50 to 100 ft (20 to 30 m) higher than the system head at the connection point to the system. A flow control or pressure control valve is used downstream of the reservoir to control the supply into the water system based on either pressure or flow control. This technique keeps conditions in the system from being modeled independently of the supplying system.

The additional advantage of using a reservoir and control valve is that flow and/or discharge pressure at the connection point from the additional source can much more closely match actual system operation. In addition, model controls can be used to regulate system supply from that source in a way that is consistent with how the actual system operates and how most users control model operation. However, if the turnouts are modeled, the extended-period results should be checked to ensure that flow and discharge pressures from the turnout are as expected. This will ensure model accuracy.

6.3.4. Control Valves

Proper control valve operation is critical to performing an accurate ESP. Typical control valves include PRVs, PSVs, flow-control valves, and altitude valves. Proper pressure or flow settings and valve elevations are important if these elements are to accurately represent the real system within the EPS model simulation.

Control valves are defined by the type of function they perform. Some types of control valves operate automatically based on pilot settings at the valve and respond to pressure conditions directly up- or downstream of the valve. These include PRVs, flow-control valves, PSVs, and pressure-breaking valves. In most modeling programs the valve control settings are input directly with the valve's characteristics but can be overridden with initial status or control statements. Other types of valves such as throttle control and general purpose valves regulate flow by simply inducting head loss on the valve based on curves that specify head loss versus flow or the minor loss coefficient of the valve specified by the modeler. Specific valve types for specific modeling software may also have additional capabilities to perform additional control functions.

Any model valve that does not operate continuously over the entire EPS will require initial valve status and/or control statements to control the operation of the valve. As with pump controls, the user will need to define the on and off set points for the valve, the parameter and element to be used as the location of the controlling point in the system (e.g., tank, pipe, or system junction point), and the initial condition to be input in order to allow the valve to function in the model as it would in the real world.

Modelers need to use caution when creating control structures for valves, as they often need a slightly different control structure than pumps in order to be made active in the model. For instance, in programs that use EPANET as the hydraulic engine, changing the status of a valve to "open" does not make the valve active but rather makes EPANET treat it as an open pipe. Instead, valves are made active in control statements in EPANET by assigning a setting value to the valve. Users should check their specific software requirements for valve controls to make sure they operate as intended in the control statements.

▶ KEY CONTROL-VALVE TYPES

- Altitude valve: Controls water supply to a storage tank based on the water level in the tank
- Pressure-reducing valve: Maintains a set pressure downstream of the valve by varying the flow supplied
- Flow control valve: Controls supply to the system based on a specified flow value
- Pressure-sustaining valve: Modulates the flow supplied by the valve to maintain a specified upstream pressure setting

There is the potential for a model to fail to converge to a solution if valve operation is based on control statements or on the built-in operation of the valve. For instance, in EPANET, multiple parallel PRVs that are hydraulically close to one another have downstream head values that they are trying to maintain. These values are within roughly 5 to 10 feet (2 to 3 m) of one another. This may generate conflicts between the valves and cause the model to fail to converge to a solution during one or more periods of the EPANET solution. This is often the case in PRV vaults with multiple valves in the same vault. If possible, the user can either adjust the valve settings to be slightly apart to reduce the conflicts or increase the number of solution iterations the model can use. Some software packages have advanced modeling parameters such as the relaxation factor to potentially assist in convergence. A full discussion of resolution of these issues is beyond the scope of this chapter. However, the modeler needs to be aware that inaccurate valve settings and valve layout can impact the model's ability to consistently converge to a solution.

6.4. SCADA INFORMATION

An EPS is designed to mimic water system operation in response to system demands over time. The more precise and detailed the information included in the model, the greater the model accuracy and the greater the potential rewards of the evaluation. It is essential to obtain two basic pieces of information in order for extended-period system analysis to be accurate. The first is an accurate description of system demand fluctuations over a defined time period. The second is a detailed description of the operating parameter changes (e.g., pump on–off) and the water system's reaction to those system demand fluctuations (storage volume changes) over the same finite time period. Is it essential that system status information be obtained every 24 hours when describing system peak day events for a water system. This information is also useful in describing static model evaluations.

However, 24-hour system status information is insufficient in developing an EPS model. Central in obtaining system operation parameter changes and reactions to the water system is a SCADA system that monitors operation parameters as they change and stores information in an easily retrievable form for use in the model. This information, broken into finite time intervals such as hourly (at a minimum), is the standard for measuring system performance versus model predictions.

▶ TYPICAL USES OF SUPERVISORY CONTROL AND DATA ACQUISITION INFORMATION

- Initial conditions: As recorded at the time, representing simulation time zero
- System controls: The times when an element status changes (e.g., a pump coming on or off) when using time-based controls
- Demand multiplier: The ratio of total measured production to modeled demands
- Calibration:
  - Tanks: Levels
  - Pumps: Flows and pressures
  - Pipes: Flows
  - Junctions: Pressures
  - Valves: Flows and pressures
  - Controls: Confirm status changes in response to set point changes

Following is the basic SCADA information often used for an EPS:

- Reservoir fluctuations over time,
- Pump run times and flow variations over time,
- System pressure fluctuations over time,
- Source flows for major supply points, and
- System control valve status, pressures, and flow rates, if measured.

6.4.1. Mapping SCADA Data to Models

In most cases, SCADA data is used in several ways for extended-period modeling analysis (Figure 6-3). These include establishing the initial status of pumps, valves, and tanks; defining the time varying patterns at system boundaries, such as tank levels or flows through pumps or inflow points; defining or confirming controls that govern pump or valve operation; and, most important, comparing modeling results with system measurements.

Prior to obtaining SCADA data for use in EPS models, it is important to know the time intervals and run durations that will be used in the model (e.g., a 24-hour run at 1-hour intervals). The next step is to assess the types and formats of the data that are available in SCADA, because each system is slightly different and SCADA systems are not always set up with modeling in mind. Many SCADA systems have historians or other types of report generators that can provide information for the models. It is important to coordinate with the operators or integrators in charge of the SCADA system to identify the format used by the SCADA. Often, the SCADA data can be exported in a file format that can be opened in a spreadsheet or database program. The data can then be reformatted to match the needs of the modeling software, as necessary. It is critical that the units in the SCADA data match the units in the model; if they do not match, the SCADA data units must be converted accordingly.

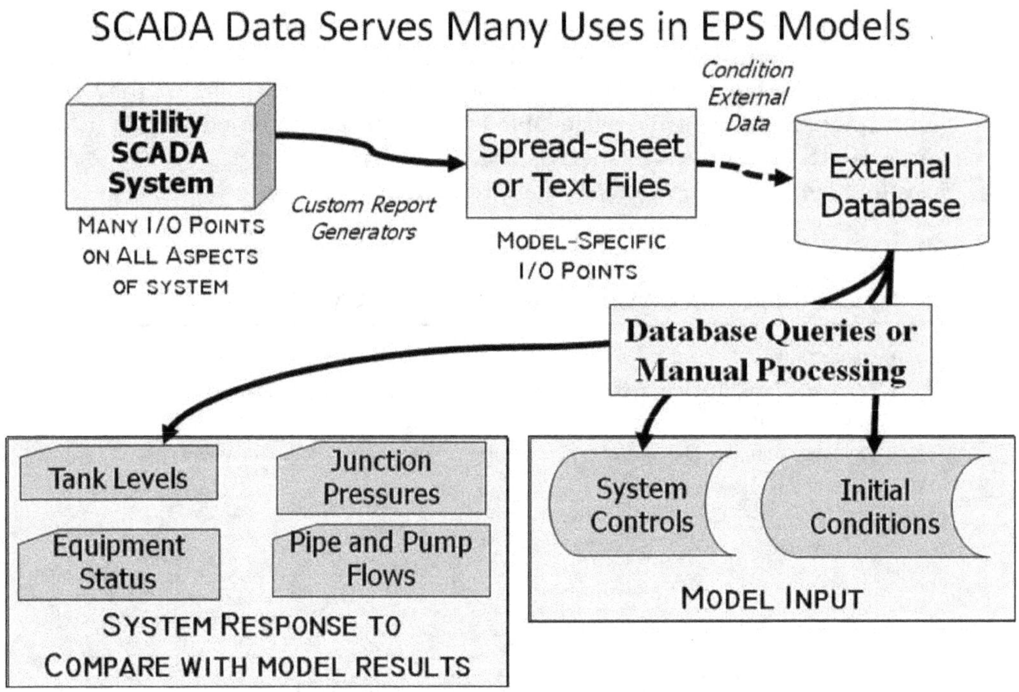

Figure 6-3 Using supervisory control and data acquisition data in extended-period simulation models

Abbreviations: EPS, extended-period simulation; I/O, input/output; SCADA, supervisory control and data acquisition.

Many important considerations for using SCADA data are discussed in chapter 2. When SCADA data are used for EPS models, the time interval of the data is one of the most important issues to consider. This is because the time intervals at which SCADA data are polled can be very frequent, usually in minutes or seconds. Even so, issues can arise when abrupt changes occur. For example, the polling interval may be 20 minutes and the pump may be off at 3:23 and on at 3:43. There is usually no way to know exactly when the pump status changed. Except in specialized cases, EPS models generally require time intervals of several minutes or hours. Furthermore, there may be considerable noise in the data recorded by field instruments, and this can cause issues. Signals may be dropped or latched due to communication problems. SCADA data may require some cleanup before it can be used. Rather than using SCADA data at intervals that match the model's intervals, it may be helpful to use more frequent SCADA intervals and derive a running average of values at the model intervals to filter out noise. In other cases, the SCADA system may only record data on change in status of the variable being recorded. In these cases, you may have to extrapolate the data for time intervals without recordings.

6.4.2. Modeling Without Electronic SCADA Data

While using electronic SCADA data certainly makes EPS modeling more convenient, lack of SCADA data does not mean EPS modeling is impossible. There is likely some way to measure flows and tank levels in place that is used by operations staff. This information may be in a different format, such as a paper circular or strip charts. Hourly data can be extracted from these charts and tabulated for use in EPS models. Generally, daily operations logs indicate when pumps were turned off or on, tank levels at the end of the day, and total water production. Although the degree of accuracy of the model results may be reduced, the data are critical for developing diurnal curves and for calibration/validation efforts. In the absence of permanent SCADA instruments, flowmeters and pressure sensors can be installed on a temporary basis, with the system information from these instruments saved on local data loggers. Stored data can be downloaded into portable computers and copied for use in the model.

6.5. MODELING CONTROLS

An EPS relies on controls within the model to govern how selected elements (typically pumps, valves, and pipes) function during the course of the simulation. Controls allow pipes to be opened or closed, pumps to be turned off or on or have their operating speeds changed, and valves to be opened or closed or to have their settings changed. Controls work when a condition occurs in the model that then causes an action by one or more of the elements in the model. Conditions that trigger controls can be based on time, on the status of other model elements, or on system conditions. Examples include turning on a pump based on a tank water level, junction pressure, or flow in a pipe.

► SIMPLE VERSUS COMPLEX CONTROLS

- Simple controls:
  - Single element controlled based on time or on conditions at a single element
- Complex controls:
  - Multiple elements controlled by conditions at a single element
  - Single element controlled by conditions at multiple elements
  - Combination of both

Modeling programs handle controls in different ways, and users are cautioned to become familiar with the software's guidelines. It is important for the modeler to carefully review the software documentation to gain an understanding of how the control statements are set up and created. Then the modeler can determine the best way to model the control statements needed to accomplish the desired model behavior in the EPS runs.

In general, controls are simple or complex. Simple controls act on a single element based on a single condition, such as turn on pump X when tank Y level drops to Z feet. Complex or rule-based controls can operate on multiple elements based on one or more conditions, such as turn on pump X1 and pump X2 when tank Y drops below Z feet and tank W is also above V feet.

The importance of understanding how rules work can be illustrated in EPANET. In EPANET there are two major types of control types – simple controls and rule-based controls – that are evaluated differently by the program. EPANET's simple controls are evaluated at the beginning of any iteration and initiate changes before the iteration is run. EPANET's rule-based controls, however, are evaluated at the end of an iteration; any status change for an element is not executed until the time equal to one rule time-step is elapsed. The differences in the control philosophy used for simple and rule-based controls result in slight differences in model output for each rule type, with rule-based controls having a slight delay in implementation. As such, the modeler is encouraged to fully understand the control structure used in the software package to ensure proper model operation of the control statements. In many cases, program interfaces for accessing simple and rule-based controls are different. Modelers are advised to consult the user's manual before attempting to use complex controls. For example, rule-based controls are necessary if multiple conditions must be met before an action is taken. This can be helpful if a pump needs to be turned on or off using tank levels that have different on and off set points for different periods of the day.

All control statements typically consist of an action statement (on, off, or a setting change) for a pipe, valve, or pump that is to be completed if certain criteria (conditions) are met at the element used for control, such as the flow in a pipe, the pressure in a junction, or the level in a tank. Figure 6-4 illustrates some examples of controls in EPANET.

Users must be prepared to gather information regarding how the real water system is operated and use this information to develop control statements within the model that can simulate this behavior. System control data are often best gathered from operations staff who understand how the system is operated and who have access to the control set points that the specific system uses. It is essential to set aside time to interview operations staff both before and after developing control statements to develop an accurate EPS model. In addition, follow-up questions to operations staff during calibration can help answer any questions that may arise and best ensure that model controls are replicating system behavior as accurately as desired.

Control Statement	Meaning
Simple Controls	
LINK 12 CLOSED IF NODE 23 ABOVE 20	Close link 12 when the level in tank 23 exceeds 20 ft (6 m)
LINK 12 OPEN IF NODE 130 BELOW 30	Open link 12 if the pressure at node 130 drops below 30 psi (207 kPa)
LINK 12 1.5 AT TIME 16	Set the relative speed of pump 12 to 1.5 at 16 hours into the simulation
LINK 12 CLOSED AT CLOCKTIME 10 AM LINK 12 OPEN AT CLOCKTIME 8 PM	Link 12 is repeatedly closed at 10 AM and opened at 8 PM throughout the simulation
Complex or Rule-based Controls	
RULE 1 IF TANK 1 LEVEL ABOVE 19.1 THEN PUMP 335 STATUS IS CLOSED AND PIPE 330 STATUS IS OPEN **RULE 2** IF TANK 1 LEVEL BELOW 17.1 THEN PUMP 335 STATUS IS OPEN AND PIPE 330 STATUS IS CLOSED	This set of rules shuts down a pump and opens a bypass pipe when the level in a tank exceeds a certain value and turns the pump back on and closes the bypass when the level is below another value
RULE 3 IF SYSTEM CLOCKTIME ≥8 AM AND SYSTEM CLOCKTIME <6 PM AND TANK 1 LEVEL BELOW 12 THEN PUMP 335 STATUS IS OPEN **RULE 4** IF SYSTEM CLOCKTIME ≥6 PM OR SYSTEM CLOCKTIME <8 AM AND TANK 1 LEVEL BELOW 14 THEN PUMP 335 STATUS IS OPEN	These rules change the tank level at which a pump turns on depending on the time of day

Figure 6-4 Examples of controls in EPANET

SCADA data can be an invaluable tool for the modeler to verify system operation as well as to back-check the key on and off set points provided by the operations staff. In certain cases, control on and off set points initially provided by operations staff may not match the actual system operation and may need to be clarified and potentially adjusted during calibration. For instance, if a pump off level for a tank is identified as 20 ft, but the pump is routinely seen in the SCADA data to turn off at 24 ft, a discrepancy exists that should be clarified with operations staff. Control on and off points for all elements controlled in a model are generally validated during the calibration phase, where differences between control values are normally easier to see.

In most cases, actual system controls can be replicated easily within the model software. However, in certain cases, the model software cannot be use to completely replicate actual system control due to software limitations. In these instances, the modeler must use good engineering judgment and develop controls that replicate system behavior as closely as possible. SCADA data can also be used to determine how best to model a behavior or to develop a control statement that best replicates model behavior. For instance, if an altitude valve generally flows at a fairly constant flow rate when open, the modeler could reasonably replicate this behavior using a flow control valve, even if the actual valve is operating as a pressure-sustaining valve. The modeler can use SCADA information to understand system operations and guide decision making on how to best replicate this behavior using control statements in the model.

Other important considerations for using controls include the following:

- Understand the units being tested for set point triggers. For example, junction set points may be based on pressure or hydraulic grade line (HGL) elevation.
- Some programs allow for assignment of priority to controls if multiple conditions occur simultaneously. The control with the higher priority will be executed. In other cases, controls are based on the order in which they appear in the input tables.
- Care must be taken to avoid conflicting controls or opposing actions based on too narrow a band of conditions. These can cause instability in the calculations.
- Pressure, head, and level controls must use appropriate data that are consistent with system controls.

6.6. EXTENDED-PERIOD MODEL CALIBRATION

EPS calibration involves simulating a series of conditions to represent dynamic operations and demand fluctuations that occur over a specific time range, commonly 24 hours or longer. As a first step, model results are plotted against time to compare with measured values, as shown in Figure 6-5. Calibration typically focuses on a 24-hour simulation because most water distribution systems follow daily repeatable variations. System components that can be adjusted during calibration include operational controls, diurnal demand patterns, piping configuration, and pipe friction coefficients. Calibration can also result in changes to model simulation settings. For example, if a time-step setting is too large such that the model is not adequately capturing pump status changes that are occurring in the actual system, the time-step setting may need to be reduced to proceed with calibration.

Calibration of an extended-period model is important because an extended-period model often serves as a base for performing advanced evaluations such as energy or water quality analysis. Discrepancies between actual data and model data can be accentuated when hydraulic results are applied to energy consumption or complex water quality reaction expressions. As with steady-state model calibration, the degree of accuracy needed

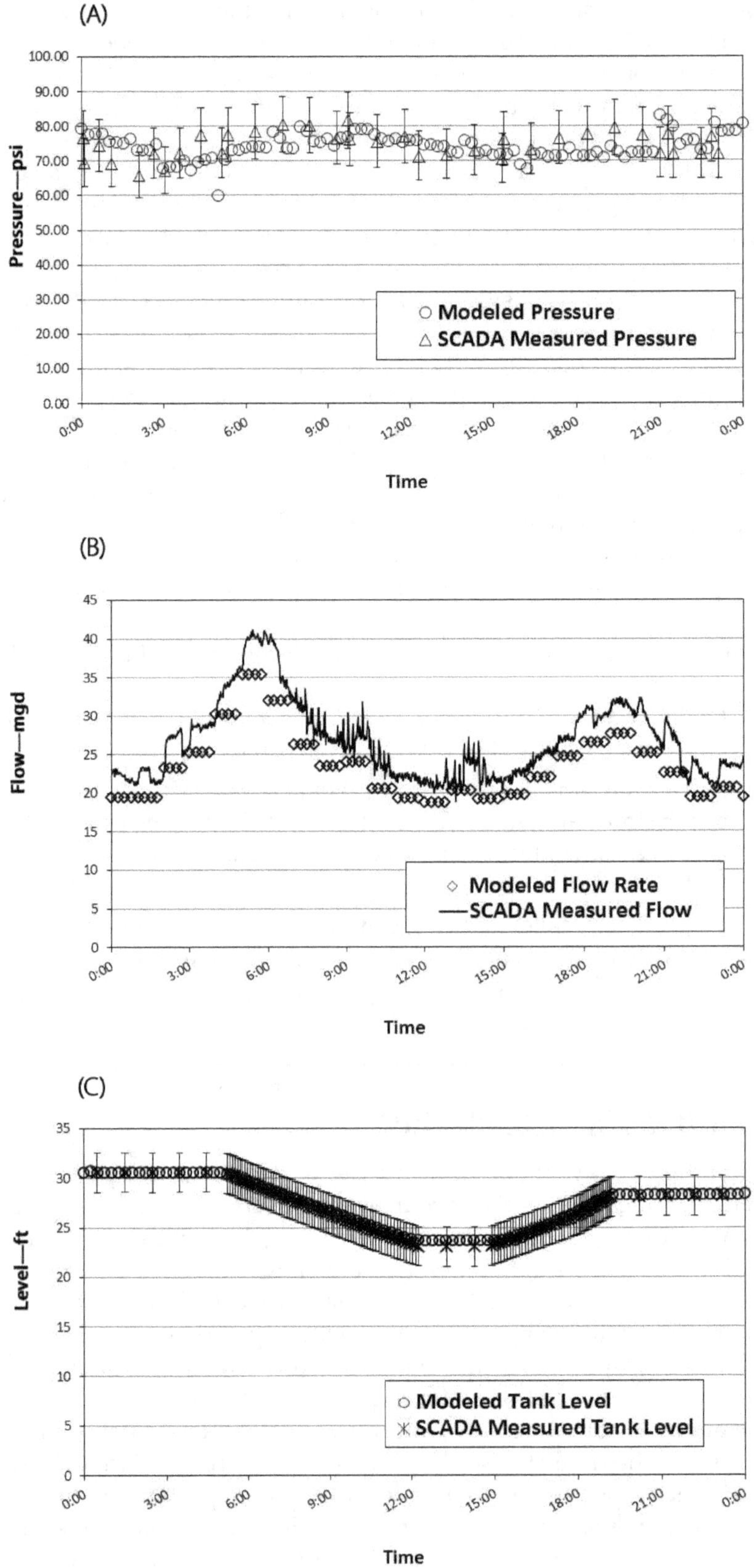

Figure 6-5 Examples of EPS calibration graphs: (A) pressure, (B) flow, (C) tank levels

Abbreviation: SCADA, supervisory control and data acquisition; EPS, extended-period simulation.

for EPS model calibration depends on the intended use of the model. For example, it may not be reasonable to expect a model that was developed for long-range master planning purposes to have the same level of accuracy as a model developed for water quality investigation. Practices for model calibration, both steady-state and EPS, can be found in chapter 4.

Steady-state models use initial system conditions such as pump status or storage tank level to establish the system-wide hydraulics, whereas extended-period models rely on changing system conditions from one time-step to another throughout the simulation period. Therefore, extended-period calibration includes the refinement of time-varying model components such as controls and patterns in addition to steady-state calibration components. The additional calibration effort also provides greater confidence that the model is simulating the system performance with adequate accuracy for its intended purpose.

6.7. SYSTEM EVALUATIONS WITH EXTENDED-PERIOD SIMULATIONS

EPS analyses are performed after the model is calibrated and are tailored based on the objectives of the analysis. Extended-period system hydraulic analyses reveal results that are similar to steady-state analyses; however, the variation in conditions during the EPS can indicate that hydraulic deficiencies may occur in different places at different times.

► TYPES OF EXTENDED-PERIOD EVALUATIONS

- Storage versus production
- Operational evaluations
- Supply management
- Pressure zone management and design
- Pump evaluations
- Vulnerability analyses
- Emergency planning
- Energy management

Modeling criteria must first be described; the actual analysis technique is then determined based on those criteria. In general, the same criteria used in steady-state analyses for hydraulic conditions are valid for an ESP. The system must maintain adequate pressures over the entire duration of the simulation. If the EPS modeling results indicate low pressures, these deficiencies could be caused by capacity in a pipe or pump that is inadequate to deliver water where it is needed. This may be due to high elevation of a service point relative to pump output or storage tanks levels or to inadequate pipeline sizes. Evaluating the resulting EPS pressures over a 24-hour period (or longer) can help identify the reasons for the deficiencies. Low pressures due to poor capacity generally occur during a period of high demand, while the more persistent low pressures are likely found with customers at higher elevations. EPS is also useful for master planning to assess future conditions. Water quality modeling, which is used for different types of extended-period analyses, uses similar hydraulic criteria in addition to the water quality criteria discussed in chapter 7. Several of the more typical analysis objectives for an EPS are discussed in the following sections.

6.7.1. Storage Versus Production

An ESP is well suited for assessing and designing storage facilities. Based on steady-state modeling, the desired minimum and maximum tank water level elevations can be selected based on the ability to provide sufficient pressures in the area served by the tank. Once these levels are set, an EPS is used to determine the range of water levels in the tank for various demand conditions including minimum day, average day, and maximum day (including maximum day with fire flows). Different tank locations can be evaluated to assess the levels of service throughout the day and to assess how well the tank will operate with the pumps used for filling. Storage tank volumes, as discussed in chapter 5, involve a number of components.

The most basic EPS evaluation involves determination of system storage adequacy relative to production and demand equalization. The main premise is that the water system's storage facilities should return to their initial conditions (levels) within a prescribed time frame. In addition, the fluctuations in tank level should not cause the tank to overflow nor should the level drop below a level that cannot sustain adequate system pressures or drop below water levels reserved for fire and emergency storage during normal operation. Be aware that there can be different minimum acceptable levels for normal demand conditions and for fire flow conditions when lower pressures are temporarily allowed.

The following examples illustrate some of the aspects involved in evaluating storage versus production. In these types of analyses, the primary objective is to ensure that initial tank levels return within a prescribed time frame, usually 24 hours – that is, the total system production should equal the total system demand over the prescribed time frame. At any given time, storage will make up for differences between demand and production-supplying water when demand exceeds production, and storing the excess when production is more than demand. Because demands are not symmetric with respect to storage or production, the combination of storage volumes plus water system piping must be sized appropriately.

The results of a real water system analysis are provided to illustrate the impact these analyses have in a real-world situation. The example system involves interaction of two reservoirs, each with a well-defined service area and each having two very distinct diurnal demand curves.

The system demand in this example includes two separate components, a residential demand and a golf course demand. Figure 6-6 represents a peak-day diurnal curve for this system. In this figure, the diurnal curve for the entire residential population (shown as a single dot dashed line) is based on the residential diurnal curve established for the residential population, adjusted to the actual residential population served. The diurnal curve for the golf course load over the same time period (shown as a double dot dashed line) is based on the course's diurnal curve and recordings of the actual instantaneous demand. In an EPS evaluation, these two loads are added to create the total system diurnal curve for the period, shown in the figure as a solid line.

Each reservoir has an on-site well supply plus a single well within the service area. The system's physical parameters are described in Figure 6-7.

In its existing configuration, the example utility includes only reservoirs 1 and 2 and wells 1 and 2. A 48-hour analysis was performed using the peak day diurnal curve to determine the adequacy of storage and production. The results of the analysis with this existing configuration (case 1) are shown in Figure 6-8. The figure indicates that the water system could not completely recover for the second day, resulting in a situation in which the reservoir could drop into the storage set aside for fire protection or even lower. In some locations in the system, the lower reservoir levels may lead to unacceptably low pressures.

The introduction of additional production would alleviate the problem with water level recovery, as shown in Figure 6-9. An evaluation was conducted using well 3 as a supplemental well to the system. Case 2 (Figure 6-9) illustrates the effect of well 3 on the system. Well 3 is set to operate when the reservoirs approach a water level 8 ft (2.45 m) below overflow. The reservoir water levels returned to a 24-hour full situation by providing the additional supply from well 3.

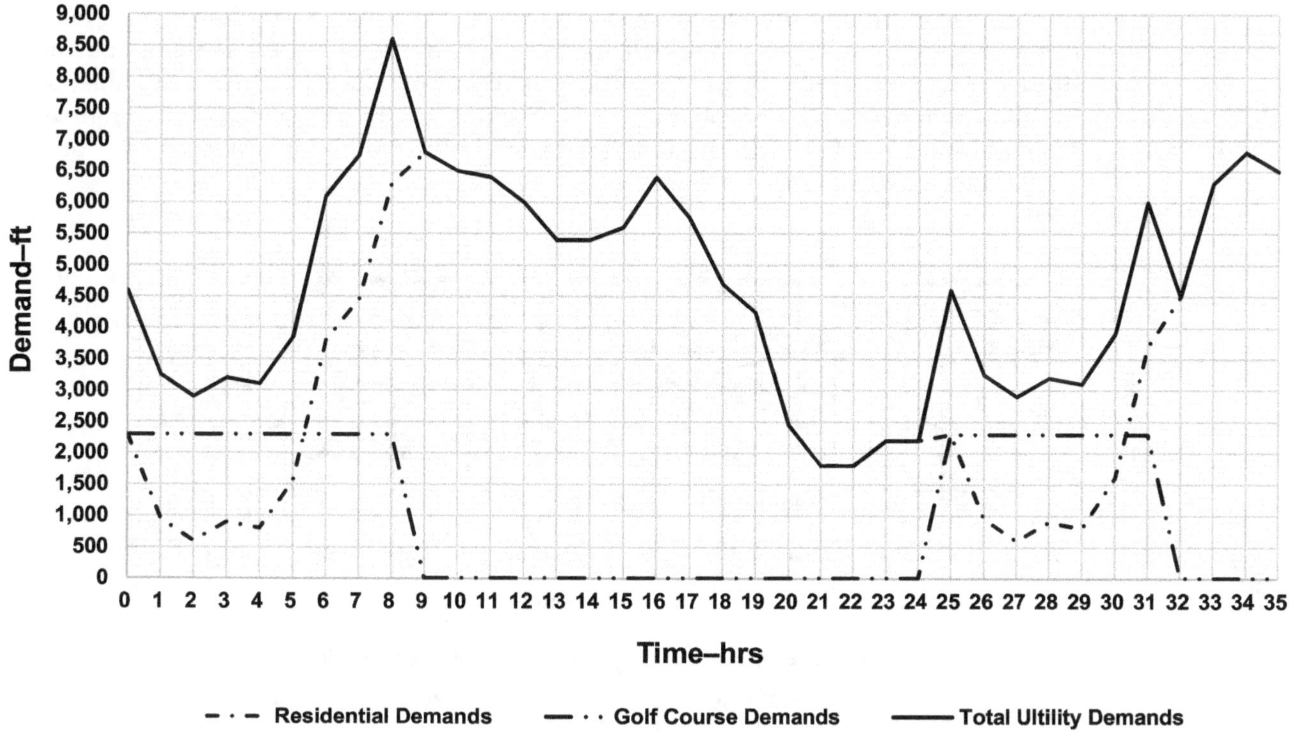

Figure 6-6 Example utility demands versus time

Reservoir No. 1		Reservoir No. 2	
Overflow elevation	5,485.00 ft	Overflow elevation	5,485.00 ft
Bottom elevation	5,453.00 ft	Bottom elevation	5,450.63 ft
Minimum water level	5,480 ft	Minimum water level	5,480 ft
Total storage	1.994 mil gal	Total storage	3.308 mil gal
Effective storage	0.312 mil gal	Effective storage	0.481 mil gal
Diameter, ft	103 ft	Diameter, ft	128 ft
Storage/ft	62,500 gal/ft	Storage/ft	96,250 gal/ft

Well no. 1	1,800 gpm @ 563 ft of lift
Well no. 2	2,075 gpm @ 563 ft of lift
Well no. 3	2,075 gpm @ 563 ft of lift
Fire protection requirements	5,000 gpm for 5 hours or 1.5 mil gal

Figure 6-7 System physical parameters for extended-period simulation analysis

However, adding production may not prevent the tanks' water levels from falling below the minimum level to maintain adequate system pressures. An evaluation using a third storage tank similar in size to tank 2 was conducted. Case 3 (Figure 6-10) illustrates the effect that new tank 3 has on the system. Not only do the storage tank levels return to their starting levels but, with the additional storage, the minimum tank level stays above what is needed to maintain adequate system pressures.

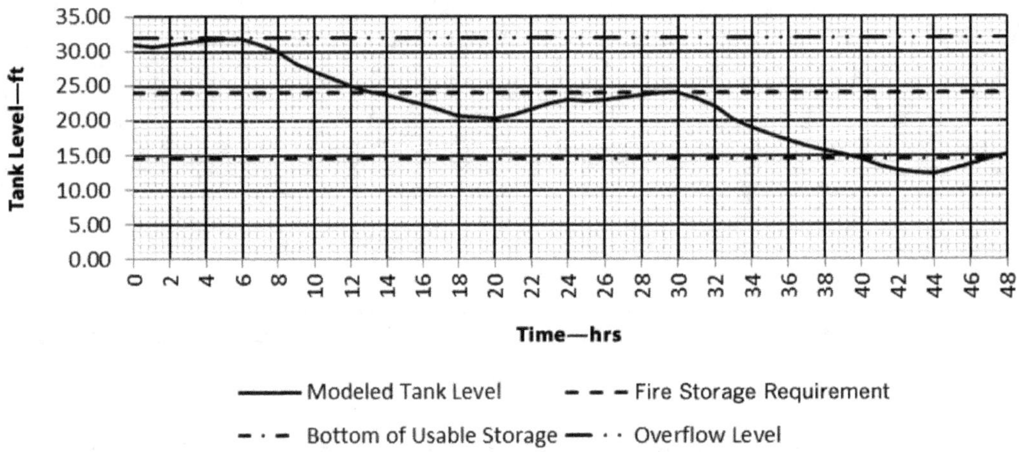

Figure 6-8 Example of storage versus production for existing conditions, case 1

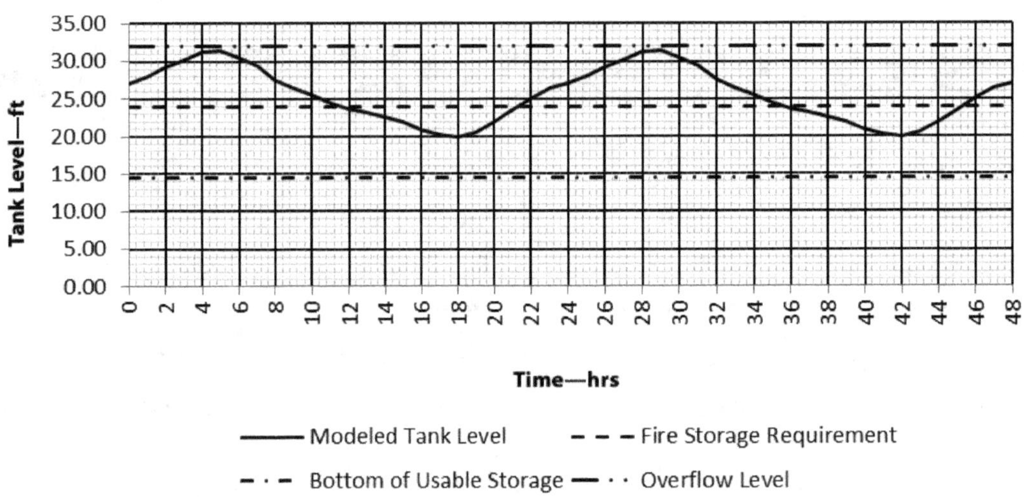

Figure 6-9 Example of storage versus production with new production from well, case 2

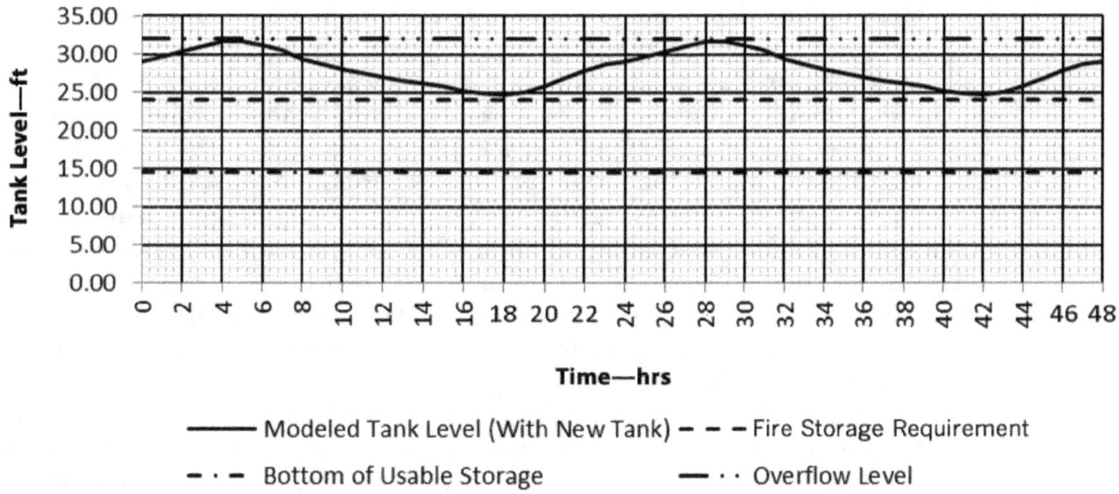

Figure 6-10 Example of storage versus production with new production (well 3) and storage (tank 3), case 3

6.7.2. Operating Inefficiencies

An EPS can help pinpoint a number of potential operating improvements within existing and future systems. For example, there are cases when a storage facility is maintained at an unusually high water level throughout maximum demand conditions and, therefore, never fully contributes significant water supply to the water system. The system boundary condition that sets the pressure gradient setting (such as a pump or pressure-regulating station) may be set too close to, or even above, the reservoir's high water level. This setting causes the pressure zone to be served primarily by the pressure-setting boundary and not the reservoir. The reservoir would stay continually full from this higher pressure setting because the HGL maintained by the valve is higher than the HGL of the water in the tank. The reservoir head is too low to provide flow to the pressure zone. As a result, the tank experiences inadequate water circulation. Lowering the valve's pressure setting would reduce flow through the pressure-setting boundary by lowering the head downstream of the valve. This also would allow much more flow from the reservoir during high demand periods. The same effect can be achieved by shutting off pumps or valves for part of the day to allow the tank to drain down.

In another example, water pumped from one pressure zone into a higher zone may flow back into a nearby intermediate-pressure zone through a pressure-regulating station. It would be more efficient to pump water directly to the intermediate zone from the reservoir where it is needed, rather than to the higher zone. By adjusting the booster pumps that move water from the low zone to the intermediate zone to run more often, recirculation between the high and intermediate pressure zones and associated wasted pump energy could be reduced.

6.7.3. Supply Management

EPS modeling can be used to assess possible water supply operations. Examples of this include when the availability of various sources changes seasonally or even when the costs of the supplies vary seasonally based on cost of treatment and/or pumping. Some utilities switch to surface water and allow groundwater to recharge; EPS analyses can be used to plan for these changes. EPS results indicate if adequate pressures are maintained over all time periods.

Many utilities have interconnection agreements with neighboring utilities, but details on these interconnections may not always be thought through completely. One utility may have an agreement for a set quantity of water. However, if the supply pressure is inadequate, that quantity of water cannot be supplied. EPS analyses would not only indicate if system pressures are maintained and if sufficient head is available to supply the water, they would also allow the total quantity of water supplied over time to be calculated.

These analyses can be especially helpful when assessing future demand conditions. If future demands exceed existing supplies, EPS analyses can be used to compare different supplemental supply alternatives. The results indicate if adequate pressures will be maintained and if system operating controls will still be valid. EPS analyses can be a key part of an overall system economic analysis where the costs of new supplies can be compared along with the hydraulic results.

If future water system improvements are needed, time-dependent analysis can be used to determine the most efficient water supply system operation under future conditions. This can include setting schedules for when supplies are active, adjusting set points to maintain stable pressures, and identifying conflicting operating rules that would need to be resolved. Analyses of different supply strategies should also be combined with evaluation of PRVs, booster pumps, and storage facilities, which are discussed in more detail in the following sections.

6.7.4. Pressure-Regulating Stations

As demands vary over time, the flow through PRVs will also change. EPS analyses are helpful to confirm that the pressure settings of these valves remain valid over time. Adjustments can be made to regulate valve settings in conjunction with evaluation of alternate supply and pump operations.

It is always recommended that the HGL associated with a regulating valve's setting be confirmed. The HGL is calculated as the valve elevation plus the pressure setting converted to feet of head. This is critical for pressure zones that are fed by multiple PRVs because the valves with the highest flows are generally associated with the valves with the highest downstream HGL values based on their settings. The flows through the valves can be better managed by coordinating the settings of parallel valves. For example, one valve may provide most flows to a zone, while the other may not be active if the HGL settings differ significantly. In other cases, a valve may provide flow to a zone only during high demand periods. This intermittent operation could lead to flow reversal within the zone, which can cause resuspension of sediments. A change in regulating valve settings can help prevent stagnation by allowing some flow to pass through a zone at all times. However, care must be taken to avoid excessive amounts of flow back through the valve to reduce energy use.

EPS analyses are helpful when investigating modification to zone boundaries. Changing a zone boundary changes the amount of demand in that zone and may change the usage patterns as well. EPS results indicate if the boundary changes can maintain pressures within acceptable ranges while verifying that the valves are still functioning as designed.

6.7.5. Booster Stations and Wells

An EPS can be used for a number of analyses of pumps, pumping stations, and pumping systems. EPS analyses can be used when selecting the right pumps or combination of pumps to meet the range of demands that occur over a day as well as when setting the controls that turn these pumps on and off. As with analysis of regulating valves, EPS analyses can be used to assess pressure zone management by adjusting operations at different booster stations that supply the zones.

One unique benefit of EPS analysis is the ability to plot variations in pump flow and head over time and to compare model results with the actual pump curves, as shown in Figure 6-11. The results can be used to determine if the pumps are running near the best efficiency point as well as to determine if there are times when the pumps are running too close to runout or shutoff.

EPS analyses can be used to evaluate pressure zones and determine whether it is better to add second-tier booster pumping to increase pressures or to increase pumping head at existing facilities with the addition of PRVs to prevent excessively high pressures in low elevation areas.

6.7.6. Storage Facilities

To determine storage capacity requirements for a water system, future needs of the entire system need to be considered as well as the needs of each individual pressure zone. The system as a whole may have enough storage, but there could be areas with excess capacity while others may have a deficit. Options to consider when evaluating storage requirements with EPS models include the following:

- Tank configurations such as tank shape or height-to-diameter ratio,
- Tank placement options within a pressure zone,
- Ability to fill tanks,
- Ability to transfer water between pressure zones,
- Elimination of physical constraints in the distribution system piping, and
- Fire storage and emergency storage requirements.

In addition to normal operating scenarios, modelers should consider reservoir operation under emergency conditions as well as system operation without tanks, such as when they are out of service for maintenance.

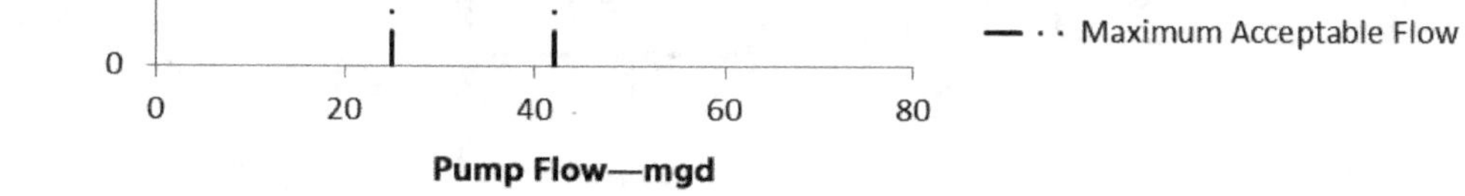

Figure 6-11 Comparing modeled pump output against a pump curve

Abbreviation: EPS, extended-period simulation.

6.8. SPECIAL TYPES OF EXTENDED-PERIOD SIMULATION ANALYSES

An EPS is useful for various applications including vulnerability analyses, emergency preparedness, energy optimization, operator training, and water quality evaluations. Some of these are discussed briefly below, while others (e.g., water quality) are presented in greater detail in other chapters.

6.8.1. Vulnerability Analysis

A key evaluation that can be performed using EPS is reliability, or vulnerability, analysis. This type of analysis includes loss of pumping, supply, or storage facilities; loss of critical transmission mains; planned maintenance outages; and pump cycling studies. To determine system response, these interruptions should be assumed to occur during critical demand periods. Under such circumstances, wells and pumping stations without standby power sources would be out of service. Therefore, the only water available would come from supplies with emergency power generators or that feed the system via gravity, system storage, or emergency interconnections with other systems. This type of analysis can be used to determine how long the system can reliably operate under these conditions. An EPS can also be used to assess recovery time from any of these events.

Vulnerability analysis is used to identify critical weak points within the system, assess the extent of those weak points and their effect on water service, and determine system modifications that provide some relief from the problem. This relief might lead to identifying the need for some redundant production, storage, or a combination. Using the example from case 3 in section 6.7.1, the loss of all well supplies results in reservoir elevations, as shown in Figure 6-12 (case 4), assuming all other assumptions stay constant. The system that was balanced previously is now unbalanced to the point that reservoir levels compromise fire storage. The analysis indicates that the system could provide at least minimum pressures for 11 hours after supplies are lost. The analysis also indicates that recovery could take longer than 24 hours, which would require operation of additional well pumps. Additional discussion of modeling for system vulnerability is included in chapter 11.

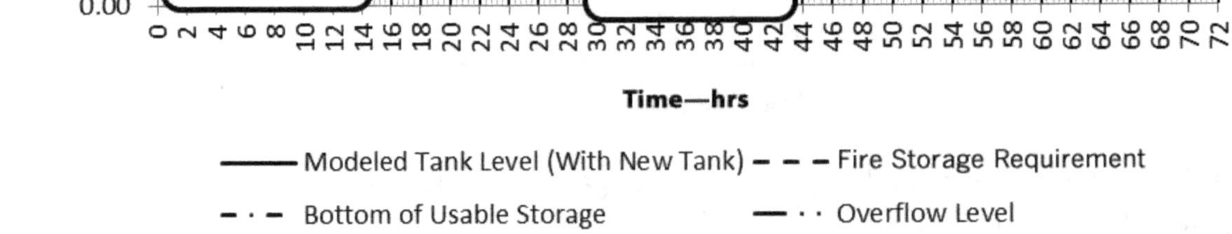

Figure 6-12 Example of storage versus production with supply interruption and recovery, case 4

6.8.2. Emergency System Operations

An extremely valuable benefit of an EPS is the evaluation of system response under emergency conditions, such as simulated major fires during a maximum day demand condition. These what-if analyses provide operations personnel with important information on how water system facilities respond when stressed. This enables advanced planning for better response in emergency situations if, and when, they occur.

Major industrial areas typically require very large fire flows. To simulate system response under worst-case conditions, analyses assume that these large fire flows occur during peak demand conditions, often in the early morning hours. For example, a simulated fire in the system results in a drop in reservoir water level elevations, as shown in Figure 6-13 (case 5), assuming all other assumptions stay constant. It is important that the tank levels stay above those required for minimum levels of service at all times during a fire.

During emergency fire flow simulations, system response should be analyzed for areas that have below-standard pressures (less than 20 psi [138 kPa]), negative pressures, and low or empty reservoir levels. For systems with multiple pressure zones, water transfers between zones should be analyzed to ensure that the whole water system responds as desired to emergency conditions. The system must have sufficient storage volume and sufficiently high tank levels to maintain pressures during the emergency. These simulations are useful tools in changing operating procedures to improve system performance.

6.8.3. Energy Optimization

Energy optimization is very important for efficient water system operation because the rising cost of energy is a concern for many water systems. Energy optimization involves a delicate balance among supply, pumping, and storage based on cost differentials among various energy sources, either by type or by time of service.

Energy use is a significant contributor to operating costs, and any potential energy savings from energy optimization can have a high return on investment. For this reason, evaluating energy requirements and developing a power management program are key steps in the development of distribution system improvements.

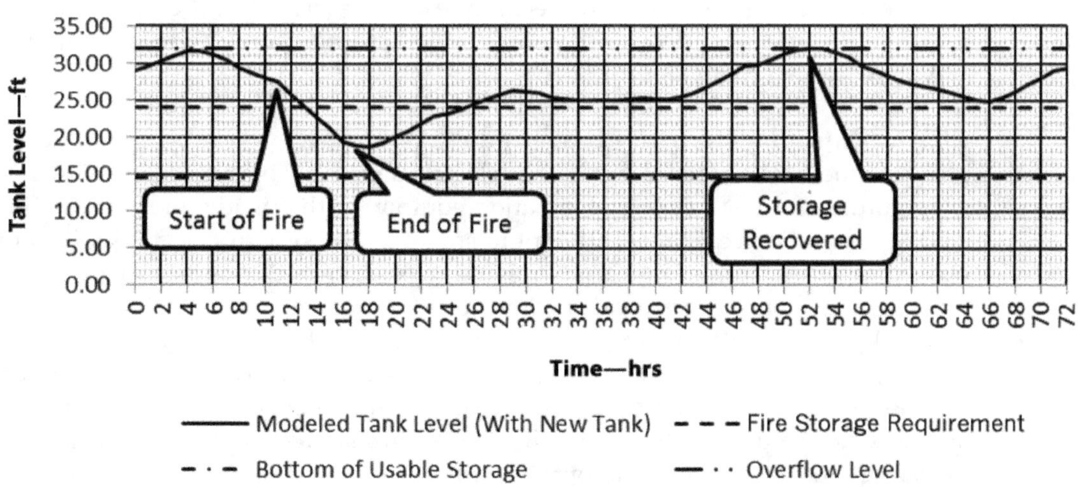

Figure 6-13 Example of storage versus production with fire demand and recovery, case 5

To assess potential energy savings, the electrical rate structure should be analyzed to determine the least expensive times to pump. Electrical rates typically consist of two components: the demand charge and the energy charge. Most modeling software packages can account for the cost of direct energy but may not always be able to replicate fully the cost associated with demand charges. Evaluating the system simply on the basis of direct energy used provides a direct comparison that can be used even if the total charge cost is slightly off.

The energy charge represents the total amount of energy used, in kilowatt-hours, over the billing period, making the reduction of overall usage a key goal. For some electric utilities, the cost for energy varies depending on the time of day. If the utility schedules reservoir filling during the lowest-cost period, additional savings can be found.

The demand charge is usually a direct cost associated with the highest instantaneous power usage, usually in kilowatts, at a facility over a given time period. In some rate structures, the demand charge is based on the billing period, but in others, the demand charge is based on the highest demand in the past 12 months. In this case, the utility pays a substantial amount for that single occurrence, even if the peak was an isolated event. Thus, reducing peak usage at each facility is another goal of energy optimization.

Some items to consider when developing an energy optimization plan include:

- Power consumption by system components,
- Alternate system operations to reduce energy consumption,
- Peak shaving through standby power,
- Water demand management, and
- Construction of additional pipes, pumps, or storage facilities to reduce overall pumping requirements.

An assessment of these items can result in many different combinations of alternatives, which can be very labor intensive. Many software packages include some type of energy optimization capabilities that can help a utility determine the pumping routine that will achieve the best energy management.

Several energy optimization techniques are described in the following sections to illustrate the effects of time-of-day rates and blending energy types to maximize water service or system capital requirements.

6.8.3.1. Managing Electricity Usage and Peaks. Electricity generation, unlike water supply, must meet or exceed instantaneous electric demand. This results in significant capital expenditures for peak electric generation equipment that is idle most of the time. Some electric generation companies attempt to smooth their system's electricity load over non-peak times, rewarding large users for off-peak electric demands and penalizing them for energy demand that coincides with the system's peak electric demand. Water system demand peaks usually mirror electric demand peaks. In fact, a significant percentage of water system demand occurs during a 12-hour peak electric demand time frame.

Pumping curtailment during periods of peak energy costs can reduce costs. EPS analyses can be used to determine if the system's storage capacity is sufficient to make pumping curtailment a viable option without causing inadequate pressures. If so, the analyses can be used to assess different curtailment periods and durations.

Water utilities that use only electric power have other options for curtailing on-peak electric consumption. Adding new storage and supply can help prevent deficits during curtailment periods. Results from more detailed analysis can be used to determine the actual number and extent of system peaks expected to occur. Also, these results can be used to balance the cost of additional infrastructure with the additional energy costs incurred in violating the on-peak pumping for those periods of the year. Supply sources at different HGLs may also have different horsepower requirements. Using supply sources that require less energy will reduce electricity costs, especially during peak energy demand periods.

6.8.3.2. Pump Drive Types. If available, one way to avoid peak electric demand rates is to supplement electric power needs using other nonelectric pump drivers. A natural gas or diesel engine connected to a well pump can provide water during on-peak electric periods and supplement electric drivers in off-peak times. In addition, peak water demand occurs during the growing season at a time when natural gas requirements for heat are traditionally at low levels. Some electric utilities may credit water utilities for running emergency generators during peak electricity demand periods and deliver the excess electricity back into the power grid. EPS analyses can be used to assess different pump schedules that include use of nonelectric drives.

Another way to improve energy efficiency is to use variable-speed pumps. Constant-speed pumps with a throttling control valve to adjust flows use the same amount of energy regardless of the flow rate. Adjusting pump speed allows operation closer to the pump's best efficiency point and will use less energy when lower flow rates are pumped. Many modeling programs have the capability to simulate system operation with variable-speed pumps based on maintaining pressures or flows.

The electrical industry throughout North America has seen substantial changes over the years. The modeler should meet with power utility representatives to review the impacts of different operating strategies on the water utility's power cost. Generally, these different operating strategies require the water utility to have additional pumping and storage capabilities to take advantage of power cost savings.

6.8.4. Operator Training

Operator training is an excellent application of distribution system modeling within a utility. Models can be used to train new staff on the water distribution system's unique operational characteristics. Models are also useful in simulating different types of system failures and in identifying the most effective response to reduce the impact on customers.

Operators are also the best resource for determining if the model accurately represents flow and pressure measurements based on the utility's SCADA system. Conversely, a well-calibrated model can give operators a better understanding of how water flows through a system and provide a tool to gauge how changes in controls will impact the system. This allows the modeler and the operator to work together to identify system deficiencies or problems in the model assumptions. Cooperation can also extend to planning for upcoming changes in operating strategies.

6.9. ADDITIONAL RESOURCES

Bhave, P.R., 1991. *Analysis of Flow in Water Distribution Systems.* Lancaster, PA: Technomic.

Cross, H., 1936. *Analysis of Flow in Networks of Conduits or Conductors.* Urbana: University of Illinois.

Eggener, C.L., & Polkowski, L., 1976. Network Modeling and the Impact of Modeling Assumptions. *Journal of the American Water Works Association,* 68:4:189–196.

Gupta, R., & Bhave, P., 1996. Comparison of Methods for Predicting Deficient Network Performance. *Journal of Water Resources Planning and Management,* 122:3:214.

Jeppson, T.W., 1976. *Analysis of Flow in Pipe Networks.* Ann Arbor, MI: Ann Arbor Science Publishers.

Jordan, R.A., Priest, M., Jain, D.K., & Jacobsen, L.B., 1999. Master Planning Utilizing H2ONET Extended-Period Simulation and Microsoft Access Generated Summary Reports. In *Proceedings of the Information Management and Technology Conference.* Denver, CO: AWWA.

Shamir, U., & Howard, C.D.D., 1968. Water Distribution Systems Analysis. *Journal of the Hydraulics Division,* ASCE, 94:1:219.

Stone, K., Barbato, L.M., & Koval, E.J., 2002. Colorado Springs Utilities: A Case Study for Extended-Period Simulation. In *Proceedings of the Information Management and Technology Conference.* Denver, CO: AWWA.

Todini, E., & Pilati, S., 1987. A Gradient Method for the Analysis of Pipe Networks. In *International Conference on Computer Applications for Water Supply and Distribution.* Leicester, UK.

Walski, T.M., Gessler, J., & Sjostrom, J.W. 1990. *Water Distribution—Simulation and Sizing.* Ann Arbor, MI: Lewis Publishers.

Wood, D.J., 1980. *Computer Analysis of Flow in Pipe Networks.* Urbana: University of Illinois.

Wood, D.J., & Charles, C.O.A., 1972. Hydraulic Analysis Using Linear Theory. *Journal of the Hydraulics Division American Society of Civil Engineers* (ASCE), 98:7:1157–1170.

Wood, D.J., & Rayes, A.G. 1981. Reliability of Algorithms for Pipe Network Analysis. *Journal of the Hydraulics Division,* ASCE, 107:10:1247–1248.

WRc Plc. 1989. *Network Analysis—A Code of Practice.* Swindon, UK: Water Research Centre.

Chapter 7

Water Quality Modeling

7.1. INTRODUCTION

A primary goal of water distribution systems is to deliver potable water when and where it is needed at adequate pressure. Ideally, there would be no change in the quality of water from the time it leaves the source until the time it is consumed. In reality, significant changes can occur as water travels through a distribution system and on to the customer's tap.

Under normal operating conditions, these changes occur for several reasons. Chemical reactions that began at the treatment plant continue in the distribution system. Some of the most important are the reactions that involve disinfectant residual, usually a form of chlorine. These reactions reduce the concentration of the disinfectant over time, limiting its ability to control pathogens and possibly producing undesirable by-products such as chloroform or taste-and-odor-producing compounds. Another example is the formation of calcium carbonate or other coatings on pipe walls. Blending of waters from different sources in the distribution system produces changes in water chemistry such as pH and chloramine speciation.

Water also reacts with the walls of the pipe through which it flows. Corrosion and subsequent buildup of tubercles and oxide coatings are typical examples. Pipe walls also support the growth of attached colonies of microorganisms or thin biofilms. When this growth is excessive, it often increases disinfectant demand, produces taste-and-odor-causing compounds, and offers protection for opportunistic pathogens. In addition to the normal chemical reactions that occur in the distribution system, water quality degradation through accidental or intentional contamination is an important concern.

Chemical reactions that occur within distribution system storage facilities may differ from the reactions that occur within the bulk water or along the pipe wall of distribution pipes. Examples include the impact of isotherms on mixing within a storage facility and a change in water temperature that significantly alters chemical reaction rates.

7.2. NEED FOR WATER QUALITY MODELING

The spatial and temporal variation of water quality in a distribution system involves very complex phenomena because of both the hydraulic mixing patterns and the water quality transformations that occur within the network. The topological complexity of pipe networks in distribution systems and the ever-changing patterns of water usage rates over the course of a day and during different seasons create myriad pathways for water to travel on its journey from treatment plant to consumer. The variation in the quality of finished water that enters the distribution system adds another variable that impacts distribution system water quality spatial variance. Individual parcels of water lose their identity as they mix with other parcels with different residence time and quality throughout the system. Thus, the quality of water that reaches a tap or fixed sampling location can vary significantly over the course of a day and between seasons. In addition, the water at two taps, even if they are in close proximity, can have significantly different quality because of the travel pathways, or flow paths, and transformations between the finished-water sources and the taps.

The complexity of flow paths and transformations makes it difficult, if not impossible, for a person experienced with a particular distribution system to determine travel times, blending, and subsequent water quality throughout the system. After careful application, the mathematical models of the hydraulics and water quality of a system can provide a method for estimating the movement, transformations, and water quality in a network. The increasing number of distribution system water quality sensors and the increasing amount of data being generated can expand a utility's understanding of its distribution system. This information can be used to modify or adjust operations, prevent or mitigate water quality issues, and effectively prepare for and implement emergency management plans.

7.3. USES OF WATER QUALITY MODELING

Water quality models are used to predict the spatial and temporal distribution of a variety of constituents within a distribution system. These predictions include the following:

- Portion of water originating from a particular source (tracing),
- Average age of water (i.e., how long since it left a particular source),
- Concentration of a nonreactive tracer compound either added to or removed from the system (e.g., fluoride or sodium),
- Concentration and decay of a residual disinfectant (e.g., chlorine or chloramines), and
- Concentration and fate of disinfection by-products (DBPs; e.g., trihalomethanes [THMs] or haloacetic acids [HAAs]).

The ability to model the transport and outcome of these parameters helps system managers perform a variety of water quality studies. Examples include the following:

- Calibrating and verifying models of the system using chemical tracers;
- Locating and sizing storage tanks and modifying system operation to reduce water age or improve various water quality constituents;
- Identifying ways to reduce water age and stagnation;
- Determining the locations and amount of mixing from multiple water sources;

- Modifying system design and operation to provide a desired blend of waters from different sources;
- Designing a cost-efficient routine monitoring program to identify water quality variations and potential problems;
- Identifying the most effective sampling locations in a distribution system;
- Determining the best location and dosage for rechlorination stations;
- Assessing and minimizing the risk of consumer exposure to DBPs;
- Assessing system vulnerability to incidents of external contamination;
- Identifying response times for various contamination scenarios;
- Finding the best combination of pipe replacement, relining, and cleaning; reduction in storage holding time; and location and injection rate of booster stations to maintain desired disinfectant levels throughout the system; and
- Performing a forensic analysis of water quality incidences in the system.

Water quality simulation capabilities include estimation of the fate, or growth and decay, of multiple constituents simultaneously. This capability is useful for evaluating DBP formation, planning corrosion control programs, controlling secondary nitrification in chloraminated systems, and simulating sedimentation in pipes to help plan flushing programs. Efforts to extend water quality models to include such features as predicting biofilm buildup and growth of microorganisms are still in the research stage (Powell et al. 2004). Water quality modeling to directly assess such distribution system problems as minimizing biofilm growth, controlling coliform outbreak occurrences, and avoiding red water problems will be handled in future advances in software.

7.4. WATER QUALITY MODELING TECHNIQUES

While there are many applications and uses for the latest water quality simulation capabilities, the three predominant uses of water quality modeling include simulation of water age, source/contaminant tracing, and constituent fate, as described in the following sections.

7.4.1. Average Water Age

Water age is the weighted average time that a parcel of water, which constitutes the flow at a node, has spent in the distribution system. Finished water that enters the network from reservoirs or source nodes has an age of zero. The age linearly increases with extended-period simulation (EPS) time until the point of consumption. Water age typically serves as a general surrogate for the measure of water quality at any point in the distribution system. In simple terms, age is treated as a reactive constituent whose growth follows zero-order kinetics with a rate constant equal to 1 (i.e., each second the water becomes a second older).

Water distribution modeling applications allow for initial water age values to be input by the user. This input reduces the EPS initialization time, that is, the time when the water age within all portions of the distribution system has reached a diurnally repeatable water age. This also allows for representation of water age that includes part of the time water is in the treatment facility before it enters the distribution system, such as within the clearwell.

7.4.2. Source/Contaminant Tracing

Source tracing tracks the percent of water from a particular source node that reaches any junction node in the network over time. The source node can be any tank or reservoir in the network. In simple terms, the source node is treated as a constant source of a nonreactive constituent that enters the network with a concentration of 100. Source tracing is a useful tool for analyzing distribution systems that draw water from two or more different raw water supplies. It can show degree to which water from a given source blends with water from other sources and how the spatial pattern of this blending changes over time. It can also be used to simulate introduction of a contaminant into the system at a given point. The contaminant may be introduced as an instantaneous slug or may be continuous over a period of time.

7.4.3. Constituent Fate

Of great importance in water quality modeling is the simulation of growth and decay, or fate, of chemical or biological constituents in the water system. Water quality modeling applications are capable of simulating zero-order (e.g., water age) and first-order (e.g. disinfectant decay or DBP fate) reaction kinetics. Tools exist that allow for simulation of more complex chemical reactions.

7.5. GOVERNING PRINCIPLES OF WATER QUALITY MODELING

Just as hydraulic models conform to the conservation laws of fluid mass and energy, water quality models are based on conservation of constituent mass. These models represent the phenomena that occur in a distribution system (Clark and Grayman 1998), which are described in the following sections.

7.5.1. Advective Transport of Mass Within Pipes

A dissolved substance travels down the length of a pipe with the same average velocity as the carrier fluid while at the same time reacting (either growing or decaying) at a given rate. Longitudinal dispersion is not an important transport mechanism under most operating conditions. This means that the modeler can assume that there is no intermixing of mass between adjacent parcels of water traveling down a pipe. The governing equation for advective transport of mass within pipes is

$$\frac{\partial C_i}{\partial t} = -u_i \frac{\partial C_i}{\partial x} + r\left(C_i\right) \tag{7-1}$$

Where:

C_i = concentration (mass/volume) in pipe i as a function of distance x and time t

u_i = flow velocity (length/time) in pipe i

r = rate of reaction (mass/volume/time) as a function of concentration

Researchers have found that pipes in low-flow zones, such as dead-end pipes, which are common in municipal water distribution systems, experience a degree of longitudinal dispersion (Tzatchkov et al. 2009). Further research is needed to assess the full extent of this issue and to develop appropriate modifications to network models.

7.5.2. Mixing of Mass at Pipe Junctions

Most water quality models assume that at junctions that receive inflow from two or more pipes, the mixing of fluid is complete and instantaneous. Thus, the concentration of a substance in water leaving the junction is simply the flow-weighted sum of the concentrations in the inflowing pipes. The governing equation for mixing of mass at pipe junctions is

$$C_{i|x} = \frac{\sum j \in I_k Q_i C_{j|x=Lj} + Q_{k,ext} C_{k,ext}}{\sum j \in I_k Q_j + Q_{k,ext}} \tag{7-2}$$

Where:

j = pipe with flow leaving node k
I_k = set of pipes with flow into k
L_j = length of pipe j
Q_j = flow (volume/time) in pipe j
$Q_{k,ext}$ = external source flow entering the network at junction node k
$C_{k,ext}$ = concentration of the external flow entering at junction node k

The notation $C_{i|x=0}$ is the concentration at the start of pipe i, while $C_{i|x=L}$ is the concentration at the end of the pipe.

Researchers have shown that under some circumstances, mixing may be incomplete at junctions of three or more pipes (Romero-Gomez et al. 2008; Austin et al. 2008). This is potentially most noticeable when there are limited pipe bends or appurtenances.

7.5.3. Mixing of Mass Within Storage Tanks

Most water quality models can model the following four types of mixing characteristics within storage tanks:

- Complete mixing,
- Plug flow (first in/first out),
- Short-circuiting (last in/first out), and
- Compartment.

The characteristics of each of these types of mixing models are described in chapter 8.

7.6. REACTIONS WITHIN PIPES AND STORAGE TANKS

While a constituent travels within a pipe or resides in a storage facility, it undergoes reaction. The rate of reaction, measured in mass reacted per volume of water per unit of time, depends on the type of water quality constituent being modeled. Some constituents, such as fluoride, do not react and are termed *conservative*. Others, such as chlorine residual, decay with time, while DBPs, such as THMs, grow with time. Some, such as chlorine, react with materials in both the bulk liquid phase and at the liquid–pipe wall boundary.

Distribution system water quality models represent these phenomena (transport within pipes, mixing at junctions and storage tanks, and appropriate reaction kinetics) with a set of mathematical equations. These equations are then solved according to an appropriate set of boundary and initial conditions to predict the variation in water quality throughout the distribution system. Even though water quality calculations are made throughout the length of each pipe, output reporting of quality is usually made only for nodes. The set of equations that comprises a typical water quality model are described in section 7.5.

7.7. COMPUTATIONAL METHODS

Water quality models rely on a well-calibrated EPS hydraulic model as the basis for performing quality simulations. Extended-period models take explicit account of how changes in flows through pipes and storage tanks that occur over an extended period of system operation affect water quality. These models predict both spatial and temporal variations in water quality throughout a network and provide a realistic picture of system behavior.

Several solution methods are available for extended-period water quality models (Rossman and Boulos 1996). They require that a hydraulic analysis be run first to determine how flow quantities and directions change from one time period to another throughout the pipe network. Usually, the time-steps used for water quality analyses are shorter than those used in hydraulic analyses, usually less than 5 minutes. Such small time-steps can result in very long run times but are necessary to achieve accurate results.

At the start of each new hydraulic time period, each pipe is divided into a number of parcels, and water is either routed between parcels (the Eulerian approach) or the parcels are moved downstream (the Lagrangian approach) over a succession of smaller time-steps. At the same time, account is taken of any reactions that occur in each parcel. At the end of each water quality time-step, the water that enters each node from inflowing pipes is blended to determine a new concentration value for that node. This concentration is released from the node into its outflowing pipes at the start of the next time period. When a new hydraulic time-step begins, the water quality routing begins again under a new set of hydraulic conditions.

Because the common Eulerian and Lagrangian water quality routing algorithms do not advect mass beyond the first downstream node within one water quality time-step, care should be taken to choose a sufficiently small water quality time-step. Of interest in this regard is the minimum travel time over all pipes in the network. If the water quality time-step is significantly larger than this minimum value, the routing of water between parcels will be artificially truncated, leading to simulation errors where the advection of mass is too slow. Thus, using highly detailed all-pipes models for accurate water quality analyses may require extremely small water quality time-steps and pose a large computational burden. Such models may need to be processed through a mild skeletonization or reduction step to eliminate unusually small pipes. Until future research produces algorithms that are not affected by short travel times in pipes, it is best to experiment with various water quality time-steps and be mindful that mild skeletonization may be advantageous.

7.8. DATA REQUIREMENTS

7.8.1. Hydraulic Data

Water quality models use the same physical and operational characteristics as are used with hydraulic models, incorporating each pipe's length and diameter and the way in which a distribution system's nodes and links are connected to one another. In addition, the water quality model relies on the flow solution of a hydraulic model as part of its input data. An extended-period water quality model uses a time history of flows in each pipe and flows to and from each storage tank as determined from an extended-period hydraulic analysis. Water quality simulation algorithms generally do not consider hydraulic data such as pipe roughness, nodal elevation, head, or pressure for water quality calculations, although the water level in tanks figures prominently in these calculations. Also, water quality kinetics, such as the pipe wall reaction rate coefficients, can, in some instances, be correlated to hydraulic data, such as pipe roughness coefficients.

7.8.2. Water Quality Data

A water quality model needs to incorporate the quality of all external inflows into the network. An example would be the concentration of chlorine leaving the clearwell of a water treatment plant or at a source of supply from an adjacent utility as well as the quantity of chlorine injected at any booster stations or storage facilities in the system. Extended-period models also need to consider varying source water quality from adjacent or consecutive systems. Source water quality data are commonly obtained from existing source monitoring records when simulating existing operations, from field data sampling with field or laboratory analysis, or they can be set to desired values to investigate potential system impacts on source quality variations.

Dynamic water quality models are supplied with a set of initial water quality values for all nodes and storage tanks throughout the network. These initial values are established in several ways. One way is to use the results of a field monitoring study, if they are available. A second method is to equilibrate the model by running it for a sufficiently long period of time from an arbitrary set of initial conditions under a repeating pattern of source and demand loadings until the model reaches a repeating pattern of outputs (dynamic equilibrium). This equilibration approach is limited by the calibration of the hydraulic model and the assumption that the demands repeat every day. The results of this approach can be used as the initial conditions for subsequent water quality analyses. The duration of the simulation for this approach varies based on the type of water quality analysis being run, system size, and the initial water quality estimates. The modeler should note an initial rise in water quality parameter followed by leveling off at the equilibrium value, as shown in Figure 7-1.

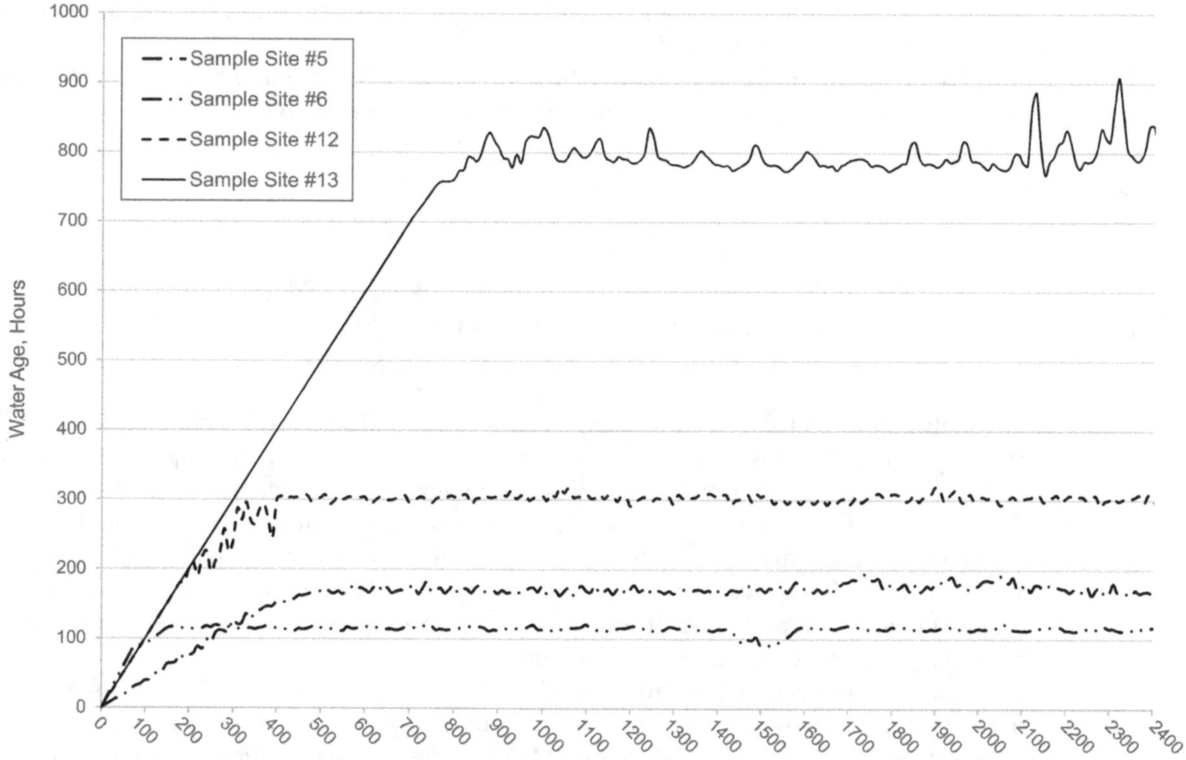

Figure 7-1 Illustration of water quality model equilibration

Abbreviation: DBP, disinfection by-product.

7.8.3. Reaction Rate Data

7.8.3.1. Chlorine Decay Coefficients. Modeling the fate of residual disinfectant is a common application of distribution system water quality models. The two most frequently used disinfectants in distribution systems are chlorine and chloramines (a combination of chlorine and ammonia). These two disinfectants have a residual that can be simulated using most modeling programs. The term *free chlorine* denotes the amount of uncombined chlorine that consists of both hypochlorous acid and hypochlorite ion. *Combined chlorine* is the amount of chloramines (primarily monochloramine) present, and *total chlorine* is the sum of free and combined chlorine. Free chlorine is more reactive than chloramines, and its reaction kinetics have been studied more extensively. Most research in this area has shown that there is more loss of chlorine observed within the distribution system than for the same amount of water stored in an inert container over the same period of time. This implies that there are several separate reaction mechanisms for chlorine decay—one that involves reactions within the bulk fluid and another that involves reactions with material on or released from the pipe wall (Vasconcelos et al. 1997). A combination of these mechanisms may be occurring as water passes through storage tanks or reservoirs. Separate reaction rates are often assigned to tanks in most modeling programs.

7.8.3.2. Bulk Decay Coefficients. The rate of free chlorine decay in the bulk fluid depends on the water's chemical nature and its temperature. Chlorine reacts rather quickly with such inorganic compounds as iron and manganese. These types of reactions are probably completed during the time the water spends in a treatment plant's clearwell. Reactions with organic matter proceed at a slower rate and typically will not have been completed by the time the water enters the distribution system. The actual rate depends on both the amount and type of inorganic and organic matter present at the entry point, among other factors such as temperature and pH.

Free chlorine in the bulk water phase is typically represented by a first-order decay rate, which can be expressed as an exponential decay over with time as indicated by

$$C = C_o e^{(-k_b t)} \tag{7-3}$$

Where:

C_o = the initial concentration at the head of the pipe

t = the time of travel spent in the pipe

k_b = a bulk reaction coefficient, with units of 1/time

Laboratory methods for measuring k_b are described later in this chapter. The k_b value can vary widely for different waters and for the same source water subjected to different treatment processes or at different times of the year. Highly reactive waters have coefficients that are greater than 1.0/day, while slightly reactive waters have coefficients that are less than 0.1/day (Hart et al. 1986; Vasconcelos 1996). For the same water, k_b can roughly double for each 10-degree rise in water temperature. It is often necessary to assign different k_b values to different pipes throughout a distribution system to account for the blending of source waters with different reaction rates.

Source tracing analyses can be used to identify which parts of the system are influenced by which source. Reaction rates may be adjusted to account for blending of the source waters. Multispecies extension (MSX) modeling (see section 7.9) may be used in multisource systems to dynamically calculate time-varying bulk decay coefficients for individual pipes based on the fraction of water emanating from different sources (Grayman et al. 2011). Bulk decay coefficients in tanks should be considered carefully because daily temperature changes in a tank's contents affect the bulk decay coefficient and because residence times in tanks can be substantial. The way the contents within a tank are mixed affects tank residence time, which in turn affects the chlorine concentration of water leaving the tank.

7.8.3.3. Wall Demand Coefficients. Pipe wall demand for free chlorine typically occurs in unlined metallic pipes or in pipes where a significant growth of biofilm has occurred. In the former case, the demand is caused mainly by corrosion, resulting in release of iron from the pipe wall. Iron oxide has been shown to act as a catalyst in chlorine decay along the pipe wall (Hassan et al. 2006). In the latter case, the organic material is associated with the biofilm layer that exerts a chlorine demand. Two approaches are used to model wall demand. With the first approach, the value of k_b is increased to account for the increased chlorine demand. The second approach introduces a second rate coefficient, k_w, so that the overall coefficient K becomes (Rossman et al. 1994)

$$K = k_b + {}^{2k_w}/r \tag{7-4}$$

The term $2/r$, where r is the pipe radius, is the pipe surface area per unit of volume available for reaction.

The wall coefficient, k_w, with units of length divided by time, depends on the pipe's surface characteristics, such as the amounts of exposed iron surface and biofilm buildup. Because older pipes tend to exhibit more chlorine demand than newer ones and because these pipes tend to have lower pipe wall roughness coefficients, k_w is hypothesized to be inversely proportional to a pipe's Hazen–Williams C-factor. In this case, the overall chlorine decay coefficient becomes

$$K = k_b + {}^{\alpha}/(r\ C_{HW}) \tag{7-5}$$

where α is a factor that is system dependent.

Both k_w and α are very site specific. In a study of portions of four distribution systems where wall demand was exerted, researchers found that k_w values ranged from 0.1 to 5.0 ft/day (0.03 to 1.5 m/day) and that α ranged from 10 to 650 between the different systems (Vasconcelos 1996). Reaction rate variations within pipes contribute to water quality variations at any specific point in the distribution system. Changing the flow path that water travels to any specific location may impact water quality at the sampling location.

Some researchers have found anomalous behavior in wall demand that is not described by the methodologies previously outlined (Powell et al. 2004). Additional research is needed in this area to refine the wall demand processes.

7.8.3.4. THM Formation Coefficients. As free chlorine reacts with organic material in water, classes of compounds known as *disinfection by-products* that include THMs and HAAs are formed. Members of these classes are suspected carcinogens, and the US Environmental Protection Agency (USEPA) has set a maximum limit of 80 μg/L of total THMs in water and a maximum limit of 60 μg/L of the total of five HAAs (HAA5s) in water. USEPA requires compliance sampling for these constituents in the distribution system. The amount of sampling required can be reduced through use of water quality modeling of the distribution system.

The kinetics of THM formation are rather slow, with reactions proceeding for many hours after water leaves the treatment plant. It is not uncommon for more than half the total THM production to occur in the distribution system. There is a maximum THM level that forms for each water, depending on the nature and amount of organic matter and on the available chlorine level present. This amount is called the THM formation potential (THMFP). DBP formation is complex and involves many reaction pathways, and a general approximation may not apply for all systems. However, under constant conditions of temperature and pH, THM formation with time for many systems is approximated using the following equation:

$$THM = THM_0 + (THMFP - THM_0)(1 - exp^{-}k_{THM}{}^{t}) \quad (7\text{-}6)$$

Where:

THM_{0o} = an initial THM concentration
k_{THM} = a formation rate constant
t = elapsed time

The constant k_{THM} is determined from a laboratory test similar to that used to determine the decay of residual chlorine that also determines the THMFP. This method is described later in this chapter.

Researchers have suggested alternative models for representing the formation of THMs in distribution systems. Speight et al. (2009) summarize some of the alternative methodologies that have been studied.

The growth of HAA5s in a distribution system initially follows similar kinetic rates as for THMs. However, the rates for HAA5s are likely to follow a more complex reaction with growth in the distribution system as HAA5s have been observed to subsequently decay in some parts of distribution systems. Simulation of HAA5s was not feasible until recently with the advent of simulation of multiple species.

7.9. MODELING OF MULTIPLE SPECIES

Recent developments in modeling software allow users to define and model complex reaction schemes between multiple chemical and biological species in both the bulk flow and at the pipe wall (Shang et al. 2008a). The USEPA developed the first MSX for the EPANET engine; it is called EPANET-MSX (Shang et al. 2008b). Examples of processes that can be modeled using MSX include the autodecomposition of chloramines to ammonia, the fate of DBPs such as HAA5, biological regrowth including nitrification dynamics, combined reaction rate constants in multisource systems, and mass transfer of limited oxidation pipe wall adsorption reactions.

These developments recognize that many constituents can be found in two different physical phases within a distribution system. These are the bulk water phase relative to the motion of water in the pipes and a surface phase relative to the fixed pipe surface. Bulk phase species can be chemical or biological components that are carried in the bulk volume of water moving at average velocity. Surface phase species are integral with or attached to the pipe wall and react with the bulk water as it passes by a fixed point on the pipe. An example of this phenomenon is the decay of residual chlorine in the bulk fluid as it reacts with natural organic matter while it also decays as it deactivates biofilms attached to the pipe wall. Other examples of bulk phase species include dissolved ions or compounds, suspended matter, and chemicals adsorbed onto suspended or colloidal particles. Other examples of surface species include ferrous iron ions on iron or steel pipes, oxidized forms of iron, calcium carbonate scale, phosphate species used for corrosion control, particulate matter that has settled out of solution or become attached to the pipe wall, and organic compounds diffusing from or to the walls of plastic pipe materials. Some of these components exist in both phases and can be transferred to or from either phase. These are cases for the use of multiple-species modeling.

Software algorithms developed to model multiple species apply these integration techniques using a choice of different methods that are applicable to different types of constituents or combinations of constituents. The user can usually specify the mathematical expressions that govern the reaction dynamics of the system being studied. This provides users the flexibility to model a wide range of chemical reactions of interest to water utilities, consultants, and researchers. For more information, refer to the specific software user manuals, for example, Shang et al. (2008b). Examples of MSX include Klosterman et al. (2009) and Helbing and VanBriesen (2009).

7.10. OBJECTIVES OF WATER QUALITY TESTING AND MONITORING

Reliable and accurate field measurements are essential to the development and calibration of a useful water quality model. Due to complex reaction pathways, field testing is necessary to establish local reaction rate coefficients for initial model development and model refinement during calibration. Field measurements serve multiple purposes when used to support water quality modeling, including:

- Model development: as a means of developing water quality reaction and transformation relationships;
- Model calibration: to develop the best values for reaction and transformation parameters used by existing water quality models;
- Model validation: to ascertain that a previously calibrated water quality model performs appropriately for the situation being studied; and
- Calibration and testing of the underlying hydraulic model.

Five general types of water quality tests and measurements are discussed in this chapter: distribution system water quality surveys, use of previously collected (historical) field data, tracer studies, tank and reservoir field studies, and laboratory kinetic studies. Frequently, multiple types of studies are conducted in tandem to fully parameterize, calibrate, and validate a hydraulic and water quality model. The extent of data collection and calibration is dependent on the purpose of the distribution system model. A model for system-wide master planning may not require the level of water quality calibration needed for a detailed, neighborhood-area quality improvement analysis.

7.11. MONITORING AND SAMPLING PRINCIPLES

Water quality field testing procedures are centered on the collection and analysis of water samples. When used for regulatory purposes, sampling and analysis must follow standards prescribed by federal and state agencies. Water quality sampling done exclusively in support of water quality modeling typically is not required to follow these standards. However, to ensure useful results, a quality assurance–quality control program that specifies methods and practices should be established and followed.

There are a variety of mechanisms for collecting and analyzing representative water samples. These mechanisms are summarized in the following sections.

7.11.1. Sample Collection

When collecting a sample, the sample must be representative of the location or facility being sampled and contamination of the sample must be avoided. There are four general methods for collecting water samples:

- In situ measurements: Instruments are inserted into the water to make direct measurements. This procedure is used for taking measurements in pipes, tanks, and reservoirs. Measurements may be taken automatically and logged, sent remotely to a receiver, or taken manually, such as by reading a thermometer inserted in water.
- Manual grab samples: A sample of water is manually taken from the distribution system for analysis. Dedicated sampling taps, hose bibs, sinks, and fire hydrants are generally used for manual collection into sampling containers.
- Automated grab samples: Automated sampling devices collect multiple samples at specified time intervals and/or in response to a signal. Samples are collected into separate containers or into a single container to collect a composite sample.
- Continuous sample collection: A continuous sidestream can be extracted from a distribution system tap for subsequent automated or manual analysis.

Care should be taken and approved sampling practices followed to ensure that the sample that is collected is representative of the desired water location. For example, when sampling from a faucet or tap in a public facility, adequate flushing is necessary to ensure the sample is representative of water in the main. In addition, excessive flushing, which could introduce atypical water into the main from other areas of the system, should be avoided. Where possible, continuous flushing of the line with appropriate discharge to local drainage should be performed to ensure that water quality is representative of that in the distribution main. Similarly, when collecting a sample from a pressurized hydrant, an understanding of the hydrant lateral and barrel length is necessary to ensure that the water that is collected is the water that is currently in the main.

Sample collection procedures must include adequate storage and preservation techniques to prevent water quality reactions from continuing in the sample bottle after collection and prior to analysis. Common practices require that there be no headspace in the sample bottle in order to prevent constituent volatization. In addition, when analyzing DBP concentrations, preservation with a quenching agent is required to prevent further reactions with the organic matter and the disinfectant after the sample is collected.

Sample collection for water quality modeling requires careful planning to ensure that all data that may influence the reaction kinetics is properly measured and recorded. Development of a field sampling plan and review by all stakeholders is recommended prior to sampling. Considerations include defining sample locations and the procedures for sampling at each location. It is also necessary to confirm that the time when samples are taken aligns with staff availability and lab capacity so that samples are analyzed within the maximum allowable holding time.

7.11.2. Sample Analysis

There are several options for analyzing field samples. The appropriate method is selected based on the specific parameters being monitored, availability of equipment, required level of accuracy, cost of analysis, required expertise, and other factors. Sampling methods include the following:

- Manual analysis in the field: Field kits are available for manually analyzing many parameters in the field. For example, chlorine residual, turbidity, pH, and temperature are typically measured in the field using such kits. Care should be taken when performing field analysis as the ambient conditions are not controlled as they are in a lab.
- Automated analysis in the field: For longer-term analysis or as part of an intensive field study, automated instrumentation may be connected to a continuous source

of flow in order to measure water quality at a prescribed frequency. Parameters that are frequently measured in the field using online monitors include chlorine, pH, temperature, conductivity, oxidation–reduction potential, turbidity, and total organic chlorine.

- Laboratory analysis: For many analyses, it is either necessary or preferred to analyze the field-collected samples in the laboratory. Proper methods must be used in collecting, labeling, preserving, and transporting the samples to the laboratory.

7.11.3. Monitoring Parameters

The parameters to be monitored depend on the specific uses of the monitoring data. When used as part of a model calibration/validation process, parameters should correspond to the specific parameters being modeled and related parameters required within the calibration/validation process. The most frequently modeled water quality parameter in distribution systems is chlorine. In this case, free and/or total chlorine residual measurements should be taken in addition to temperature measurements. For modeling studies of DBPs, the specific DBPs should be monitored. In the case of tracer studies, the specific tracer being used (e.g., fluoride) or a surrogate parameter, such as conductivity for measuring sodium or calcium chloride, or UV-254 for measuring organic content, should be measured. Other water quality parameters that are sometimes measured in conjunction with a modeling study include bacteria counts, organic compounds, and specific chemicals.

7.11.4. Monitoring Locations

Monitoring locations are selected based on a number of factors including ultimate use of the data, availability and cost of equipment and resources, and accessibility of alternative sites. For online equipment, additional factors include the availability of electricity, protection of the environment, and a drain for sidestream water. Various mathematical methods have been used in selecting monitoring locations that best fit a set of stated criteria. These methods have been used to site monitors and sampling locations for both water security purposes (e.g., through use of the US EPA-developed software Threat Ensemble Vulnerability Assessment Sensor Placement Optimization Tool) and for routine water quality sampling (Grayman 2008).

7.12. WATER QUALITY SURVEYS

Water quality surveys encompass a wide range of water quality field data collection and analysis activities. Surveys range in time and complexity from the collection of data at a few stations within the distribution system to intensive multiday or longer-term data collection activities at many stations that involve significant labor and equipment resources. Though the scale of these activities varies significantly, many planning and implementation activities for a survey are similar.

There are five major steps that apply to all water quality surveys:

1. Develop a detailed sampling plan and make other preparations for sampling such as procuring materials (e.g., sampling bottles, reagents, colorimeters), developing forms, obtaining preapproval for staff overtime, and so on.
2. Perform the field sampling study.
3. Analyze post-sampling data.
4. Prepare a report that documents the field study.

5. Use the field data in the development (parameterization) of the distribution system model.

When preparing the sampling plan, the distribution system/water quality model is applied to the sampling area using the operating conditions likely to be in effect during the field study. Based on predicted results, sampling locations and sampling frequency are established. In some cases, alternative operating conditions are simulated to determine their effect on the planned sampling strategy. Even though the model is not fully calibrated and, in fact, is only an approximation of the actual system, the information gained by predicting the system's behavior under the expected sampling conditions is frequently helpful. The following issues should be addressed when developing a sampling plan, preparing for sampling, conducting the actual sampling campaign, and analyzing the post-sampling data (Clark and Grayman 1998).

7.12.1. Sampling Locations

Sampling sites should be carefully and accurately delineated in the plan. Examination of results from the preliminary modeling runs is an important factor in selecting general sampling locations. Samples are taken from permanent sampling taps, public buildings, businesses, residences, and water utility facilities. The locations are selected to reflect places where water quality data are desired, including water sources. Other selection criteria include accessibility during the entire study, required flushing times, safety issues, liability, and the degree to which the sample site is representative of some portion of the distribution system. Customer plumbing greatly affects water quality, so a site close to the water main is preferable. There are differing opinions on the use of hydrants for samples. While hydrants, with proper flushing, provide a convenient method for obtaining samples from a main, stagnant water in the hydrant or excessive biofilm and corrosion products typical of hydrant risers can confound the water quality data.

7.12.2. Sampling Frequency

Samples are taken manually or using automated collection or analysis instruments. Sampling frequency is generally governed by the availability of personnel, the number of samples to be analyzed in the laboratory, and most importantly, the expected temporal variation in water quality in the study area. For manual sampling, a sampling circuit is generally established in which samples are taken sequentially by a sampling crew. The circuit should be clearly marked on a map. The time it takes to sample at a single station and over the entire circuit is estimated by performing a preliminary test of the circuit or by estimating the times associated with each aspect of sampling (e.g., time to flush the sampling tap, time to take and analyze a sample, time to travel between sites). The duration of a sampling program varies depending on study goals, travel time across the network, and whether or not a tracer is used in the study.

7.12.3. System Operation

The water system operation (e.g., plant or well flow rates, tank levels, pumps) can impact the distribution system water quality. During a sampling study, system operating conditions are viewed as a controlled set of variables rather than a random set of occurrences. This requires close cooperation between the sampling team and the system's operations group. System operation during the sampling event either reflects normal operating conditions or reflects a desired set of operational conditions. Operational changes affect the time required to establish a steady-state water quality condition and, therefore, the length of the sampling study.

7.12.4. Preparing Sampling Sites

Prior to actual sampling, sampling sites should be evaluated and prepared. Preparation includes installation of sampling taps, if needed, calculation of required flushing time, notification of owners, and marking sites for easy identification. Valves and appurtenances are checked to ensure that they are in good working order.

7.12.5. Sample Collection Procedures

Procedures for collecting samples should be specified in the plan. These procedures include required flushing times (generally 2 to 6 minutes), methods for filling and marking sample containers, listing reagents or preservatives to be added to selected samples, methods for storing samples, and data logging procedures. The sample is only as good as the collection methods; therefore, USEPA, AWWA, and other resources should be consulted. In addition, the water quality model is only as good as the sample analysis results.

7.12.6. Collection of Ancillary Data

Use of sampling data in conjunction with a model requires extensive knowledge of the system. Preparations should be made to collect information on flows, pump operations, valve settings, and any other information required by the model during the sampling study.

7.12.7. Analysis Procedures

Samples taken in the field are analyzed at the sampling site, at a field laboratory located in the sampling area, or at a centralized laboratory. Procedures for each type of analysis should follow commonly accepted practice and/or *Standard Methods for the Examination of Water and Wastewater* (APHA 2017) and be specified in the plan.

7.12.8. Logistical Arrangements

Logistics refers to the practical arrangements that are necessary to organize an event successfully, especially an event that involves many people or many pieces of equipment. Logistical arrangements include the coordination of lodging for nonresident participants, provision of meals, transportation, and all the other minor details associated with any field study.

An important element of logistics is personnel scheduling. A detailed personnel schedule should be developed for the sampling study. Ideally, the schedule includes crew assignments and a work schedule for each study participant. The personnel scheduling should include coordination with utility staff and outside personnel such as consultants and equipment providers, as appropriate.

Other logistical arrangements include coordination related to safety issues and ensuring that all necessary equipment is available. These items are discussed separately and in more detail in the following section.

7.12.9. Safety Issues

The following safety-related concerns should be addressed in preparation for the sampling campaign: notification of police and other governmental agencies, public notification, notification of customers who will be directly affected, issuance of safety equipment such as flashlights and vests, use of marked vehicles and uniforms that identify the participants as official water utility employees or contractors, and issuance of official identification cards or letters that explain participation in the study. Whenever possible, samplers should not work alone, especially at night or in dangerous areas. If sample collectors will be using hose bibs from homes that are not part of the routine monitoring programs, homeowners should be notified.

7.12.10. Data Recording

An organized method for recording all data should be devised, including design of a data recording form. Data are recorded in ink using military (24-hour) time, and the individual data sheets are numbered sequentially and transferred to a central location at frequent intervals. A date–time–location identifier should be included on both the recording sheet and on the sample containers' labels.

7.12.11. Equipment and Supply Needs

Equipment includes field sampling equipment, safety equipment, laboratory equipment, and similar equipment. Expendable supplies include sampling containers, reagents, marking pens, batteries, and similar supplies. As part of the sampling plan, the needs and availability of equipment and supplies are identified, and alternative sources for equipment are investigated. Some redundancy in equipment should be planned because equipment malfunction or loss is possible.

7.12.12. Training Requirements

Training of sampling crews is essential and should be specified as part of the sampling plan. Even though crews are composed of experienced workers, it is important that all crews be trained in the use of a consistent set of procedures. Topics include sampling locations, sample collection and analysis procedures, data recording procedures, and contingencies. All crew members should read the sampling plan prior to the training sessions.

7.12.13. Contingency Plans

The old adage that "if something can go wrong during a sampling study, it will" serves as a basis for contingency planning. Contingency planning includes equipment malfunction, illness of crew members, communication problems, severe weather, unexpected system operation, and customer complaints. The sampling crew should brainstorm what could go wrong during the training period and assist in developing contingencies. Backup sampling sites are part of a contingency plan.

7.12.14. Communications and Coordination

During the sampling study, key personnel are often distributed throughout the area as follows: samplers at single stations or riding a circuit; the laboratory analysts in a permanent or field lab; operations personnel at central control stations; and the study supervisor at any of many locations. In order to coordinate actions during the study or to respond to unexpected events, some means of communication is needed. Alternatives include radios (normally available in utility vehicles), cellular phones, walkie-talkies, and the low-tech solution of a person circulating in a vehicle. Ideally, a study is coordinated through a central location where a study supervisor is located. The supervisor functions as the central coordinator, receiving information from the field, laboratory, and operations center and making decisions and disseminating information back to the field. Decisions on changes in sampling schedules or operating procedures are made out of this center based on full knowledge of all aspects of the ongoing study.

7.12.15. Calibration and Review of Analytical Instruments

A specific plan for reviewing and calibrating all instruments is necessary, and this task must be carried out prior to the study. When multiple instruments are used to measure the same parameter, each instrument should be calibrated against a standard and compared. A plan to check the calibration of instruments during the study is also essential.

7.12.16. Preparation of Data Report

Immediately following the sampling campaign, the data are examined and transferred to a computer format using spreadsheet or database management software, and a comprehensive data report is prepared. The data form should have been constructed with data input in mind (i.e., rows and columns to match spreadsheet format). The purpose of the data report is to contain all pertinent information, organized in a usable manner, relating to the field study and subsequent lab analysis. This includes information on the sampling study, study area, and results of the lab analysis.

7.13. USE OF HISTORICAL DATA

Historical distribution system water quality data may be available, usually resulting from regulatory sampling required by the Total Coliform Rule, the Disinfectants and Disinfection Byproducts Rule, the Lead and Copper Rule, and other federal or state regulations. The efficacy of using such historical data in support of water quality modeling depends on many factors, including the following:

- Assurance that the data have been correctly collected and reported,
- Knowledge of the exact location and time that the data were collected, and
- Knowledge of the state of the distribution system operation (i.e., boundary conditions) at the time of data collection.

Unlike field studies that are specifically designed in support of water quality modeling, the modeler has no control over practices associated with past field studies. In many cases, complete information is not known concerning the historical data. As a result, subsequent use of the data in the calibration or validation of a water quality model may lead to erroneous results.

Application of advanced analysis techniques such as data mining or advanced statistical analysis can be completed on sufficiently large data sets to gain insight into water quality trends and causation of quality variations. Care should be taken to ensure the results are statistically valid and reliable when using advanced data analysis methods. Multiple studies have successfully applied tools such as Bayesian statistics to estimate reaction rate coefficients (Cooper 2011).

7.14. TRACER STUDIES

Tracer studies in distribution systems involve observation of the movement of a substance within the distribution system. The information gained from a tracer study is used to characterize travel times (velocity) and treated water source blending within the system and to aid in the testing and calibration of a model, especially regarding water quality applications.

Tracer studies have been successfully conducted at many test sites throughout the United States and in other countries (USEPA 2006a). In the past, tracer studies were used frequently in support of the initial distribution system evaluation requirement of the Stage 2 Disinfectants and Disinfection Byproducts Rule (USEPA 2006b). More recently, tracer studies have been conducted by utilities interested in model calibration for EPS and studying their system's overall water quality. Because tracer studies generally require a significant amount of resources, a detailed study plan should be developed and the costs and resource requirements evaluated prior to performing a tracer study.

7.14.1. Tracer Study Preparation

Advanced in-depth planning should be completed to successfully conduct a tracer study for the purposes of extended-period model calibration. The following items should be well established before starting the tracer study: selection of the tracer substance; tracer removal or injection method and dosage; measurement equipment, method, and locations; and system operations during the study period. Proactive communication of the tracer study timeline and appropriate specifics should be shared throughout the appropriate departments of the utility, including customer service, water treatment, distribution operations, and engineering. Any local or state regulatory agencies should be notified of the tracer study and its details, as required.

Tracers include constituents that are normally added to a distribution system, such as fluoride, which are turned off for some period and traced; nontoxic, conservative substances such as sodium chloride, calcium chloride, or lithium chloride, which are injected at a known concentration; and naturally occurring substances and water characteristics such as fluoride, hardness, conductivity, or UV-254, which vary between different sources. The tracer substance that is chosen is unique to each system depending on baseline raw water source quality, measurable difference in conservative constituents between treated water sources, public health considerations, regulatory requirements, operational feasibility, and/or budgetary constraints of removing or injecting substances intended for tracer purposes. For chemicals that are normally added to the water supply (e.g., fluoride) and considered for use as a tracer, data collected when that chemical is discontinued and resumed can be used for two separate events and analyzed as part of the tracer study.

Budgets are often impacted by measurement protocols and, in turn, by the selection of tracer substance. Substances such as sodium and calcium chloride are indirectly measured by a change in ionic strength. This is indicated by specific conductance, which is measure using battery-powered and relatively robust continuous conductivity probes. Consequently, there are initial costs for logistics and instrument setup. However, field measurements are, for the most part, obtained in an autonomous and continuous manner. As a different example, fluoride requires relatively expensive ion-specific probes or individual samples grabbed by field personnel who travel on sample circuits. These two types of field studies have very different logistical and resource requirements and may yield data that are of different qualities. The National Sanitation Foundation should approve all injected tracer substances for use in treated water systems.

Removal of an added constituent should be evaluated to determine if its absence can be measured precisely enough throughout the system against source water background concentrations (e.g., fluoride). The method and process of injection should be planned, including the injection point and required equipment (e.g., pumps, chemical storage, mixing). Typical seasonal demand calculations should be completed during the planned tracer study period to confirm that only the required dosage of tracer substance is injected to develop a measurable concentration in the system. Injection of the required tracer dosage can save tracer substance costs while reducing other potential effects of the tracer on the system and its consumption.

Equipment used to monitor and measure the tracer throughout the distribution system should be carefully selected. The equipment should be sensitive enough to detect the range of possible concentrations of the selected tracer. Manual and automated sampling equipment must be calibrated prior to the study for optimal results. A detailed field test and sampling chain-of-custody plan should be established for in-field or laboratory tests of manual grab samples. Any automated sampling equipment should be secured and checked and batteries replaced as needed throughout the tracer study. All clocks should be synchronized on the sampling equipment prior to the study.

Areas of the system to monitor during the tracer study are selected for a general validation of modeled travel times and for verification of treated water source blending in zones and/or in specific areas of hydraulic or water quality interest. Measurement locations should be placed along the normal flow path of water through the system as well as in areas around the extremities of the system. Areas where the tracer could be diluted beyond detection should be avoided unless the areas are chosen in order to verify treated water source blending zones.

System operations should be discussed and modified, if needed, during the tracer study period. All important operating decisions should be noted for replication in the model, including facilities that are out of service during the tracer test. Generally, normal operations should be maintained to best represent the system during the intended model calibration period. System-wide supervisory control and data acquisition (SCADA) data collection and quality should be verified for use in the model simulation of the tracer study. Any equipment adjustments or calibrations to the SCADA system assist in establishing valid boundary conditions during the tracer simulation.

7.14.2. Tracer Study and Model Calibration Process

The underlying concept behind tracer studies for use in the model characterization/calibration/testing process is as follows: When modeling a conservative substance, there are essentially no water quality parameters that need to be adjusted. In other words, if the model hydraulic parameters, including initial boundary and controls throughout the tracer period, are correct and if the initial baseline and loading concentrations for the tracer substance are accurately known, the water quality model should provide a relatively accurate estimate of the spatial and temporary concentration of the tracer substance throughout the modeled network. The use of the water quality model and conservative tracer substance as a means of calibrating the hydraulic model is based on this relationship.

Using water quality modeling, the calibration process is summarized as follows:

1. An appropriate conservative tracer is identified for a distribution system. Factors that affect this decision include cost, baseline source and treated water quality, analysis requirements, local or state regulations, system operational prerequisites, and ease in handling the tracer.
2. An ideal time period is selected for the tracer study based on the specific use of the model calibration. For example, if nitrification increases or disinfectant residual decreases within the distribution system during the fall season, this time period should be targeted for the tracer test. With this approach, the system hydraulics that occur during the time period of interest can be adequately captured and validated through the tracer study.
3. A controlled field experiment is performed in which the conservative tracer, such as sodium chloride, is injected into the system for a prescribed period of time; a conservative substance that is normally added, such as fluoride, is shut off for a prescribed period; or a naturally occurring substance that differs between sources is traced.
4. During the field experiment, the concentration of the tracer is measured at the previously selected locations in the distribution system along with other hydraulic

parameters that are required by a model, such as tank water levels, pump operations, and flows. The tracer study is typically conducted in conjunction with a water quality survey. Depending on the tracer selected, measurement of the tracer is completed manually using in-field grab sample analysis, completed in the laboratory with samples collected in the field, or involves the use of automated, continuous instruments such as electrodes or chemical analyzers, along with data loggers. The measurement of the tracer should continue until a detectable concentration has reached the furthest extremities of the system.

5. Standard methods are used to adjust the parameters in the model to represent distribution system operations including storage, pump and valve boundary conditions, and controls. The water quality model is used to simulate the conservative tracer over the entire tracer time period in order to validate the measurements in the field to the model results, including travel time and propagation extents in the case of blending zones.
6. If the model adequately represents the observed concentrations, the calibration of the model for the demand and operational conditions being modeled is likely to be good. Significant deviation between the observed and modeled concentrations indicates that further model calibration is required. The model demands should be validated near the sample locations, especially if the locations are near dead ends or at system extremities where demands can play a significant role in travel time. Various statistical and directed search techniques are used in conjunction with the conservative tracer data to aid the user in adjusting the model parameters so as to better match the observed concentrations.

7.14.3. Tracer Study Simulation With Real-time Modeling

Tracer studies with real-time or simulated real-time modeling are being conducted, and becoming more common, for enhanced model calibration for water quality modeling. This approach combines traditional tracer studies with the increased representation of system conditions through real-time modeling methods. Real-time modeling uses SCADA data to represent the boundary conditions in the model based on what is being measured in the system, including tank levels, pump status and flow, and valve status and settings. Once the SCADA data streams are connected to the appropriate model elements, real-time modeling software can automate the historical simulation of the tracer study, potentially enhancing the model's ability to represent actual system conditions and reducing the costs of simulation-related data integration and management. Additional discussion on real-time modeling is provided in chapter 11.

Automated tracer measurement equipment can be remotely connected to the real-time modeling software through the SCADA system or a temporary data acquisition system to observe and simulate tracer movement through the system in real time. This is an advantage for tracer simulation as system operations adjustments or tracer monitoring equipment maintenance could be performed, if needed, depending on the real-time observation of the tracer movement across the distribution system. An alternative to this true real-time approach is to perform a retrospective real-time simulation using archived SCADA and tracer data. The outcome of this retrospective real-time analysis for model validation using tracers should be identical to real-time modeling but without the opportunity for benefits from adjustments made during the tracer test period.

The following steps can be taken to combine real-time modeling with tracer studies:

1. Establish SCADA tag and model element mapping to all level, flow, pressure, status, and setting locations where available. Validate that the SCADA-to-model data

connections are working through chosen real-time modeling software. The Real-Time eXtension (RTX) to the EPANET programmer's toolkit (EPANET-RTX) was designed by the USEPA, in part, to support such analyses.

2. Connect live data feeds from automated water quality monitors, if available, to the model. Observe tracer movement throughout the system during the tracer test period. Alternatively, simulate this tracer movement in the model software from archived data using retrospective real-time methods.
3. Determine where and when tracer travel time deviates from field measurements and track down potential causes in the model. Adjust model parameters, as needed, to improve the correlation between the tracer travel time and blending areas. The EPANET-MSX (Multispecies Extension) can be used in tracer modeling to simulate the tracer movement and concentration (if applicable) from different sources and the blending areas that occur between these sources.
4. Use the improved real-time or simulated real-time model parameters as inputs for the basis of an extended-period model simulation for water quality analysis purposes.

7.15. TANK AND RESERVOIR FIELD STUDIES

Field studies of distribution system tanks and reservoirs may be used both to characterize the flow and mixing behavior within the facility and to assist in the calibration or validation of mathematical models of the storage facility. Field testing may be categorized as water quality studies, tracer studies, and temperature studies. Water quality studies provide data on the temporal and spatial variation of water quality parameters within the storage facility and in the inflow and outflow. Tracer studies provide information on the mixing behavior in the storage facility. Temperature studies are used to gather information on how the temperature may vary at different locations and depths within the facility and over both short-term and long-term periods of time. The three types of studies may be performed by themselves or in tandem to develop a better understanding of how the reservoir behaves.

To understand the mixing phenomena in a tank or reservoir, a conservative tracer may be added to the inflow; the resulting concentration is measured over time at various locations within the facility and in the outflow. When a tracer study is performed for distribution system reservoirs and tanks and to understand the internal mixing phenomena, it is best to measure the tracer at available locations within the facility in addition to the outflow. Substances that are used as tracers should be conservative (not decay over time) and must conform to local, state, and federal regulations. Fluoride, calcium chloride, sodium chloride, and lithium chloride are the usual candidates for such studies.

Temperature variations in a tank or reservoir can significantly affect the mixing characteristics in the facility. Water temperatures may be measured manually from a sample tap, or water may be drawn from different depths using a pump or sampling apparatus. Alternatively, an apparatus composed of a series of thermistors and a data logger may be placed in a reservoir or tank to measure temperature at preset intervals at different depths. Figure 7-2 illustrates the results produced by three thermistors placed in a large reservoir at different elevations for a 16-day period. As illustrated, the tank displays a slight degree of stratification with a maximum temperature difference of 1.5°C that forms and then dissipates over a 5-day period.

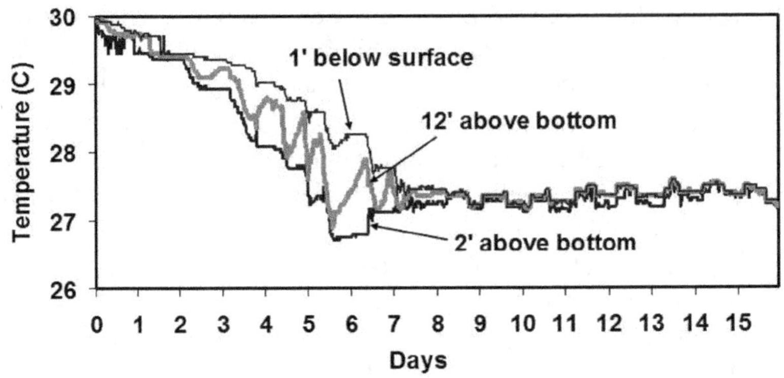

Figure 7-2 Example results from thermistor study showing temperature variation in tank

7.16. LABORATORY KINETIC STUDIES

Laboratory tests are available for estimating the kinetics of two types of reactions represented in distribution system water quality models – bulk chlorine decay and THM formation.

7.16.1. Chlorine Decay Bottle Test

The chlorine decay bottle test measures the rate of chlorine decay that occurs within the bulk water without any influence from the pipe wall. The test is generally conducted on finished water leaving a treatment plant or on the water that enters the zone of the distribution system being modeled. Waters from different sources undergo separate tests. Figure 7-3 lists the steps taken to conduct the test. Test results provide an estimate of the bulk decay rate coefficient, k_b, at the single point in time that the samples were collected. Additional data that provide insight into general water quality, such as natural organic matter (e.g., dissolved organic carbon), pH, ultraviolet absorbance, and temperature, should also be collected. The additional data can support user modification of the bulk decay rate coefficient as finished or source water quality varies.

Similar tests can be conducted for chloramine decay. In this case, the duration of the test is likely to be much longer than for chlorine because of the lower reactivity of monochloramine.

7.16.2. THM Formation Test

A similar bottle test can be used to estimate rate parameters for the growth of THMs. This is a modified form of the standard simulated distribution system test (APHA 2005). In that test, instead of a single sample being incubated for three days, multiple samples are incubated together, each for a different period of time. The test is run long enough for the THM levels to plateau to a constant level. This level becomes the THMFP, and the formation rate, k_{THM}, is estimated by plotting the natural logarithm of THM at time t minus THM at time zero against time and computing the slope of the straight line fit through these points. Alternatively, nonlinear least squares are used to estimate k_{THM} directly from the equation:

$$\text{THM} = \text{THM}_0 + (\text{THMFP} - \text{THM}_0)\,[(1 - exp\,(-k_{\text{THM}}t)] \qquad \text{(Eq 7-7)}$$

where THM is the measured THM value at time t and THM_0 is the measured value at time zero.

This procedure should be followed to generate data that can be used to estimate the rate of chlorine decay in finished drinking water.

1. Split a single sample of finished water into 24 to 36 amber bottles of 250 ml size or larger. Fill the bottles so they are headspace-free and cap them.
2. Place the bottles in a small bucket or wire cage and immerse them in a water bath in a sink with running water so that the initial water temperature of the sample can be maintained.
3. Starting from time zero, periodically remove three bottles at a time and analyze the contents for free chlorine. Discard bottle contents after taking measurements.
4. For each group of bottles analyzed, record the time (in hours from the start of the test), the free chlorine concentration of their contents, and the temperature of the water bath.

An ideal schedule for analyzing the bottles would provide at least 10 sets of chlorine values that cover a range from 100% down to about 25% of the initial chlorine concentration. A typical schedule might analyze bottles at 0, 0.5, 1, 2, 3, 6, 12, 18, 24, 36, 48, 60, and 72 hours. However, some adjustments might have to be made if the intermediate results indicate that chlorine is decaying either much more rapidly or much more slowly than anticipated.

After the test data are generated, a bulk decay coefficient can be estimated by plotting the natural logarithm of chlorine concentration versus time, and fitting a straight line through the points. The use of three separate chlorine readings at each time can be used to eliminate obvious outliers. The slope of the line is the bulk decay coefficient in units of 1/time (e.g., if time is in hours then the coefficient has units of 1/hours). When fitting the line it is best to force it to pass through the initial measurement at time zero. Alternatively, one could use a nonlinear curve fitting routine, available in many commercial spreadsheet and curve plotting software packages, to estimate K_{bt} directly from the equation

$$C = C_0 \exp(-k_{bt})$$

where C is the chlorine concentration measured at time t and C_0 is the measured concentration at time zero.

Figure 7-3 Protocol for chlorine decay bottle test

7.17. WATER QUALITY MODELING AND TESTING CASE STUDY

The following case study describes a landmark water quality modeling study conducted in 1993. Though the study was conducted in the relatively early days of distribution system water quality modeling, the methods are still highly relevant. At the end of the case example description, there is a brief discussion on how more recent developments in testing and modeling could be used to improve this historical study.

The study illustrates how water quality data collected from an intensive short-term sampling program are used to calibrate both an extended-period hydraulic model and a dynamic water quality model of a distribution system. It also demonstrates some of the complexities involved in modeling chlorine decay in multisource systems where both bulk and pipe wall demands are present.

7.17.1. Background

At the time of the study, the North Marin Water District (NMWD) served a suburban population of approximately 53,000 people in the northern portion of Marin County, Calif. In 1993, as part of an AWWA Research Foundation (now Water Research Foundation) project, a water quality sampling and modeling study was conducted to characterize and model chlorine decay in the distribution system (Vasconcelos 1996).

NMWD was served by two water sources: the Russian River, via an aqueduct used throughout the year, and Stafford Lake, which provided water from approximately May through October. Russian River water entered the system with a chlorine residual in the range of 0.3 to 0.4 mg/L and a THM concentration in the range of 10 to 20 µg/L. Stafford Lake had a high humic content. Following conventional treatment and prechlorination, the water left the clearwell with a chlorine residual of approximately 0.5 mg/L and THM levels that sometimes exceeded 100 µg/L. Sodium hydroxide was added in the treatment process, resulting in average sodium levels of 23 mg/L, compared to sodium levels in the Russian River of about 9 mg/L. The significant difference in sodium concentrations provided an easy method for identifying the water source as it blended within the distribution system.

The NMWD system was divided into a series of zones and satellite systems. For the sampling and modeling study, only the primary zone (zone I) was represented, with transfers to other zones represented as external demands. Zone I contains three tanks and several pump stations used to lift water into higher-pressure zones. The export to these zones is represented in the model by demands assigned to nodes located at the point of export.

7.17.2. Sampling Study

A 42-hour sampling study was performed in July 1993 to gather data for use in the assessment and modeling aspects of the project. Prior to conducting the study, a network model was applied to determine the projected response of the system and to aid in selecting sampling stations and frequency. A skeletonized representation of the distribution system network is presented in Figure 7-4, which shows the location of the sampling stations. Flows from the two sources and tank water levels were recorded every 15 minutes by a SCADA system.

During the sampling study, samples were taken manually at approximately 2-hour intervals at 16 stations. Each sample was analyzed for temperature, sodium, and free and total chlorine. Selected samples were also analyzed for a wide range of parameters in order to characterize the water and to study THM formation. Bottle decay tests were run to determine the chlorine bulk decay coefficient for the two water sources and for an even mixture of the two sources. All sampling results were stored in a database management system for later graphical and statistical analysis.

7.17.3. Hydraulic Validation

Following completion of the sampling study, the initial modeling task was to test an existing hydraulically calibrated model of zone I of the NMWD system for the conditions encountered during the study. The model was first updated to reflect recent hydraulic modifications in the network. The time history of SCADA-recorded tank levels and pump station flows was used to compute total demand in the network for each hour of the sampling period. Total demands were disaggregated to individual network nodes by prorating them against demands used in the original model calibration.

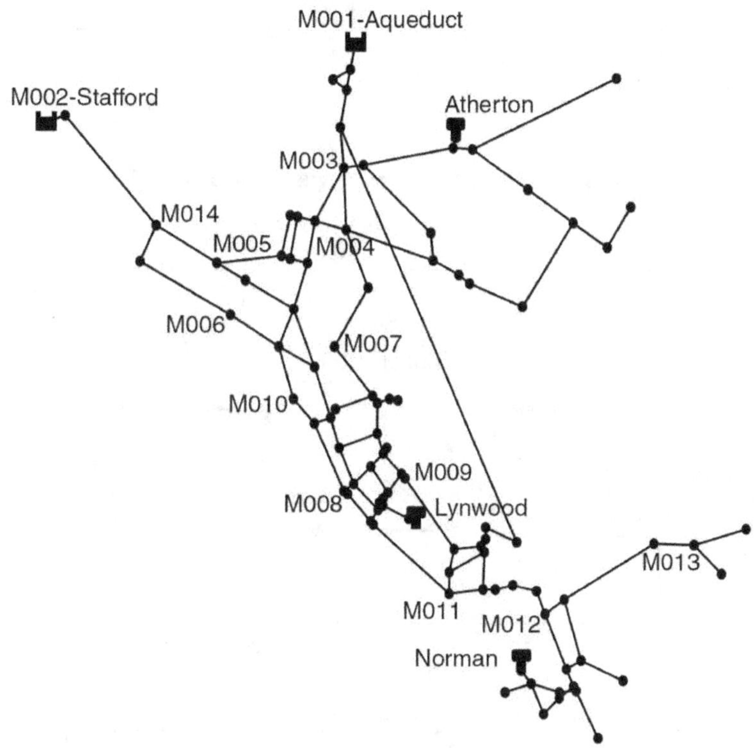

Figure 7-4 Skeletonized representation of zone I of the North Marin Water District

Model testing was done by comparing predicted and observed sodium concentrations over time throughout the system because of the blending of water from the two sources. Initial sodium levels at all nodes in the model were assigned by interpolating from the initial samples taken at the monitoring stations. During the sampling period, the Stafford Treatment Plant operated only during the day for 8- to 9-hour periods, resulting in widely fluctuating sodium concentrations throughout the system. Figure 7-5 illustrates the observed and modeled sodium concentrations at three stations: M003, which received Russian River water at all times; M005, which received Stafford Lake water when that treatment plant was operating and Russian River water at other times; and M012, which received blended water. The study relied on visual comparison (as opposed to statistical analysis) of the modeled and observed sodium concentrations, along with a comparison of the storage tank water elevation trajectories, to determine that an acceptable level of hydraulic calibration had been obtained. For systems that do not have multiple sources of water that contain significantly different water quality "signatures," similar calibration is performed by the addition of a conservative tracer, such as fluoride.

7.17.4. Water Quality Calibration

The sampling results were used to provide an understanding of chlorine decay kinetics in the NMWD system and to develop a calibrated chlorine model of the system. The two water sources were very different in terms of their chlorine decay kinetics, as illustrated by their bulk chlorine reaction coefficients:

Russian River: 1.32/day
Stafford Lake: 17.7/days
50:50 blend: 10.8/days

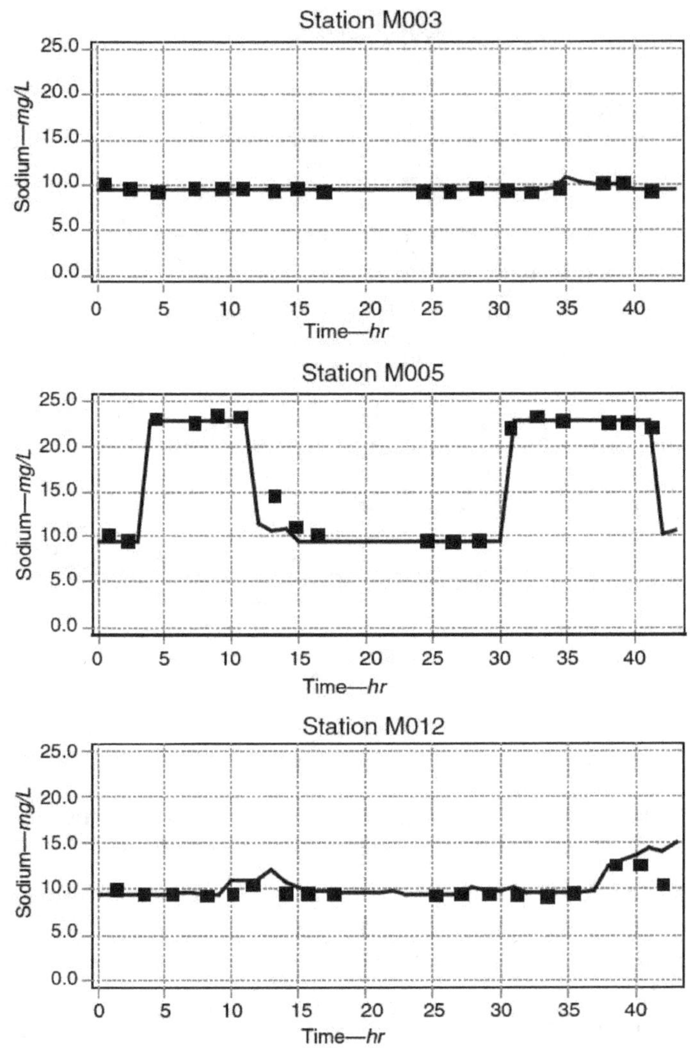

Figure 7-5 Comparison of observed and modeled sodium concentrations in the North Marin Water District system

This complicated the modeling of chlorine decay within the network because most water quality models use a single bulk decay coefficient for each pipe that is applied to all water flowing through the pipe regardless of its origin.

To resolve this situation, a model run was made to determine what the average contribution of flow to each of the monitoring stations from Stafford Lake was over the 42-hour sampling period. The results, shown in Figure 7-6, show that the Stafford source mainly serves the system's western edge. This suggested dividing the system into two zones, as shown in the figure. The pipes in the eastern zone, which receive mostly Aqueduct (Russian River) water, were assigned the bulk decay coefficient associated with the Aqueduct (Russian River) water. Because, on average, the water in the Stafford zone pipes was close to a 50:50 blend of Aqueduct (Russian River) and Stafford water, these pipes were assigned the decay coefficient determined for the 50:50 blended water. The main line connecting Stafford to station M014 was assigned the bulk coefficient for 100 percent Stafford water because no Aqueduct (Russian River) water ever travels through this pipe. As with sodium, initial chlorine concentrations throughout the network were assigned by interpolating from the initial sampled values at each monitoring station. Inputs of chlorine from the two sources were kept at their recorded values throughout the sampling study.

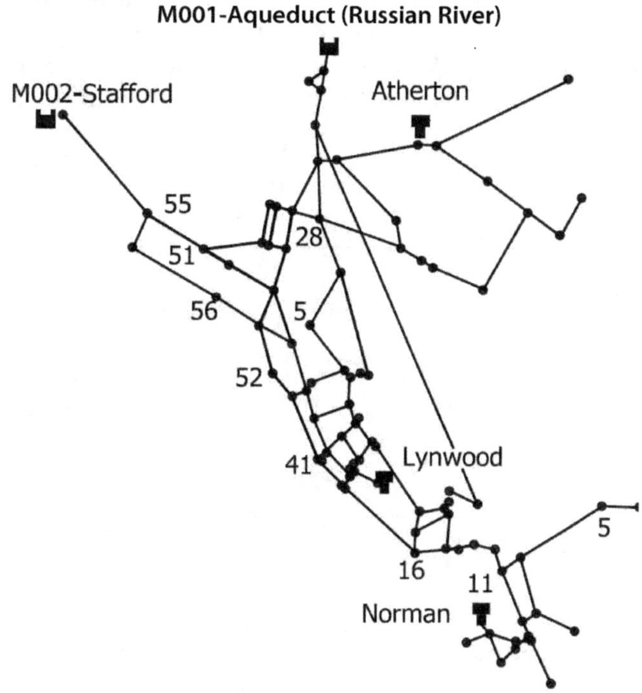

Figure 7-6 Average percent of Stafford Lake water in the North Marin Water District system

Initial runs of the model were made with no pipe wall demand for chlorine. When simulated chlorine concentrations were compared with measured values, it became apparent that chlorine levels in the Stafford zone were overpredicted. This result implied that the pipes in this portion of the system were exerting an additional chlorine demand from the pipe wall. This was logically consistent with the age and material of these pipes (40- to 50-year-old mortar-lined steel or cast iron) when compared to the newer asbestos–cement pipes found in the southerly and easterly portions of the system.

A wall demand was then introduced into the model for the pipes in the Stafford zone. The wall coefficient was systematically varied over a range of values until the overall mean absolute error between observed and predicted chlorine values at all sampling locations was minimized. This occurred at a wall coefficient of 5 ft/day (1.5 m/day), producing a mean absolute error of 0.05 mg/L. Time series plots comparing the model with observed chlorine levels for several of the monitoring stations are presented in Figure 7-7.

7.17.5. Conclusions

This exercise demonstrated that a distribution system water quality model could be calibrated to reproduce conditions observed within the North Marin system during the July 1993 sampling period. Additional testing of the model, using data collected from other operating periods, was needed before the model could be applied within the district.

In the period since the case study was performed, there have been some significant improvements in available testing and modeling methods. Boccelli et al. (2004) and Sautner et al. (2005) describe the use of continuous monitors for measuring both conductivity and chlorine. Conductivity can be used to measure the presence of chemicals such as sodium chloride and calcium chloride that are injected as tracers in a distribution system or natural tracers such as sodium that differ significantly in the two sources in NMWD.

There have been recent advances in water quality modeling using the MSX that can lead to more accurate chlorine modeling when there are two sources with differing bulk decay coefficients (Grayman et al. 2011). In place of the "zoned" approach used in the original NMWD study to assign bulk decay coefficients to parts of the system, MSX can be used to dynamically calculate the fraction of water coming from each source to each pipe at each time-step and to calculate the resulting bulk decay coefficient for the blended water.

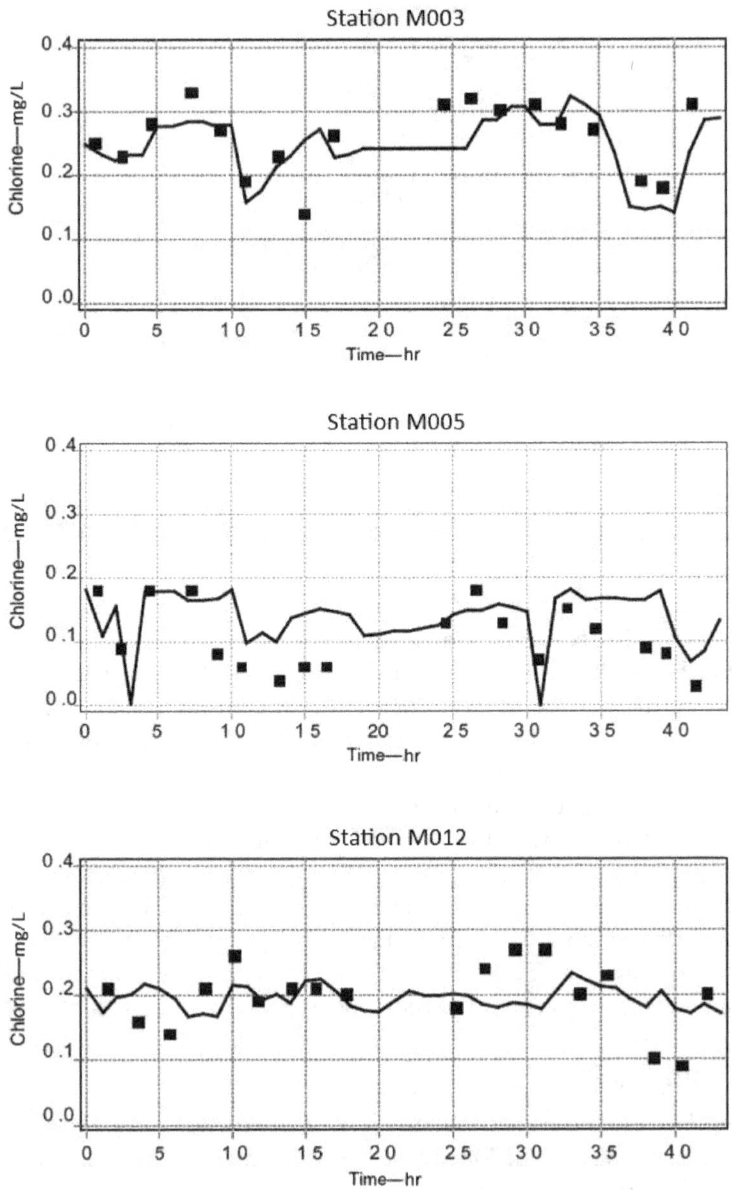

Figure 7-7 Comparison of observed and modeled chlorine residual in the North Marin Water District system

7.18. REFERENCES

APHA, 2017 (23rd ed.). *Standard Methods for the Examination of Water and Wastewater.* American Public Health Association, American Water Works Association, Water Environment Federation. Washington, DC: APHA.

Austin, R.G., van Bloemen Waanders, B., McKenna, S., & Choi, C.Y. 2008. Mixing at Cross Junctions in Water Distribution Systems—Part II. An Experimental Study. *ASCE Journal of Water Resources Planning and Management,* 134:3:295–302.

Boccelli, D.L., Shang, F., Uber, J.G., Orcevic, A., Moll, D., Hooper, S., Maslia, M., Sautner, J., Blount, B., & Cardinali, F., 2004. Tracer Tests for Network Model Calibration. In *Proceedings of the 2004 World Water and Environmental Resources Congress, American Society of Civil Engineers* [CD ROM document]. Salt Lake City, Utah.

Clark, R.M., & Grayman, W.M., 1998. *Modeling Water Quality in Drinking Water Distribution Systems.* Denver, CO: AWWA.

Cooper, J.P., 2011. Use of IDSE and Chlorine Monitoring Data to Predict Chlorine Decay and Trihalomethane Formation. In *Proceedings Annual Conference and Exposition.* Denver, CO: AWWA.

Grayman, W., Kshirsagar, S., Rivera-Sustache, M., & Ginsberg, M., 2011. An Improved Water Distribution System Chlorine Decay Model Using EPANET MSX. In *Modeling of Urban Water Systems,* Monograph 20, W. James, ed. Guelph, Ont.: Computational Hydraulics International.

Grayman, W.M., 2008. Designing an Optimum Water Monitoring System. In *Wiley Handbook of Science and Technology for Homeland Security.* J.G. Voeller, ed. New York: John Wiley & Sons.

Hart, F.L., Meander, J.L., & Chiang, S.M. 1986. CLNET—A Simulation Model for Tracing Chlorine Residuals in a Potable Water Distribution Network. In *Proceedings Distribution System Symposium.* Denver, CO: AWWA.

Hassan, K.Z.A., Bower, K.C., & Miller, C.M. 2006. Iron Oxide Enhanced Chlorine Decay and Disinfection By-Product Formation. *ASCE Journal of Environmental Engineering,* 132:12:1609–1616.

Helbing, D.E., & VanBriesen, J.M. 2009. Modeling Residual Chlorine Response to a Microbial Contamination Event in Drinking Water Distribution Systems. *ASCE Journal of Environmental Engineering,* 135(10):918–927.

Klosterman, S., Hatchett, S., Murray, R., Uber, J., & Boccelli, D. 2009. Comparing Single- and Multi-Species Water Quality Modeling Approaches for Assessing Contamination Exposure in Drinking Water Distribution Systems. In *Proceedings, World Environmental and Water Resources Congress.* Reston, VA: American Society of Civil Engineers.

Powell, J., Clement, J., Brandt, M., Casey, R., Holt, D., Grayman, W., & LeChevallier, M., 2004. *Predictive Models for Water Quality in Distribution Systems.* Denver, CO: Awwa Research Foundation.

Romero-Gomez, P., Ho, C.K., & Choi, C.Y., 2008. Mixing at Cross Junctions in Water Distribution Systems—Part I. A Numerical Study. *ASCE Journal of Water Resources Planning and Managements,* 134:3:284–294.

Rossman, L.A., & Boulos, B.F., 1996. Numerical Methods for Modeling Water Quality in Distribution Systems: A Comparison. *ASCE Journal of Water Resources Planning and Management,* 122:2:137–146.

Rossman, L.A., Clark, R.M., & Grayman, W.M., 1994. Modeling Chlorine Residuals in Drinking Water Distribution Systems. *Journal of Environmental Engineering, ASCE Journal of Water Resources Planning and Management,* 120:4:803–820.

Sautner, J.B., Maslia, M.L., Valenzuela, C., Grayman, W.M., Aral, M.M., & Green, Jr., J.W., 2005. Field Testing of Water Distribution Systems at U.S. Marine Corps Base, Camp Lejeune, North Carolina, in Support of an Epidemiologic Study. In *Proceedings, ASCE/EWRI Congress.* Anchorage, AK.

Shang, F., Uber, J.G., & Rossman, L.A., 2008a. Modeling Reaction and Transport of Multiple Species in Water Distribution Systems. *Environmental Science & Technology,* 42(3):808–814.

Shang, F., Uber, J.G., & Rossman, L.A., 2008b. *EPANET Multi-Species Extension User's Manual.* USEPA Office of Research and Development—National Homeland Security Research Center. EPA/600/S-07/021. Revised October 2008. Washington, DC: USEPA.

Speight, V., Uber, J., Grayman, W., Martel, K., Friedman, M., Singer, P., & DiGiano, F., 2009. *Probabilistic Modeling Framework for Assessing Water Quality Sampling Programs.* Denver, CO: Water Research Foundation.

Tzatchkov, V.G., Buchberger, S.G., Li, Z., Romero-Gomez, P., & Choi, C. 2009. Axial Dispersion in Pressurized Water Distribution Networks—A Review. In *Proceedings International Symposium on Water Management and Hydraulic Engineering.* Ohrid, Macedonia.

USEPA, 2006a. *Water Distribution System Analysis: Field Studies, Modeling and Management—A Reference Guide for Utilities.* Washington, DC: USEPA.

USEPA, 2006b. *Evaluation Guidance Manual for the Final Stage 2 Disinfectants and Disinfection Byproducts Rule.* Washington, DC: USEPA.

Vasconcelos, J.J., 1996. *Characterization and Modeling of Chlorine Decay in Distribution Systems.* Denver, CO: Awwa Research Foundation.

Vasconcelos, J.J., Rossman, L.A., Grayman, W.M., Boulos, P.F., & Clark, R.M. 1997. Kinetics of Chlorine Decay. *Journal of the American Water Works Association,* 89:7:54–65.

Chapter 8

Storage Tank Mixing and Water Age

8.1. INTRODUCTION

This chapter provides a summary of various methods of modeling distribution system tanks and reservoirs and the advantages and disadvantages of each. It also describes the most important concepts of flow mechanics inside tanks and reservoirs. Some systems distinguish tanks from reservoirs by the vertical location of the storage facility (e.g., tanks are elevated and reservoirs are in-ground), by size (e.g., tanks are small and reservoirs are large), purpose (tanks are for treated water and reservoirs for raw water), or by shape (e.g., tanks are regularly shaped and reservoirs are irregularly shaped). In this chapter, the words *tank* and *reservoir* are used interchangeably because both refer to facilities that are used to store water and have similar purposes in a distribution system.

Distribution system storage tanks (and reservoirs) store a significant portion of the total volume of water in a potable water system. Maintaining water quality in these storage tanks is necessary to reduce the loss of disinfectant residuals, prevent nitrification, minimize disinfection by-product (DBP) formation, and reduce biological growth. Maintaining acceptable water quality in distribution system storage tanks depends on the quality of water entering a tank, the water mixing characteristics in the tank, and the rate of turnover. The quality of the water entering tanks is primarily determined by treatment plant process performance, disinfection practices, and design and operation of the distribution system. Operational and design aspects of an individual tank affect mixing characteristics and the change in water quality in the tank. Adequate mixing is needed to prevent the formation of stagnant zones in tanks that, over time, can lead to the release of poor-quality water to the distribution system supplied by the tanks. In some cases, insufficient mixing leads to thermal stratification and dead zones of water inside storage tanks that may exacerbate water quality deterioration and freezing in colder climates.

8.2. TYPES OF TANKS AND RESERVOIRS

Tanks and reservoirs can generally be classified into two broad types: those that provide storage for fire flow requirements, emergency needs, and peak demands and those that are used for disinfection to achieve desired contact time. In most cases, a tank or reservoir serves one of these purposes, but sometimes it can be used to serve both. The tanks and reservoirs in distribution systems are generally used to serve fire flow, emergency needs, and peak flow demands, whereas the tanks and reservoirs at water treatment plants (frequently referred to as *clearwells*) are generally used for disinfection credit (although some of them are also used for peak demands). For any type of storage tank, the change in water quality entering and exiting the tank is primarily governed by the mixing processes and residence time (or volume turnover) within the tank. For distribution system storage tanks, the intent is to minimize water quality deterioration by maximizing overall mixing inside the tank and reducing water age. For water treatment plant storage tanks, the intent is to maximize disinfection by promoting plug-flow conditions and prolonging the time the disinfectant spends inside the tank. This condition is generally achieved using baffles to direct water in a plug-flow manner. However, baffles can sometimes introduce pockets of recirculation zones within a tank and therefore need to be evaluated and designed properly.

8.3. BACKGROUND

Physical and mathematical models have been used to study the mixing behavior in distribution system tanks and reservoirs. A summary of some of these studies is provided in an AWWA Research Foundation (now Water Research Foundation) report (Grayman et al. 2000). Some of the major findings of this report are represented in this section.

8.3.1. Tank Designs Should Achieve Good Mixing

There are multiple theoretical schemes that describe the ways in which water flows through a storage tank, including:

- Completely mixed state,
- Plug-flow manner,
- First-in, last-out, and
- Two-compartment mixing.

When comparing two tanks of the same size and with the same inflow–outflow characteristics, the one that operates under plug-flow conditions will generally lose more disinfectant than the one that operates under mixed-flow conditions because the rate of disinfectant decay is concentration dependent. Therefore, achieving a completely mixed state should be a primary goal in water storage tank design.

8.3.2. Fill Time Versus Mixing Time

Ideally, the typical time required to fill a tank exceeds the time required to achieve complete mixing. For a tank operating in a fill-and-draw mode, mixing occurs primarily during the fill cycle. As a result, if the tank is relatively well mixed by the end of each fill cycle, mixing problems such as stratification and dead zones are unlikely to occur. For a wide range of tank designs, experimentation has shown that the mixing effectiveness is primarily dependent on

- Volume of water in the tank at the start of filling,
- Inlet diameter,
- Filling flow rate, and
- Filling time.

8.3.3. Develop Turbulent Inlet Jet to Promote Mixing

Mixing a fluid requires an energy source. In distribution system storage tanks, this energy is normally introduced during tank refilling unless mechanical mixers are used. As water enters a tank, jet flow occurs due to high-velocity vectors at the inlet, ambient water is entrained into the jet, and circulation patterns are formed that result in mixing. To have efficient mixing, the jet's path must be long enough to allow for the entrainment and mixing process to develop. Therefore, the inlet jet should not be pointed directly toward nearby barriers such as walls, the tank bottom, or deflectors. The rate and degree of mixing depend primarily on the size of the tank and the momentum of the incoming jet. If the turbulent jet is insufficient to ensure good mixing, more complex inlet configurations or mechanical mixing may be required.

8.3.4. Avoid Conditions That Inhibit Complete Mixing

Conditions that inhibit complete mixing and promote stratification should be avoided. Whenever there is a temperature difference between the contents of a tank and its inflow, the potential for poor mixing and stratification exists. Temperature differences result in what is called a *buoyant jet*. A positively buoyant jet occurs when the density of the incoming water is less than the density of the ambient water (this may occur during winter), while a negatively buoyant jet occurs when the density of the incoming water is greater than the density of the ambient water (this may occur in elevated storage tanks during the summer). Incoming water with excessive buoyancy (either positive or negative) relative to its momentum will lead to ineffective mixing and could produce stable stratified conditions within the tank. Based on this relationship, tall tanks and tanks with large-diameter inlets have a greater tendency toward stratification. If significant temperature differences are experienced, increasing the inflow velocity is an effective strategy for reducing the likelihood of stratification. Standpipes, in which the water depth exceeds the diameter of the tank, are typically more susceptible to stratification.

8.4. FACTORS THAT AFFECT WATER QUALITY IN TANKS

8.4.1. Hydraulic Residence Time

Hydraulic residence time (HRT) is the average amount of time that water spends within a storage tank. This is frequently referred to as the tank *water age*. However, HRT is a more precise term because it differentiates between the total water age within the water system (i.e., the combination of residence time within the distribution system and in tanks) and the water age associated only with the time spent in the tank. Excessive HRT in a distribution system storage tank can make it difficult to consistently maintain an adequate disinfectant residual inside the tank and can lead to excessive DBPs, regardless of mixing conditions. This can result in inadequate residuals and elevated DBPs downstream of the tank when the water flows into the distribution system. Acceptable values for HRT depend on both the water's inherent quality (e.g., water with higher amounts of organic material results in faster consumption of disinfectant) and the relationship of the tank to other parts of the distribution system (i.e., if a tank feeds into other tanks, the total HRT through the system can quickly become excessive).

Acceptable HRT is system specific because of differences in the quality of the water entering the distribution system, disinfection strategy, and water temperature among different systems. A bulk water disinfectant decay test that indicates disinfectant residual as a function of time can be conducted to determine the criteria for desirable, acceptable, and undesirable categories.

The HRT can be calculated as follows (Mahmood et al. 2005):

$$\text{Theoretical average residence time} = \frac{V_{max}}{V_{max} - V_{min}} \times \frac{1}{N} \tag{8-1}$$

Where:

V_{min} = typical minimum cycle volume
V_{max} = typical maximum cycle volume
N = number of fill–draw cycles per day

8.4.2. Fill Time

Mixing primarily occurs during the filling cycle. The ability of a water storage tank to achieve complete mixing can be assessed by comparing the actual average fill time to the mixing time theoretically required for a complete mix to occur. For good mixing, actual average fill times should exceed the theoretically required time for complete mixing. For a given filling flow rate, this can be expressed in terms of fill time or volume turnover per fill cycle. Based on detailed laboratory tests, the fill time required to achieve complete mixing can be defined using the following empirical equation (Grayman et al. 2000; Roberts et al. 2006):

$$\text{Mixing time (hours)} = \frac{k}{3{,}600} \times V_{min}^{2/3} / M^{1/2} \tag{8-2}$$

Where:

M = momentum = U × Q
Q = average fill rate (cfs)
U = average inflow velocity (ft/s)
V_{min} = average minimum daily volume (ft^3)
k = mixing coefficient
= 10 (for H/D <1)
= 10 + 3.5 (H/D − 1) (for H/D >1)
H = tank height (ft)
D = tank diameter (or equivalent diameter for noncylindrical tanks) (ft)

The mixing time equation assumes isothermal conditions (i.e., the temperature of the inflow is the same as the temperature of the ambient water in the tank). As a result, if the water temperature of the inflow significantly differs from the water temperature in the tank, the mixing equation may not be applicable, and mixing may be inhibited and lead to stratification. The mixing equation applies to tanks that operate in either simultaneous inflow–outflow mode (tank inflow and outflow through different pipes) and in fill-and-draw mode (tank fills and empties through the same pipe). However, for the simultaneous inflow–outflow mode, there is no representative fill time that can be used to compare to the mixing time. Therefore, professional judgment must be applied to determine the adequacy of the calculated mixing time.

8.4.3. Inlet Momentum

Inlet momentum is a key factor for mixing in storage tanks. Generally, the higher the inlet momentum, the better the mixing characteristics in a storage tank. Inlet momentum is defined as:

$$\text{momentum} = \text{velocity} \times \text{flow rate}$$

Average inlet velocity and inlet momentum are calculated for a given tank based on the average filling rates. Higher momentum can be achieved by increasing the inflow rate or by increasing the velocity. In many situations, it is not practical to increase the flow rate into tanks because of system hydraulic limitations. Pumping water into the tanks may increase the flow rate, but this approach is typically not desirable due to increased operational complexity and cost. It is generally more feasible to increase the inlet velocity by decreasing the inlet diameter. Prior to decreasing inlet diameter, the resulting increase in head loss needs to be considered.

8.4.4. Inlet Pipe Location and Orientation and Tank/Reservoir Geometry

The location and orientation of the inlet pipe relative to the tank walls can have a significant impact on mixing characteristics. For example, when the height of a tank is much larger than the diameter or width, a horizontal inlet pipe at the bottom of a tank is more likely than a vertical inlet to result in incomplete mixing of the water in the tank. In this type of tank, the horizontal orientation of the inlet pipe would most likely result in the water jet hitting the opposite vertical wall of the tank before sufficient time has elapsed for mixing, resulting in a loss of inlet momentum.

A tank's mixing characteristics also depend on the initial water depth in the tank. When the inlet orientation is horizontal and the initial water depth is very high, the inlet momentum may not be able to mix all the water in the tank completely during the fill cycle.

8.4.5. Volume Turnover

Turnover is the percentage of water that is exchanged on a daily basis between the tank and the distribution system. Increasing the daily turnover can lower the residence time in the tank and improve mixing. There are no firm standards on the amount of turnover that is needed.

8.4.6. Seasonal Variation

Several factors that influence the quality of water in tanks are affected by seasonal variation. These include changes in water usage and system operations, changes in water and air temperature that may lead to stratification or affect mixing, and changes in temperature that may affect reactive processes in the tank. To account for seasonal variation, water quality within tanks should be assessed under different seasonal conditions.

8.5. TYPES OF TANK MODELING

8.5.1. Physical Scale Modeling

Tanks and reservoirs can be modeled by building a geometrically scaled model and observing water movement and flow characteristics. However, a physical model has to be scaled properly; dimensional analysis is used to achieve this goal. The scale range for most studies is typically between 1:10 and 1:100. Generally three conditions need to be satisfied according to the dimensional analysis:

- Geometric similarity: the model should be the same shape as the actual tank but scaled by a constant scale factor.
- Kinematic similarity: the velocity at any location in the model and the actual tank should be proportional to each other by a constant factor.
- Dynamic similarity: all forces in the model flow should be proportional to all forces in the actual tank flow.

To achieve dynamic similarity, it is necessary to establish dimensionless parameters and to match these parameters between the modeled tank and actual tank. The common dimensionless parameters used in tank hydraulics are:

- Froude number, $F_r = V^2/gL$, ratio of inertial force to gravitational force;
- Reynolds number, $Re = VL/\nu$, ratio of inertial force to viscous force.
- Weber number, $W = \varrho V^2 L/\sigma$, ratio of inertial force to surface tension force.

Where:

V = velocity of water
g = gravitational acceleration
L = characteristic length such as diameter of tank or depth of water
ν = water viscosity
σ = water surface tension
ϱ = water density

Sometimes it is difficult to match all of the dimensionless parameters between a physical model and the actual tank. In this case, emphasis should be placed on matching the parameters that correspond to the predominant forces. In the case of tanks, the most important dimensionless parameter is the Froude number. For inlet and outlet pipes, the most important dimensionless parameter is the Reynolds number. When tanks and reservoirs are modeled according to Froude number similitude, the following relationships govern:

- Geometric similarity: length = L, area = L^2, volume = L^3
- Kinematic similarity: velocity = $L^{0.5}$, time = $L^{0.5}$, flow = $L^{2.5}$
- Dynamic similarity: force = L^3, mass = L^3

Roberts et al. (2006) applied a unique tank scale modeling technique called *three-dimensional laser-induced fluorescence* (3DLIF). Two fast scanning mirrors drive a laser beam from an argon-ion laser through the flow in a programmed pattern. A small amount of a fluorescent dye is added to the effluent; the laser causes the dye to fluoresce, and the emitted light is captured by a video camera. Mirrors are used to sweep the laser beam and to create multiple vertical slices through the flow. After computation of tracer concentrations, the data can be converted, by image processing techniques, into three-dimensional images of the flow field. The 3DLIF system can obtain millions of sampling points, leading to great insight into the hydrodynamics of turbulent mixing in tanks.

Physical scale models can provide accurate information about water flow mechanics inside a tank. However, depending on the tank's geometry and inlet–outlet, it may become challenging to achieve similarity of the dimensionless parameters. Larger-scale models are preferable, but building large physical models is more expensive. Building a model may also be impractical for many utilities that want to study mixing within a water storage tank. Therefore, other methods of modeling need to be explored.

8.5.2. Empirical Modeling

Tank and reservoir mixing can be studied using empirical models that can be used as stand-alone tank models or integrated into network simulations. Several types of empirical models have been developed for tanks and reservoirs:

- Completely mixed,
- Plug flow (first in/first out),
- Short-circuiting (last in/first out), and
- Compartment.

Figure 8-1 illustrates the types of empirical models previously listed. Though each theoretical mode of behavior is an idealistic representation of flow behavior in tanks that is not actually fully achieved in practice, a properly applied empirical model that is based on specific tank design and operational conditions may provide a good representation of a tank's mixing behavior. These types of empirical models are also referred to as *systems models, input – output models,* and *black box models.*

Completely mixed models assume that the water in the tank mixes instantaneously and completely and remains in this state indefinitely. The concentration throughout the tank is a blend of the tank's current bulk water and the newer water entering the tank. The concentration at the outflow of the tank is equal to the concentration in the completely mixed tank. This is the most commonly used empirical model and is the default mechanism used by most network modeling software. Depending on tank geometry, inlet–outlet configuration, and flow rates, it is possible to approximately model many distribution system tanks as completely mixed. Generally, tanks with high momentum jets and high-volume turnover achieve predominantly mixed flow behavior.

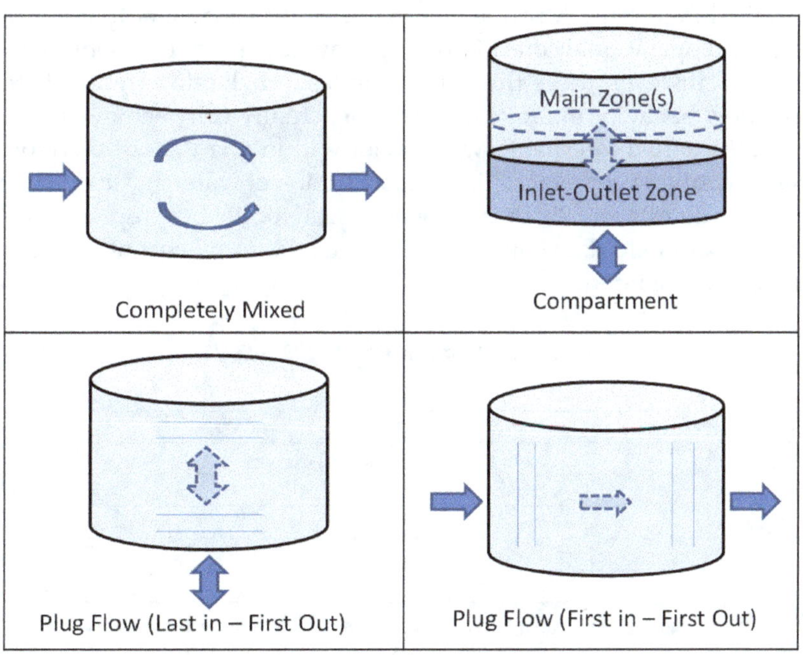

Figure 8-1 Schematic representation of the types of empirical models

Plug-flow models assume that the water does not mix at all and moves as parcels through the tank from the inlet to the outlet. This type of model represents first in–first out (FIFO) behavior for water flow, meaning the first parcel of water to enter the tank is also the first parcel of water to leave the tank. Tanks that operate in simultaneous inflow–outflow mode with baffles to guide the flow of water through the tank are most likely to achieve the plug-flow FIFO behavior. Treatment plant clearwells and contact chambers most commonly follow this type of behavior. Most tanks in the distribution system do not fit this type of model because they are not typically baffled and usually operate in fill-and-draw mode.

Short-circuiting models assume that water entering a tank is bypassing a significant portion of the tank volume by traveling toward the outlet preferentially. This type of model represents LIFO behavior for water flow. Depending on the tank geometry, inlet–outlet configuration, and flow rates, it is appropriate to model some distribution system tanks as short-circuiting type. Generally, tanks with low flow rates, stratified conditions, low-volume turnover, and a common inlet–outlet or an outlet close to the inlet may exhibit short-circuiting behavior. This type of model may be applicable to tall and narrow standpipes in the distribution system.

Compartment models are hybrid representations that assume that the entire tank volume does not fit into one of the model types previously indicated but may be represented by two or more compartments that are each completely mixed and interact with other compartments. The volume of each compartment may vary or can remain fixed (e.g., to represent stratified conditions). New water entering the tank completely mixes with the stored water in the first compartment, and when the first compartment fills, excess water is sent to another compartment where the water completely mixes with the stored water in that compartment. There has been only limited application of compartment models due to the lack of guidance and experience on model selection and parameterization.

Empirical models are most frequently used to model water age or disinfectant concentrations in a tank over an extended period of time. Figure 8-2 illustrates the output from an empirical model used to model water age in a tank for 1 year.

Empirical models offer a very fast computational solution for modeling water age and disinfectant residuals in the *outflow* from a tank and can therefore be easily used in network simulations. A properly applied empirical model that is based on specific tank design and operational conditions may provide a good representation of the tank's mixing behavior. However, these theoretical modes of behavior are idealistic representations of actual flow behavior in tanks that are not actually fully achieved in practice. They do not predict the actual internal flow behavior within a tank and, therefore, cannot be used to assess the mixing characteristics or distribution of water age *within* the tank. Field data can be used to determine which type of empirical model best represents a tank, but care must be taken with field data collection, and the additional cost of field sampling and analysis should be considered.

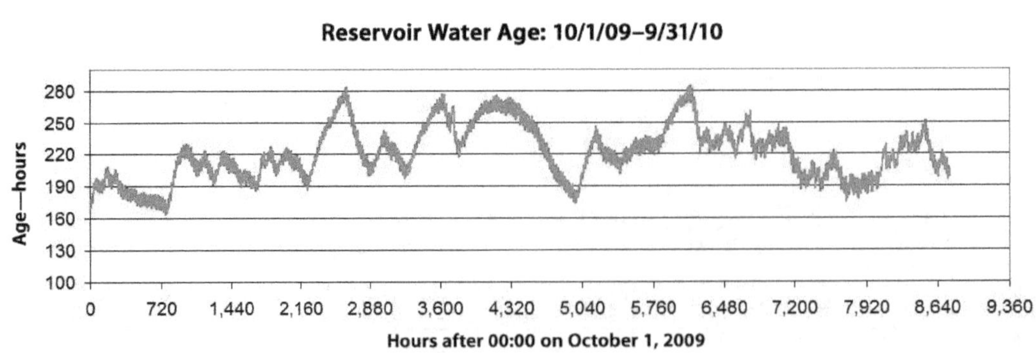

Figure 8-2 Tank water age calculated using an empirical model assuming complete mixing

8.5.3. Computational Fluid Dynamics Modeling

Mixing in tanks and reservoirs can be studied using computational fluid dynamics (CFD) modeling techniques that apply the fundamental equations of fluid mechanics as governing equations. These fundamental equations are:

- Conservation of mass,
- Conservation momentum, and
- Conservation of energy.

Within the framework of a CFD model, a tank's geometry, including inlet and outlet features, is represented by a grid or mesh that consists of cells or control volumes with which the equations are solved. Gridding may be characterized by both the density (number of grid cells and size of the individual cells) and the quality of the gridding. There are various ways to generate grids. The resulting quality of the gridding, the type of grids, and the grid density can influence both the results and the computational time associated with solving the equations.

Appropriate boundary conditions need to be specified for the walls, air–water interface, inlets, and outlets to obtain accurate CFD solutions. For example, tank walls can be specified as having zero velocity both normal to the wall and tangent to the walls. However, if the water is allowed to move along the surface, the tank water surface can be specified as a slip wall with zero stress to allow tangential movement of water along the surface. The inlets and outlets can be specified with known velocity or known pressure crossing the inlet–outlet. If reverse flow is observed at the outlet boundaries, the computational domain needs to be extended outside of the confines of the tank.

Appropriate solution algorithms should be selected based on the type of problem. For example, if the water surface remains relatively constant, a tank can be modeled using a single-phase approach that involves modeling of the water only. However, if the water surface level varies significantly, a multiphase approach that involves the air–water interface is more appropriate. If the flow is turbulent, an appropriate turbulence closure scheme also needs to be used.

As a CFD simulation progresses, the user should monitor the convergence of the solutions, defined as *residuals*. Convergence occurs when the residuals are small. The residuals should decrease with time. The magnitude of various residuals at any given time provides information about how well the solution is approximating the actual tank hydraulic behavior as defined by the Navier–Stokes equations.

Because CFD modeling is capable of generating millions of data points that represent the spatial and temporal behavior of flow within the tank, graphical methods for visualizing the output data are important. Figures 8-3 through 8-6 provide examples of how CFD modeling results can be displayed to evaluate water movement and water age inside a tank for various types of scenarios. Figure 8-3 (tall tank) and Figure 8-4 (elevated tank) illustrate the effects of thermal differences using tracer and section views. These figures show that mixing conditions can be exacerbated for aboveground storage tanks during the summer when inflow is cooler than the bulk water in the tank, especially if the tanks are made of steel. Figure 8-5 illustrates the effect of operational or design changes using velocity values and isometric views. Figure 8-6 illustrates different methods of displaying water age distribution inside a tank using pathlines; one shows pathlines by water age values and the other shows percent of the total tank volume with specific water age ranges. Figure 8-7 shows flow paths of an unbaffled and baffled storage tank, illustrating the impact of baffles in reducing recirculation zones significantly but also the potential to create new areas of recirculation zones.

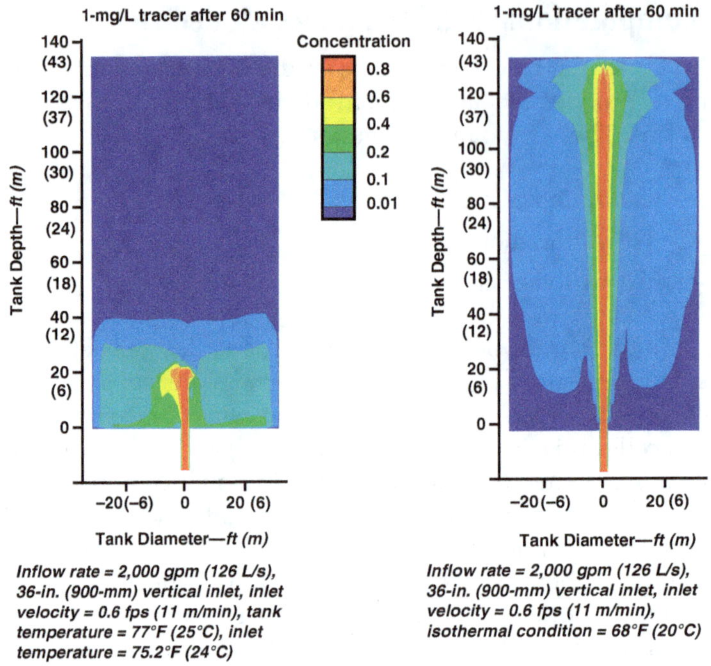

Figure 8-3 Effect of thermal differences for a tall tank

Figure 8-4 Effect of thermal differences for an elevated tank

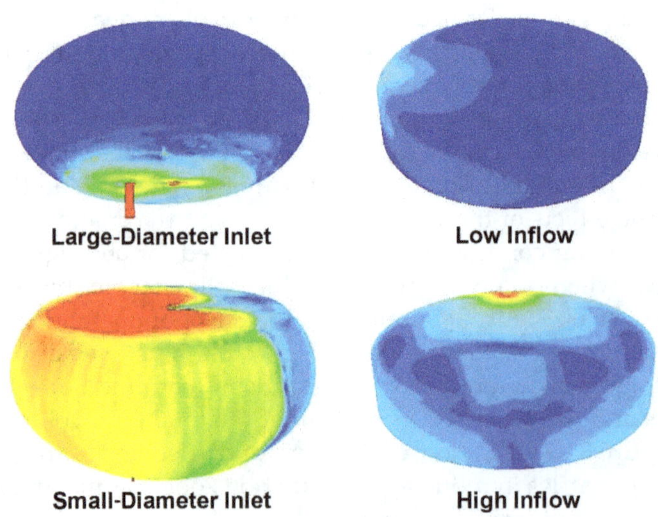

Figure 8-5 Effect of operational and design changes

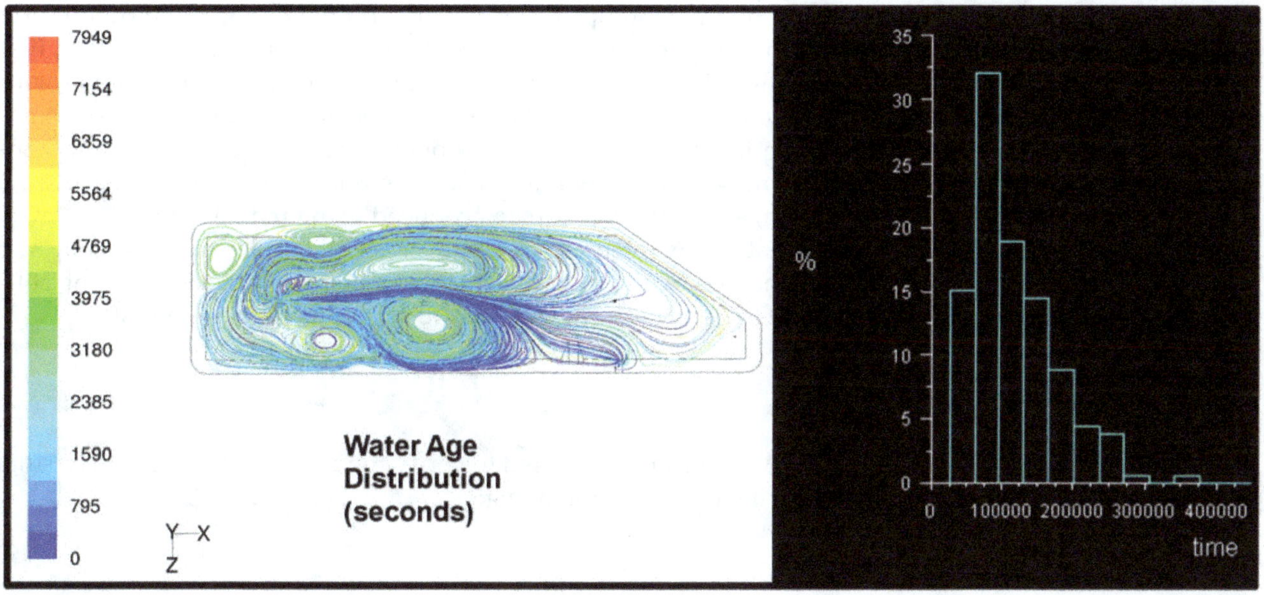

Figure 8-6 Water age distribution

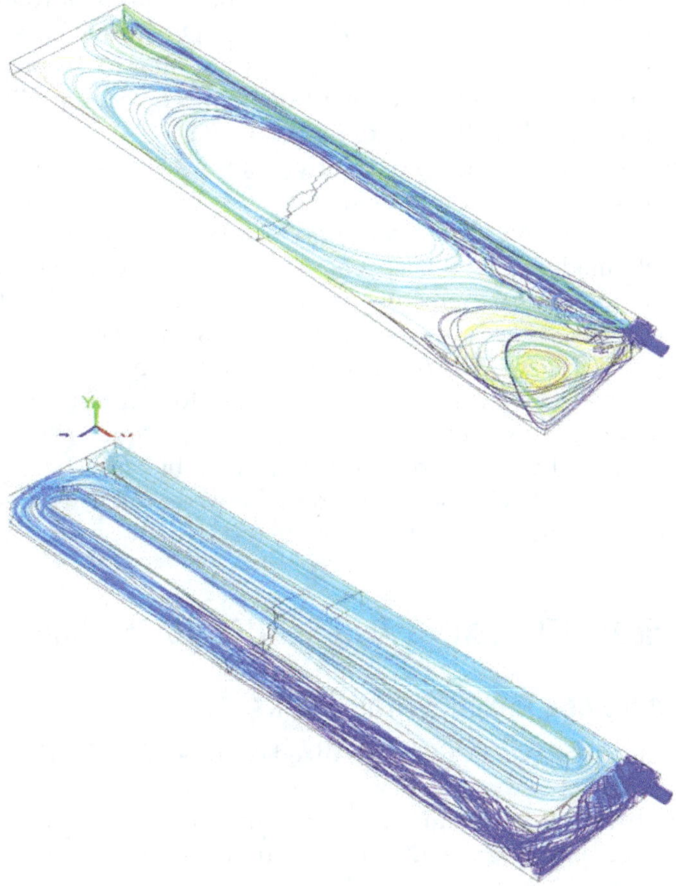

Figure 8-7 Flow paths in unbaffled and baffled storage tanks

The output from CFD modeling provides insights on how a tank could be represented in a distribution system (e.g., completely mixed, compartment, last in–first out, first in–first out). CFD modeling, if applied properly, can provide detailed information with regard to the hydraulic behavior of tanks and reservoirs, insights for the design of mixing mechanisms and inlet piping, and improvement of operations. In addition, three-dimensional images of important parameters of interest such as velocity, temperature, pressure, constituent concentration, and water age can be produced using model results. However, CFD modeling is computationally intensive and requires a good understanding of physics and fluid mechanics principles by the modeler. As with other commercial modeling software, CFD modeling software has a capital outlay and an annual operating cost.

8.6. TANK MODEL VERIFICATION

Several types of data can be collected and used to verify the accuracy of modeling results for tanks and reservoirs, including the following:

- Tracer data,
- Temperature data,
- Water quality data, and
- Hydraulic data.

Field data may be collected as part of a special study or routinely measured in a tank using permanent monitors that report via a SCADA system. The procedure for data collection to verify the results from tank modeling is similar to the procedure for data collection to calibrate network models. The collection of tracer, water quality, and hydraulic data has been discussed in other chapters of this manual.

Tracer studies that use conservative tracers such as fluoride or calcium chloride can be used to both verify mixing models and to directly calculate the HRT distribution of the water inside a tank. The tracer tests can be simulated using physical scale models and CFD models; modeling results are compared to the field data to verify or refine the model.

Temperature measurements may be taken at different levels within a tank to determine whether the tank is stratified. These measurements can also be used to verify or refine physical scale models or CFD models of the tank.

Water quality parameters can also be used to verify results from physical, empirical, or CFD modeling. However, many water quality parameters such as chlorine residual are not just impacted by tank hydraulics but also by biological or chemical reactions. Hence, care must be taken to choose the appropriate water quality parameter for model verification and for accurate representation of the reactive processes.

8.7. STRATEGIES TO PROMOTE MIXING AND REDUCE WATER AGE

8.7.1. Improving Mixing Within Tanks

The consistent achievement of well-mixed conditions in distribution system storage tanks minimizes the potential for development of stratified conditions and dead (stagnant) zones. Stratification and stagnant zones within distribution system storage tanks can lead to wide ranges of water age, DBP levels, and microbial activity within the tanks. Slugs of poorer-quality water present in the higher stratified layers or stagnant zones within a tank can be released into the distribution system when larger than normal drawdowns occur in the tank, resulting in transient occurrences of low residuals and high DBPs within the distribution system.

The goal of achieving acceptable mixing within distribution system storage tanks is relevant to all system-wide water quality management strategies. Table 8-1 identifies typical alternatives for improving the mixing and turnover characteristics of distribution system storage tanks. These improvements are characterized as physical or operational modifications. The table also identifies the relative ease with which these modifications can typically be made.

8.7.1.1. Inlet and/or Outlet Modifications. Physical modifications in inlets and outlets for improving mixing within the tanks may include the following:

- Changing the orientation of the inlet pipe to increase mixing paths and to reduce energy dissipation by reservoir walls or floors,
- Decreasing the inlet diameter to increase the jetting action, and
- Separating inlets and outlets to reduce short-circuiting.

The primary advantage of inlet and/or outlet modification is that it is a one-time physical modification that typically does not result in increased annual operational (energy or maintenance) costs. Generally, the head loss for modified piping in distribution system tanks does not exceed more than a few feet, even at relatively high velocities, which is small compared to typical system pressure heads. Nonetheless, a reduction in the inlet or outlet diameter should be checked hydraulically to ensure that the hydraulics are not significantly impaired.

There are a variety of commercially available inlet–outlet systems as well as the possibility of custom-designed systems to improve mixing in storage tanks. One frequently used modification is to combine the separation of inlets and outlets with installation of "duck-billed" inlets that effectively result in changing inlet diameters in response to varying inflow rates.

8.7.1.2. Mechanical Mixers and Pumps. Several mechanical approaches that require an energy source at the reservoir to provide the necessary mixing energy are discussed in the next section. Variations on these general methods are available, or they may be custom designed and meet the mixing objectives just as effectively.

Table 8-1 Example modifications to improve tank mixing characteristics

Variable Impacting Mixing and Water Quality	Type of Improvement	Desirable Characteristic	General Ability to Control/Implement
Inlet diameter	Physical	Smaller diameter for higher velocity	Good
Inlet orientation	Physical	Toward maximum length of water	Good
Separate inlet and outlet	Physical	Avoid short-circuiting	Good
Active mixer	Physical	Increase mixing	Good
Booster chlorination	Physical/ Operational	Boost residual	Good
Initial water level	Operational	Lower water level at beginning of fill cycle	Moderate
Filling flow rate	Operational	Higher flow rate during fill cycle	Moderate
Temperature difference	Physical/ Operational	Inlet temperature ≥ tank temperature	Low/No independent control

8.7.1.3. Pumps Outside Reservoir. If the energy available within the incoming flow is insufficient to provide adequate mixing, energy can be added using a pump. Water can be drawn from the fill/draw line or from inside the reservoir to a pump located just outside the reservoir and pumped through a nozzle system to provide effective mixing. The sizing and layout of the nozzles and recirculation flow rate should be determined using the methodology described in the previous sections. Pump capacity should be based on the hydraulic requirements to meet these conditions.

The advantages of this approach include the following:

- Operation is straightforward using on–off control of a pump.
- Access to mechanical and electrical equipment for inspection and maintenance can be relatively easy since all equipment is located outside the reservoir.
- Installation of the necessary mechanical and electrical equipment can be straightforward if there is space available.

The primary disadvantages of this approach include the following:

- The mechanical and electrical equipment required for pumped recirculation adds capital, maintenance, and operating costs.
- Additional indoor space for mechanical and electrical equipment could be required, further increasing the capital, maintenance, and operating costs.
- Large reservoirs could require high-capacity pumps to provide effective mixing, and installation of large equipment could be impractical.

8.7.1.4. Recirculation Pumps Inside Reservoir. Individual pumping units can be installed inside the reservoir to provide the energy required for mixing. These submersible "pumps" can have a number of configurations, but in all cases, the objective of producing a jet to promote mixing is the same. Therefore, the only penetration through the reservoir wall is the electrical power and control cabling, which can be installed above the high-water level. Commercially available systems can include devices with and without a casing or enclosure around the impeller. These systems can be mounted on rails that extend the height of the reservoir or on stands located on the reservoir floor.

The advantages of these systems include the following:

- Small reservoir wall penetrations can be installed above water for power and control cabling.
- Minor modifications to existing reservoirs can be installed while the reservoir is in operation.

The primary disadvantages of this approach include the following:

- The mechanical and electrical equipment required adds capital, maintenance, and operating costs.
- Mechanical and electrical equipment located inside the reservoir must be removed or the reservoir taken out of service for inspection and maintenance.
- Additional electrical and control equipment and indoor space are required for mixers located inside the reservoir.
- Unless the mixers are affixed to the tank bottom, they can "walk," or move, from their initial position.

8.7.1.5. Mechanical Mixers. Reservoirs can be effectively mixed using mechanical mixers that include paddles or an impeller mounted on a vertical shaft that is rotated by an electric motor, usually through a gearing system to reduce the speed. These are distinct from the relatively small "pumping" systems discussed previously in that the effective diameter of the rotating paddles or impeller is larger relative to the tank diameter. This type of mixing is commonly used to mix chemicals in dedicated mixing chambers. The mixer generates a vertical flow that recirculates to mix relatively large volumes of water.

The drive units are typically located at the top of the tank or reservoir. So, depending on climatic conditions, some protection of the drive units may be required. For larger tanks, multiple units that are strategically located can effectively mix large, relatively shallow reservoirs.

The advantages and disadvantages of mechanical mixers are similar to the approaches discussed previously. Addition of mechanical equipment increases capital, operating, and maintenance costs. However, in this case, most of the equipment that requires inspection and maintenance is located outside the reservoir.

8.7.1.6. Aeration. If one of the process requirements includes aeration, reservoirs can be mixed using the momentum of bubbles rising through a column of water. Like the jet mixing systems discussed above, the upward momentum transferred to the water by the buoyancy of small or large air bubbles is redirected radially when the free surface is reached and a toroidal circulation is developed. These systems would provide more effective mixing in taller, narrower reservoirs where the bubble travel is long relative to the tank width or diameter.

Aeration systems require blowers and the associated electrical equipment outside of the reservoir as well as distribution piping and diffusers inside the reservoir. Also, the reservoir head space must be well ventilated. Aeration mixing systems are not typically recommended unless there is an important process objective, as the operating costs can be higher.

Aeration is also an effective way to remove contaminants such as trihalomethanes, which are regulated DBPs. A review of Henry's law coefficients for constituents within water is recommended prior to implementing aeration for water quality purposes other than mixing.

8.7.2. Reducing Water Age in Tanks

The ability to reduce the average HRT of stored water typically requires significant modifications to system operations. These may include changes to pressure gradients, inventory management, and operating set points (e.g., pump station operation and level set points). These modifications can result in increased operational complexity. Potential impacts to customers and the assurance of the adequacy of reserves for fire flow and emergencies are important considerations to address as part of any implementation strategy.

In addition to the example physical modifications to tanks listed in Table 8-1, operational changes are often needed to reduce HRT in distribution system storage tanks. Possible operational changes include the following:

- Increasing the daily drawdown in the tank;
- Increasing the number of fill-and-draw cycles each day and performing deep cycling;
- Manually draining the tank at set intervals or as needed to force refilling with fresh water;
- Installing tank booster pumps to pump water out of "locked out" tanks and back into the distribution system;

- Installing new water mains to eliminate hydraulic bottlenecks that prevent a tank from contributing to the supply of water during peak hourly demand times;
- Demolishing tanks that are too low, too small, too old, or otherwise no longer needed; and
- Seasonal modifications to tank usage that include lowering water levels in the winter or taking selected tanks out of service in the winter.

It should, however, be recognized that the ability to implement operational modifications may be limited by the following considerations:

- Control of flow rates during tank filling may be needed to minimize the potential for low pressure in the distribution system;
- Changes in operating protocol may be necessary for booster stations and other tanks to achieve turnover while maintaining adequate pressure system wide; and
- Increase in energy costs associated with extended operation of pumps.

8.7.3. Sampling Locations Within Tanks

While improved mixing and reduced HRT may improve water quality and reduce DBP formation within storage tanks, sampling within storage tanks can help provide early warning of impending water quality problems if the sampling ports are placed at appropriate locations. Water quality parameters such as disinfectant residual and temperature can provide information about mixing and water quality conditions within a tank. Tools such as CFD modeling that provide three-dimensional flow patterns inside storage tanks are useful in locating poorly mixed and well-mixed zones and thus can help determine appropriate sampling locations.

8.8. REFERENCES

Grayman, W.M., Rossman, L.A., Arnold, C., Deininger, R.A., Smith, C., Smith, J.F., & Schnipke, R., 2000. *Water Quality Modeling of Distribution System Storage Facilities.* Denver, CO: AwwaRF.

Mahmood, F., Pimblett, J.G., Grace, N., & Grayman, W.M., 2005. Evaluation of Water Mixing Characteristics in Distribution System Storage Tanks. *Journal of the American Water Works Association,* 97(3):74–88.

Roberts, P.J.W., Tian, X., Sotiropoulos, F., & Duer, M., 2006. *Physical Modeling of Mixing in Water Storage Tanks.* Denver, CO: AwwaRF.

Chapter 9

Model Maintenance

9.1. INTRODUCTION

An up-to-date model, representative of the current system conditions, can yield benefits to a range of utility departments, including planning, engineering, operations, customer service, and management. Depending on the pace of system changes, the types of changes, and the utility objectives for model applications, model maintenance may require frequent and/or regularly scheduled updates. These updates result in a model with data that represents current system facilities, accurate hydraulic parameters and model controls, and a reasonable level of calibration. Using a model that is reflective of current conditions increases the confidence in results when the model is used in planning and operational decision making.

Model maintenance is an essential task to leverage the initial model investment and to sustain and build on the model's benefits. Historically, many models were developed and used with minimal planning for model maintenance. Models were recreated periodically from scratch and were viewed as a tool to be used only for major planning studies.

Today, models are used frequently to evaluate operational, design, energy, vulnerability, water quality, and many more issues in addition to the more traditional planning applications (for ideas on the latest model applications see chapter 11). As models increase in complexity and level of detail, the development of a new model becomes increasingly costly, and the investment in model maintenance becomes more attractive. Furthermore, existing models include a range of information such as pump curves and controls or pipe roughness factors that should be retained in the updated model.

Performing updates of an existing model rather than fully reconstructing a new model can result in significant time and cost savings. Regular model maintenance can also support ongoing model applications and help identify field–model data discrepancies when validating the model, resulting in a continuous return on the investment with the modeling tool.

Model maintenance can range from manually updating pipe and facilities data periodically to an automated update of some or all of the model elements from various data sources (e.g., geographic information systems [GIS], computer-aided design [CAD], as-built drawings, spreadsheets, databases, and operator input). Model maintenance activities are varied but will usually consist of the following categories:

- Updating hydraulic network data (pipes, nodes),
- Updating system facilities (e.g., pump stations, tanks, wells, interconnections),
- Updating system demand averages and associated diurnal curves,
- Updating system operational controls and settings, and
- Performing model validation and recalibration.

Before updating a model, its current and potential future applications should be identified (see section 1.4 for examples of different model applications). The desired model applications will dictate the model's level of detail, its required level of accuracy, and the data needs. Finally, desired model applications will determine the necessity and the frequency of model updates. The model should be updated when the existing model does not meet the utility's needs.

After the model application(s) has been determined, decisions on topics such as the appropriate model scenarios and the model calibration goals naturally follow. By planning the modeling objectives first, a utility can also determine necessary steps for model maintenance.

This section discusses the key processes needed to maintain model data accuracy and adequacy for the intended uses. In addition to the ongoing or scheduled model data updates, model validation and recalibration, and other tasks used to keep the model up to date, model maintenance can include other tasks such as maintaining the modeling software license, updating the model to the current modeling software version, educating the utility on the benefits of the model, and others. While these tasks are important, this section focuses primarily on model data maintenance.

9.2. REASONS FOR MODEL MAINTENANCE

While most water distribution systems have many of the same types of assets, each water system's size, configuration, and operations are unique. Similarly, utilities can differ significantly in the extent and the accuracy of asset information available and the frequency of the updates for this information. The water system and model applications should be carefully considered when planning for model maintenance.

A model must be maintained to keep up with system changes to support current and future model applications and provide useful information to support system operation and management. This need for model updates and regular maintenance should result from ongoing use of the model. Once model results used in decision making start diverging from the system they represent, the model needs to be updated. This diversion can occur because the system has changed since the model was updated and/or calibrated or because the model is to be used to simulate new system conditions. In either case, the model should be updated before it can continue to provide useful information.

9.2.1. Change Drivers

A considerable driver for model maintenance and the determining factor for model maintenance frequency is the current and proposed model application. The model should be updated only if the updates will significantly improve model results and support the decisions that are based on these results. The benefits of the updated model results should outweigh the cost of model maintenance. If a model does not have any current or proposed model applications, it does not need to be maintained.

Another important driver for model maintenance and its frequency is the rate of system changes, the speed at which adequate information about this change is collected, and the level of impact these changes have on model results. If the system is not changing, the utility does not have adequate information about system changes, or the model is used in such way that system changes are not affecting results, the model does not need to be updated to match these changes.

9.2.2. Change Type

Two types of changes can drive a model update: changes in the model and real-world system changes. The first change is user driven, and it is relatively easy to plan if the model needs to be updated to support the new model applications.

System changes, however, can happen without the modeler's knowledge. A well-calibrated model used for its originally intended purpose can, over time, become less representative as the system changes. Regular model validation is needed to ensure that the model can be used to represent a changing system. For more information on model validation, see chapter 4.

9.2.2.1. Model Use Changes. Application of the model is driven by the user. When a need arises to use the model for new applications such as the development of an extended-period simulation model from a steady-state model, the steps identified in chapter 2 would be used to further develop the model. Since there is a need for a new model use, it should be relatively straightforward to define the value of the updated model, as well as to identify the data needs and steps required to update the model. Model use changes are typically conducted only a few times during a model's life cycle. Typically, no regular maintenance schedule is needed to account for these changes.

9.2.2.2. Data Corrections. When erroneous data are identified in the model during normal use, the model should be updated immediately to correct this. The source data that were used to create the model should also be updated. Updating the original data source ensures that all utility users are accessing the up-to-date, validated information. It will also prevent the old, incorrect data from being reimported into the model during any future model updates.

After each data correction, the modeler should determine if the correction is significant enough to affect model results and warrant model recalibration. For example, if an erroneously small pipe was used in the model during the calibration, resulting roughness coefficients might not represent the true system condition. When the model results are significantly changed by the error correction, or several error corrections, the model needs to be recalibrated. One way to determine if the model results have changed is to check and see if the last calibration results were different after the change(s). It should be noted that even if a change is not large enough to affect model results, a series of small changes can eventually become significant enough to affect more results. For this reason, the model should be compared to the last calibration data or validated to recent field conditions, not just to the model results from immediately before the change.

9.2.2.3. New Data. New data in the source asset database can represent three system data changes: new data about new assets that were not originally in the database, updated attributes of existing assets, and data about asset(s) that were removed from the system. When new data are available, the model should ideally be updated to reflect these changes. If GIS is frequently updated with new or revised data, it might not be feasible to update the model every time GIS changes. Instead, the GIS changes should be checked on a routine basis to determine the extent of the changes, which may drive an update. For such situations, scheduled model maintenance, as discussed in the following section, might be more appropriate. If model maintenance is an ongoing process that is repeated regularly, the utility may consider automating model maintenance, as discussed in section 9.7. Similar to data changes, model updates that are driven by new data would require model validation, determination if the model is still accurate, and possibly model recalibration to ensure model accuracy.

9.2.2.4. Planned System Changes. The other change that may require a model update is a planned system change (i.e., a change that is planned in the future and thus is not yet reflected in the asset database). An example would be phased planning scenarios where a utility would want to assess the hydraulic impact on its system from proposed pipeline projects scheduled 5, 10, or 15+ years from today. Such changes might affect future scenarios or proposed new alternatives. What is unique about these changes is that they typically do not require model validation or recalibration. Similar to model use changes, planned system changes often do not require model maintenance plans.

9.3. MODEL MAINTENANCE PLAN

A model that more accurately represents the real-world system will be able to provide utility professionals with more confidence in the model results that are used to support the planning and operational decision-making process. A model that is routinely maintained and represents the necessary assets (per the desired model applications) will aid in improving confidence in the decisions developed and evaluated using models. Today, many water utilities have an asset repository, typically stored within a GIS platform, that is continually being updated with water asset data and corresponding attributes. This asset repository, along with other data sources (e.g., elevation data, demand data), is then routinely used to aid in the model building and updating process. A model maintenance plan is a documented series of steps or a work flow that is followed to systematically build and/or update a model from these data sources. This type of plan is discussed further here.

9.3.1. Planning for a Sustainable Model

A model that is planned and developed with the intent of being periodically updated and/or refined can provide a sustainable and long-term benefit to the utility. In addition to providing the benefit of having more confidence in a model that is continually updated and evaluated, regular model maintenance can also prove to be less resource intensive to maintain the model investment. By contrast, if a model is planned and executed as a one-off disposable product, it will likely need to be completely rebuilt before its next significant use, resulting in duplicated effort and possible loss of knowledge about the model. The method and data used to construct and calibrate the model initially should be documented, as this will help guide future applications, model update frequency, and the model maintenance approach. If a model was constructed and calibrated to evaluate operational and water quality scenarios, the level of detail and accuracy required for future updates of that model would be greater than if the model had been built solely for master planning purposes.

It is important to understand the data structures, development and maintenance procedures, and available tools when developing a sustainable model. As previously mentioned, an all-pipes model best facilitates automated maintenance procedures and may provide the most efficient option for a model maintenance plan.

The complexity of model development and maintenance from GIS requires that sound procedures be outlined and followed. Many development and maintenance tasks can be performed within the GIS environment. Other tasks can be performed using tools within some modeling software. Many tasks can be performed in either environment, and the actual tools used and the order in which they are applied depends on the type of model constructed, the quality and format of the data source, the skills and availability of staff, and the modeling software used. Additional details on GIS integration can be found in chapter 2 and in section 9.6.

Tight integration between GIS and the model allows for consistent tracking of physical changes and prevents the problems created when the model elements do not exactly match the real-world configuration. When errors are encountered in model data, they should be relayed to the GIS system administrator (person responsible for entering and updating data in GIS) for correction. Many utilities have formal methods for requesting data verification or correction to the GIS. However, less formal methods such as export of model data into a spreadsheet format can also be used. Regardless of the method for data exchange between the model and GIS, it is important that modelers and GIS professionals cross-check the validity of the data and use both the GIS and model systems when troubleshooting data challenges.

Updating an all-pipes model from the GIS can require effort. This can be accomplished more easily if the GIS has a way of tracking which elements have been modified. Various utilities have proposed different methods for addressing this issue, with solutions ranging from a GIS tracking field called "date last modified" (MM/DD/YYYY), which lets the modeler check modified fields within the GIS, to using available commercial comparison tools to identify changes between data sets.

Example GIS-to-model update protocol steps are summarized below:

1. Save an archive copy of latest model.
2. Perform quality assurance on new GIS data.
3. Assign relevant attributes to new GIS data (e.g., elevations, roughness coefficients).
4. Import new GIS water assets to model.
5. Remove superseded water assets from model.
6. Update model with new data.
7. Create/update hydraulic facilities (e.g., pumps, control valves) in model.
8. Reassign model demands.
9. Update scenario-specific data.
10. Validate new model results.

Maintaining a model with a one-to-one relationship to the GIS will require periodic updates. A clear plan needs to be identified for how often updates will occur and how extensive the updates will be. However, it is important to realize that updating a model from the GIS does not simply involve importing new elements into the model that are in the GIS. Any time the model is revised, the modeler may need to revisit demand values, demand allocations, and operational controls and may want to validate and/or recalibrate the model. In addition, the modeler must identify what model scenarios will be included within the updated model. This will require verification of model data sets and model operation to verify that the updated model is operating as expected. Figure 9-1 provides a general overview of a sustainable model maintenance process.

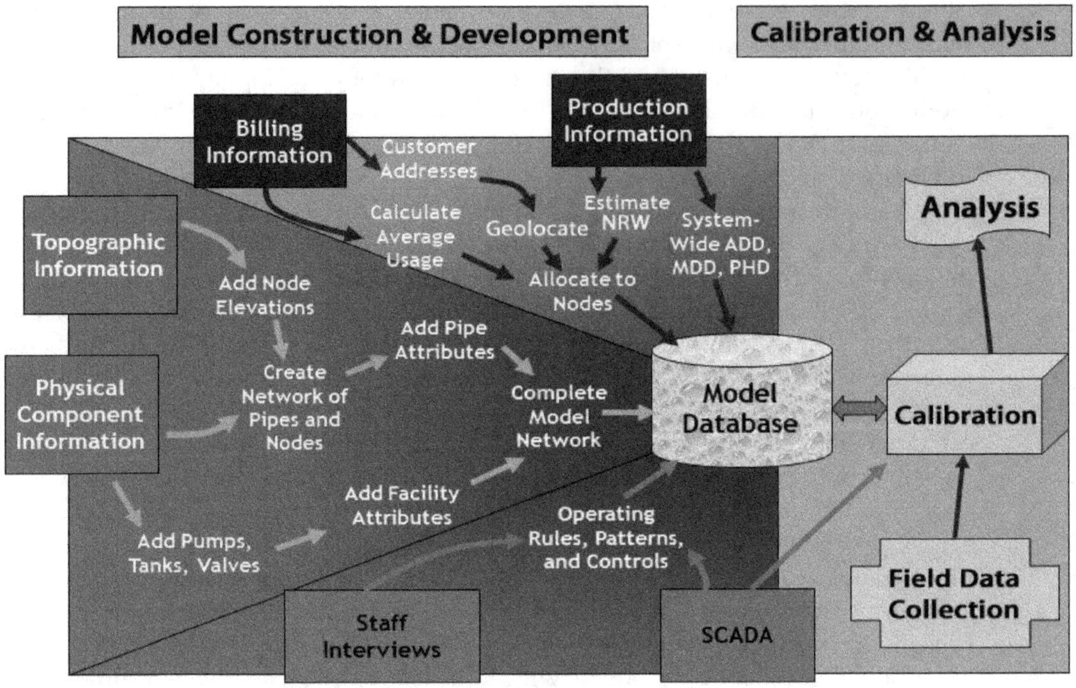

Figure 9-1 Overview of an example modeling process

Abbreviations: ADD (average day demand); MDD (maximum day demand); PHD (peak hourly demand); SCADA, supervisory control and data acquisition; NRW (non-revenue water).

9.4. MODEL UPDATE FREQUENCY

The frequency of model updates may be decided based on the rate at which changes are occurring in the system, the types of changes occurring, the intended application(s) of the model, and the model update process in place or planned. The frequency of an update may also be influenced by the data corrections found in the model, the technology used, and the personnel who will perform the update. If the process is relatively automated or a model maintenance plan has been established, the utility may decide to update more frequently. However, if the update process is labor intensive, the utility may elect to update less frequently, with only significant hydraulic changes that may have occurred within the water system acting as triggers for a model update.

Significant changes in pipeline repair and replacements, changes in major customers, and growth or decline in the population served all lead to changes in demand distribution, impacting overall system hydraulics. These types of changes would cause disparities in the model results versus the actual system and may warrant a model update. In addition, changes including addition, removal, or modifications to major facilities such as pumps, tanks, and control valves along with corresponding operational adjustments would likely have an impact on system hydraulics, thus prompting a model update.

Current modeling platforms allow separation of models into constituent alternatives that are used to build model scenarios, as discussed in chapter 2. Careful consideration should be given to determine which scenarios and alternatives should be updated during model maintenance. Data corrections might be better incorporated into existing alternatives, while new data might better fit into new alternatives.

The following sections provide more detail regarding the various changes that drive model updates.

9.4.1. Ongoing Model Verification and Changes

9.4.1.1. Data Corrections. Any error or inconsistency found in the data should be corrected in the main system of record (e.g., GIS) that the model is being updated from. Utility staff who maintain the GIS data should be in frequent communication with the staff responsible for modeling to relay any significant changes caused by data corrections. These corrections may warrant a model update ahead of the model maintenance schedule. Periodic discussion with the personnel responsible for system operations is also recommended. Both operators and maintenance personnel may change settings in the system (e.g., a valve closure or a new pump control setting). If the change is permanent, that information should be relayed to the modeler. These permanent changes should be reflected in the model and carried through to what-if simulations for the existing system. Modeling results should be reviewed to see if these corrections have a significant impact on the results.

9.4.1.2. Existing System Changes. With a model maintenance plan, once the GIS is updated with actual system changes, the model can be readily updated to reflect the same changes. The model could be edited directly to reflect these changes, but this creates a situation where the data are updated in two separate systems, causing duplicated effort and an opportunity to introduce errors into the data. Editing in the main system of record and then periodically importing into the model is more efficient and less error prone.

9.4.1.3. Future Planning Model Changes. Developing what-if modeling scenarios can be useful in understanding the impact of future real-world system updates. These system updates can be edited directly in the model because these proposed changes are typically not reflected within the GIS data. Various commercial modeling software packages provide the functionality to model these proposed infrastructure changes as separate modeling scenarios with the ability to compare and contrast with the existing (present-day) system model simulation. When proposed assets are modeled, it is recommended that the modeling software tools (network domains or topological alternatives) and/or data attributes (e.g., different labeling notation) be used to designate these future assets as being separate from the existing system assets.

9.4.1.4. Major System Changes. Significant system facility changes are defined as the changes that would inevitably affect the existing model's results. These include such modifications as adding a tank, retiring a pump station, or building a new pump station. These changes should be updated in the model once they are functional in the system. Modeling with incorrect or missing facilities may seriously compromise the results. Facility data such as booster pump stations and their related attributes (e.g., location, pump type, pump curve) usually do not reside in the GIS; therefore, this data can be updated directly within the model.

9.4.1.5. Demand Updates. Customer demand changes in the system may have a significant impact on the model results, prompting a model update. Addition or exclusion of a large user (e.g., a large industrial customer) will affect model results in that area. As these large users tend to have a relatively significant impact on system hydraulics, periodic verification of these large user demands is recommended. Also, an increase or decrease in the water usage per capita could be a driver for a model update. Last, when the system grows due to new subdivisions, annexations, or expansion, the corresponding demands should be input into the model to effectively analyze the complete system.

9.4.1.6. Model Use Changes. Model use, meaning what applications the model is being used for, dictates the time frame and the level of model updates. For example, the application of long-range planning (such as master planning) requires less detail than the application of water quality analysis. By starting out with an all-pipes model that is built and updated from GIS data, water utilities have a model that can be used for many different applications. Regular model updates of the entire system provide more confidence when the model is used for operational and planning decisions, as the model is always up to date based on actual system conditions. With a model maintenance plan, keeping a model current becomes very efficient as the documented work flow shortens the learning curve each time a model update is required.

9.4.2. Periodic Model Update

The traditional approach has been to update models on a periodic or as-needed basis. With this approach, the model is updated when a water system master plan is updated or a major system upgrade is taking place. However, with expanding capabilities and increased practical applications of the model, the update process can be undertaken with more regularity. The model update typically includes adding new facilities constructed since the model was first developed (or last updated), updating model demands, and validating (or recalibrating) the updated model to reaffirm hydraulic results are in agreement with the field performance. GIS and modeling tools support the identification and addition of new physical facilities to the model. Updating repaired or rehabilitated facilities in the model may require a database link or spatial reference to facilitate the update. If neither exists, the process may involve the manual task of translating hydraulic and/or geometry attributes from different data sources to the model. After an update, modeling scenarios for future proposed assets may have to be rerun to provide updated and more accurate results for those scenarios.

9.4.3. Model Rebuild

Because of the improved quality of GIS data and the modeling software construction and updating tools available, some utilities may consider reconstruction of the entire model directly from GIS as a viable model maintenance strategy. This eliminates the challenges associated with updating repaired and rehabilitated facilities and ensures that the model reflects the current GIS data. Depending on the extent of this approach, information specific to the model that is not typically stored within the GIS database (such as roughness coefficients and demand data) may be maintained in the GIS database to aid in the effectiveness of the model update and rebuild process. This approach to model maintenance requires a thorough planning stage prior to implementation: GIS and engineering departments need to agree on methods of data storage, naming conventions, possible revisions to the GIS data model in order to comply with their model requirements, internal updating of work flow processes, and other potential utility-wide decisions that involve model results.

9.4.4. Model Recalibration

Any time the model is updated or changed, calibration data should be revisited, if possible, to see if the model results still compare well to the field data. New calibration field data are necessary if the model changes are substantial and could directly affect the area being calibrated. For instance, a new pump station would require the pressure zone associated with that pump station to be recalibrated with new field data. New pipes in the system might also require recalibration, and new field data during the model updates would be helpful.

This new field data could also be used to identify any unintentionally closed valves or any other system anomalies that may have occurred since the last model calibration.

For model corrections, reuse of older field data might be sufficient. Time and resources permitting, new field data for changed areas provide the ability to validate and, if necessary, recalibrate the model. Keep in mind the purpose of the model and the level of calibration necessary to achieve the desired confidence in the model's capabilities.

When using a relatively old model, a full calibration may not be needed if the study involves a narrow scope. For example, an inquiry as to capacity for a new land development may not require a full calibration. However, checking the calibration with a hydrant flow test near the tie-in point can provide confidence in the system capacity for that particular use.

9.5. CHANGE NOTIFICATION

A model that is used by different departments within the utility has the most potential to provide continuous benefits to the utility. Some departments might use the model directly, others might have their own (simplified or specialized) version of the model, and most will typically rely on model results produced by the modeler. Once a model is updated, it is important to notify all model users and model results users that the results might have changed. Similarly, a modeler might not always be aware of the system changes that might require a model update and should be notified of such changes.

A utility should set up a change notification protocol to ensure the model will be adequate and representative for all intended uses and that the most up-to-date model results are used in the decision-making process.

9.5.1. System Change Notification

A system change notification occurs when the model needs to be updated based on a system change or new information. This can be as simple as new information collected by operations or a new subdivision coming on line. The model is then updated based on these changes; how each organization handles this can vary widely. Each utility with planned model updates should have a protocol that describes how the modeler will be notified of the system or data changes and how and when the model will be updated to account for these changes.

It is not uncommon for large organizations to have multiple modelers and departments working on a model. Therefore, the change notification could come from multiple sources. How each organization handles this will vary based on the organization and planning. The key principle is to set up a plan to enable the model to be updated while avoiding confusion or duplicate work. There are a number of factors to consider:

- Where is the update coming from?
- Who is the final decision maker on model updates?
- Do other departments need to be notified?
- Is the update already part of an existing scenario/model?
- Has communication/notification been established for this process?

The process needs to address these factors as well as any factors specific to the organization. Once completed, the master model should have a defined model maintenance process.

9.5.2. End-User Notification

Once a model is updated and verified, the model users should be notified about the model change. The notification should clearly indicate what has changed and which model results are expected to be different. Model users can rerun relevant simulations and update the results they are interested in. Model users who have derivative, simplified, or specialized versions of the model will have to decide when to update their version of the model. They can also choose to keep track of model changes for a future model update.

Finally, the users who do not run the model themselves but rely on model results in the decision-making process will also have to be notified of the model changes and be provided with the updated model results. The simplest approach to such notification would be to have a centralized location where the model results are accessed, such as GIS or intranet. The modeler would update the results at the central location and notify interested users that new results are available. For this to happen, a list of model result users would have to be maintained. Alternatively, model results could be time stamped, and it would be the user's responsibility to ensure that the most current model results are used in the decision-making process.

It is undesirable if decisions are made based on static reports of model results and no documentation exists on who should be notified that the results have changed. To avoid a situation in which decisions are based on outdated model results, the utility should either establish distribution protocols with centralized sharing of results or document how and where results are used to allow model change notification.

9.5.3. Model Change Tracking

As a model is being developed and updated, it will likely be managed and used by different people. Understanding why and how a model has changed over time can significantly help a new model user to understand the model. Model changes can be tracked on two levels: micro-changes, which are small, ongoing model changes, and macro-changes such as model update or recalibration.

9.5.3.1. Micro-Change Tracking. Micro-changes are changes to individual model elements that occur during regular, ongoing model use. They can be data corrections (e.g., single pipe diameter change), individual element updates (e.g., a pressure-reducing valve setting change), or local model calibration (e.g., an isolation valve assumed to be open instead of closed to match fire hydrant test results). Micro-changes should be tracked within the modeling file. Such changes can be documented using data flags or notes that are available in most commercially available modeling software platforms.

It is important to have a protocol to identify the way micro-changes will be tracked and documented. The documentation should help future users find the change logs and understand the codes used to track changes. At a minimum, the person who made the change, when, and why should be documented.

Documenting micro-changes allows the utility to periodically check the cumulative effect of all micro-changes that occurred since the last model validation or calibration. Most modeling software has tools to assist in this process. These checks can be used to schedule model updates or request model recalibration. They will also support the utility's trust in model accuracy and representativeness.

9.5.3.2. Macro-Change Tracking. Macro-changes are major changes of the whole model, such as a major model update or comprehensive model recalibration. Typically, major model changes will be differentiated in the model file using the "scenarios and alternatives" features of the modeling software platform. Scenario and/or alternative names should include sufficient information to notify model users of the change they represent. Examples of good scenario names are "2017 Recalibration" or "Downtown Redevelopment."

Since macro-changes are major changes, they will typically be accompanied by a report documenting what was done. Additionally, the modeler should add scenario notes to explain the change in more detail. Model maintenance protocol should specify where in the model the corresponding document file is referenced, for example, in the scenario notes. The scenario notes might be a good place to document this.

9.5.4. Model Backup Strategy

As a model changes over time and through use, it is important to be able to recreate old model results. This is particularly relevant if the model is used to provide regulatory support or is used in legal proceedings. Any time the model results are delivered to a regulator, the corresponding model version should be archived for this purpose.

The model should also be archived for backup purposes. A model archive should include necessary information to recover the model when needed, such as archive data, modeling software, and version used. Using model versioning within the model file without creating a separate backup file can provide a model archive for regulatory purposes but it might not provide sufficient backup protection. Model archiving and backup should be coordinated with the utility information technology department. Today's modeling files, particularly models that include results, can be excessively large, and their backup needs to be carefully planned.

9.6. DATA SOURCE INTEGRATION

Data source integration is a vital part of any water model, and all sources need to be considered. Data for the model, as discussed in chapter 2, can come from multiple sources such as a customer information system (CIS), GIS, CAD, operational protocols, and similar sources. Coordinating these resources becomes an integral part of the water system model. Some aspects to consider include the following:

- Where is the data stored?
- Who is the owner of the data?
- Is the data within a firewall?
- How often is the data updated/modified?
- What are the formats of the data? Can/should it be modified?

This section explores some of these questions; however, all should be considered when managing a model.

9.6.1. Organizational Coordination

Coordination inside your organization is the key to any successful model. To avoid time delays caused by the lack of cooperation among utility departments, the first priority should be to establish communication and delineate data responsibilities within the utility. Organizational coordination can be as simple as getting members of each department who control the data in the same room or it might require a series of meetings and development of data management and exchange protocols. Every organization is set up differently and will require unique solutions to this problem.

Investment by key personnel is vital to model maintenance in order to ensure continuous benefits through the sustained use of the model. Clear and specific benefits such as improved flushing routines, better understanding of valve operation and impacts, and input data validation and vetting should be explained and demonstrated to key personnel when requesting their support for model maintenance. For example, the GIS department may have reservations about allowing other departments to directly access asset data for fear that data may be misinterpreted, changed, or erased. However, once they learn that more accurate data in the model reflects more accurate GIS and asset management, they may be more willing to cooperate.

Not everyone needs to be involved, but the more investment by the organization in the process, the higher level of accuracy and benefits can be achieved by the model. For example, while the CIS billing data is vital for model demand development, it might not always be necessary to have someone from the CIS group involved in the model maintenance process. However, the model could be used to show where billing problems or water losses are occurring, so the CIS could benefit directly if they are involved. Often, current CIS data are already contained in the GIS and can be accessed by the modeler. Even if this is true, communication with CIS personnel is helpful for keeping track of changes that could have a significant impact on the model (e.g., changes related to large users).

9.6.2. Identification Management

Managing the model identification key (ID) with respect to the source data ID is the first obstacle when updating a model. The model ID should be based on the original data source ID whenever possible. It should be noted that generic IDs created by GIS software might change as more data are added. If the model ID does not match the GIS ID, the model ID should be included in the GIS attribute data.

For a single data source scenario (multiple databases will be discussed later in this section), the simplest approach is to have the model ID directly match the GIS ID. This approach might not be possible if the GIS database uses multiple IDs or if there is not a 1:1 relationship between the GIS and the model. When the model is created by hand, the modeler has to determine a unique ID. Today's modeling software tools automatically manage unique ID creation for new model objects for the modeler.

Regardless of the ID source, determining the element IDs is a crucial step in the process. Each ID should be unique for every element type. This reduces confusion relating to the naming convention and enables easier data management. The ID should be nonchanging and not controlled by any software such as auto-generated GIS IDs. Instead, the ID should be controlled by GIS personnel and/or the modeler. Importing duplicate IDs into a model can cause serious errors that are not readily recognizable. Old IDs should not be reused but rather retired. Reusing IDs can create connections in the model that may not actually exist and are not easy to find.

A problem with ID management is what to do when data for the same element comes from multiple data sources. For example, pipe topology and location data might come from GIS, but material and diameter data might come from maintenance records. In such instances, the utility would need to determine a unique asset identifier for this asset, and the identifier should be used as the model ID. This approach assumes that both GIS and maintenance databases would maintain the accurate asset ID. If the utility does not maintain the unique asset ID, the best choice for model ID might be the ID contained in the database from which the model is updated most frequently.

9.6.3. One-to-One Element Mapping

In an ideal world, the model represents a one-to-one match with the utility's GIS database, complete with matching unique identification numbers for each element. Some additional entities might be added to the model in order to satisfy the software requirements.

While one-to-one element mapping can streamline the model maintenance process, importing unnecessary elements into the model, such as hydrant laterals, can complicate the modeling process. For example, dead-end pipes without demands might interfere with water quality result interpretation. Thus, all-pipes models should include the elements that should be used in the model, and the GIS database should be developed with separate modeled and nonmodeled elements.

9.6.4. Topology Management

Topology management becomes more important as the accuracy of the model increases. Most modeling applications give the user the ability to specify spatial coordinates and topology. Instead of relying on X, Y coordinates from an unknown or archaic source, this allows the user to spatially orient the model and data to the real world. At this point, all topologies should be in the same coordinate system or should be transposed to the same coordinate system so that a higher level of accuracy exists and the model data can easily be managed and used throughout the organization.

Similarly, most modeling software platforms have tools that are used to apply elevation data based on ground models. Governmental agencies frequently provide free or low-cost access to flyover (ground model elevation) data, which is getting increasingly more accurate – for example, 20-ft (6-m) contours being replaced with 2-ft (0.6-m) contours.

9.6.5. Connectivity Management

Another concern during model updates from external data sources is how to handle element deletion or pipe splits and merges. Typically, elements that were removed from a source database or elements that were marked as obsolete are removed from or disabled in the model. However, when an element is changed, such as a pipe split by a lateral connection, the model update needs to be handled carefully. The simplest approach to this situation would be to make the original element obsolete and to create two new elements in the source database. The model update can then automatically adjust the model network. However, if the old source data element ID was recycled and used for one of the new elements, the model update might not work as expected. In such cases, the modeler should manually inspect model updates for pipe splits, which can be very time consuming. Assigning new IDs and retiring old IDs is a more efficient way to handle the model data when pipes are broken for new connections.

9.7. AUTOMATED MODEL UPDATE

As discussed in this chapter, the frequency of a model update may be influenced by the types of data that are accessible, the complexity of the integration technology used, and the ease of the model update process. The model update frequency tends to increase if the organization uses software technology that makes the process relatively automated. Subsequently, confidence in the model results is typically improved since the model and the data behind the model are routinely being validated.

Automated typically refers to a process or work flow that is enhanced by software routines to minimize the number of steps required. This automated approach also provides a level of data integrity, unlike a more manual approach that may introduce errors into one or more of the various steps. Various software technologies are available to help make the model update process more automated. These vary from custom database applications to commercial modeling software that provides tools to aid in the model update steps. The general concept behind these tool sets is that they connect or extract data from a source (e.g., GIS, database, spreadsheet), load the data into the model (both geometric and attribute data), and then provide tools to clean up and conform the data to the modeling software platform. These are the core functions in automated model updating. These functions can be extended by having tools that provide demand condition updates, automated SCADA data boundary condition updates, and built-in validation tools (to compare the newly updated model to current system data). This type of comprehensive automation is covered in Section 11.8. As with all other model updates, consideration should be given to which scenario or alternative should be updated and whether new alternatives should be created for the new data.

Once the model has been updated from the GIS database and the data have been validated, there remain a series of steps necessary to complete the model update, including incorporating information that is not typically included within the GIS database. These steps include reviewing the newly updated model and establishing what hydraulic properties need to be applied. For example, are there any new facilities such as pumps, tanks, or control valves? If so, facility-specific data – which may include elevation data, facility settings (e.g., pump curves, valve set points, and minor loss coefficients), and corresponding operational controls – would need to be updated. Some modeling software packages provide the ability to update model elements by importing the data from previously defined data sources. Depending on the number of new facilities (i.e., there is not data enough to justify an automated procedure), the modeler may decide to perform these data updates as a manual step. Elevation data as well as model demands would need to be applied to any new nodes introduced during the update. Several modeling software packages offer automated tools to extract elevation data from various sources (e.g., digital elevation models, contour layers) as well as load demands from available usage data. Last, model attributes such as pipe roughness or pressure zone designation can be applied manually or by using automated tools to spatially assign these attributes and/or import them via a database exchange procedure.

As this section emphasizes, automated model updating is not a single button-click process but rather is a work flow, with each step having potential tools to make it more efficient to execute. These automated tools significantly reduce the time and effort that it takes to update models, especially when compared to performing each task manually. Documenting these work flows (ideally in a step-by-step approach) is valuable for making these procedures repeatable and sustainable by modelers. The benefit is having a greater level of confidence in making system operation and planning decisions based on a modeling tool that is relatively up to date with the latest field conditions.

9.8. RETURN ON INVESTMENT IN MODELING

A well-maintained model that accurately represents the existing distribution system conditions can provide many benefits to the utility. The model can interpolate between known system information, such as predicting operating pressures in areas where no pressure gauges are available based on known tank levels and pump statuses, or extrapolate into unknown conditions, such as predicting system response to a main closure. Having additional information about system operation and performance helps utility managers and system operators better understand and manage the system and its facilities. Considering this, it is surprising how rarely models are used after the original planning study for which they were developed.

The reasons for the lack of model reuse are varied and many. The most common reason is that the model is not representative of the current system. "If the model is better maintained," the reasoning goes, "it would be used more often." However, many utilities have well-defined and strictly followed model maintenance plans and still do not reuse the model to its full potential. Even immediately after the planning study, when the model still represents the system with acceptable accuracy, model reuse is often limited.

This section attempts to explain the reasons for limited model reuse, proposes how to overcome the obstacles, and provides the continued return on investment (ROI) on the original model development effort.

9.8.1. Obstacles to Model Reuse

The most common reason for not reusing a model is that the model is not representative of the current system. While it is true that a model that does not represent the system cannot provide much benefit to the utility, it is often the case that this is not the key reason for the lack of model reuse. The true reason for not reusing a model is the lack of understanding of its limitations and potential uses.

Models were traditionally developed for engineering departments to support planning studies. Since planning typically follows the capital investment plan cycles, often after the model is developed, the planning department does not need it until the next planning cycle, and the model may go on the shelf until the next planning application. If, in a rare instance, an operator or a manager asks a question of the modeler, chances are that the model might not be set up to answer this specific question. The logical conclusion for the operator or the manager is that the model is not representative of the system and cannot support operational or management decisions.

However, the reality is that most models can be used to answer most questions about the system, and often they can be used to answer all operational or management questions with limited investment in model maintenance. The trick to having the model ready when the operational or management question is asked is to educate potential model users about the existing model limitations and the investments needed to update the model.

Thus, the following reasons are the most common for not reusing a model:

- Lack of understanding of the existing model by potential model users,
- Lack of understanding by potential model users of the level of investment needed to ensure that the model can be used to answer their questions, and
- Inability of nonmodelers to easily access model results.

The solution for the lack of model reuse is to educate potential users about the model and provide access to model results to nonmodelers. Once potential users understand the advantages and limitations of the existing model, model maintenance will come naturally to ensure that the model can continue to support the utility decision-making process. Instead of thinking about model maintenance as a cost, after realizing the ROI the model provides to the utility, the utility will consider model maintenance as an investment in better utility management and operation.

9.8.2. ROI on the Model

Model reuse can be ensured by following these two steps: Educate potential model users about the existing model, and provide simple and convenient access to the desired model results by those who can benefit from them.

9.8.2.1. Education. Utility-wide education about the model should start during the model development process. Ideally, all potential model users should be involved early in the process when the future model applications are determined. If this is not possible, future users should be involved during model calibration and validation. It is impossible for the modeler to know every fact about the system. Model result review and other involvement by system operations personnel and other potential model users can help ensure the model's representativeness. This involvement would make the end user more invested in model results and could open the door to new model applications. This is true for both the GIS database and the model. The more people who review and correct both systems, the better the model and GIS database will be.

The best way to gain trust in the model and ensure the model can represent the system under a variety of conditions is to validate the model to past observed system conditions. System operators can provide the modeler with examples of observed system conditions such as a main failure or tank being taken off line for maintenance. The model scenario developed to simulate these conditions should be able to replicate the observed system response. Testing the model under such conditions will help validate the model. Also, system operators and managers will learn about what the model can do and see how well the model can represent the system.

During model build and after the model is completed, workshops should be held to educate potential model users about model advantages and limitations. The workshop should be used to identify the ways that the model can be used to support system operation and decision making. Once opportunities for the model applications are identified and if the model is capable of providing the answer, a simple way to access needed information should be provided. Alternatively, the steps needed to upgrade the model for new uses and associated costs should be defined.

The modeler needs to understand how model results will be used, the level of expected result accuracy, and the conditions in the system that the model needs to represent. In this way, the modeler can determine whether the current model is accurate enough to provide the desired answer(s). The modeler must also understand the value of desired model answers and be able to justify model maintenance or upgrade costs to provide needed answers in the future. Many operations departments have weekly "tool-box" meetings where the modeler and operators can educate each other.

9.8.2.2. Access. Once opportunities for the model application are defined, a standardized model reporting method should be developed and agreed upon to ensure the data provided by the modeler will be useful and properly interpreted by nonmodelers. Ideally, model results would be obtained directly from the model. However, it is unreasonable to expect every potential user to be able to access modeling software and/or to fully understand model limitations. Instead, model results should be provided to end users who cannot access the model easily. Often model results can be provided as static reports, such as a GIS database layer with available fire flows at each system hydrant. Other results might require new model runs, such as an assessment of a main failure. Finally, in the most complex cases, an interactive approach with collaboration between the modeler and the operator might be needed, for example, to identify the best approach to taking a tank off line that might involve different alternative solutions.

If model results are provided as static reports, a protocol should be developed to update model results and notify potential users when the model changes, as discussed in section 9.5.

Once the model becomes an integral part of the utility decision-support system, model maintenance will be driven by the model application. Thus, any cost of model maintenance will be justified by the improved current application or additional model use benefits. Most importantly, continued model use will lead to increased trust in model results by all involved parties, which will, in turn, support additional model uses.

9.8.3. Model Maintenance and Model ROI

Developing models is expensive and time consuming. The cost of maintaining a model is marginal compared to the cost of the original investment in model development. This is particularly true if model maintenance is driven by ongoing or additional model use. The continued ROI on the model, achieved through ongoing model use, justifies model maintenance.

Another benefit of ongoing model use is the increased confidence in the model results. When a model is used to predict actual conditions in a new system, one of two outcomes is possible. If the model is representative and accurate, model predictions will closely match the observed system conditions. On the other hand, examined system conditions might be beyond model capabilities, and model results will diverge from the observed parameters. Since the second case should lead to model update and/or recalibration, both cases provide an opportunity to improve trust in the model. This is obvious in the first case when model predictions match the observed system.

However, when the model does not match the system, even more useful information can be gained. Since the hydraulics equations used in the model definitely match system hydraulics, the source of the mismatch between model result and the system must be in the model assumptions. For example, a model can fail to predict system response due to a closed valve in the system that is not included in the model. Identifying a closed valve that should be opened will not only increase trust in the model, it will also provide an opportunity to improve the system. In many cases where the model does not match the system, the reason can be traced to open or closed valves not represented correctly in the model. Alternatively, data collected in the field might be inaccurate. Sometimes, however, the mismatch between the model and the system is due to the lack of model accuracy. In this case, the new system data should be used to improve the model through additional calibration. Additional calibration should result in a model that will be able to predict system response and should increase trust in the model.

Therefore, ongoing model use supported through education of the potential model users, combined with simple access to the model results, will provide the continued ROI on the model and support continued model maintenance. Since the model is already developed, the cost of sharing model results is negligible. Continued model use will provide ongoing ROI on the original model development investment, which will in turn justify continued model maintenance.

9.9. CASE STUDIES

9.9.1. Case Study: Model Maintenance and Updating From Four Distinct Data Sets

Background: The Albemarle County Service Authority's (ACSA) jurisdictional area surrounds the city of Charlottesville (City). The University of Virginia (UVA) is a City customer and an ACSA customer. The City and ACSA sell water to UVA and the public. The Rivanna Water and Sewer Authority (RWSA) is considered a wholesaler.

It was decided at the onset of model building to use prefixes on node and pipe names to indicate ownership. Before using prefixes, there were duplicate identifiers in the data, which made determining ownership at connection points confusing. Node and pipe names must all be unique for a successful model rebuild.

Most of the participants update their GIS database based on when appurtenances are installed in the ground, rather than waiting for as-built data. Inspectors locate the constructed assets using global positioning system (GPS) equipment as the assets are constructed. Either GIS personnel or engineers, in ACSA's case, then digitize the constructed water pipes, valves, and other appurtenances. Nodes are drawn by each entity into its respective GIS database, rather than letting the software automatically create nodes. Pipes are drawn in the GIS database between the nodes using snapping connectivity, and each is given a unique identifier. Beginning and ending node names are maintained as attribute data for each pipe in the pipe layer in the GIS database (see Figure 9-2 for an example). Prefixes are included in the default editing template, so only the unique number has to be entered when the nodes and pipes are drawn into the GIS database layers.

Each entity is responsible for the connectivity of its respective system and the interconnectivity between the systems. For the model build, nodes were automatically generated in the model. Once the nodes were generated in the model, they were exported to the GIS database to allow for connectivity for future model updates. For subsequent updates, all new pipes have new nodes created in the GIS database and imported into the model. The City used this approach. Each entity had to review the nodes to make sure it connected the correct pipe segments. When a new connection is made between systems, coordination between the GIS database editors allows the unique identifiers to be appropriately determined, and duplication is avoided in the entities' GIS database. Normally, a node that breaks a pipe for a new connection belongs to the owner of the new pipe, and the new connecting pipe ends with that node name. When pipes are "broken" by a connection that requires a new node, each new pipe gets a new unique identifier and the previous pipe's identifier is retired. Both old node and old pipe unique identifiers are no longer used. When pipes are abandoned in the field, they are moved to an abandoned pipe layer in the GIS database, which is maintained for utility locating purposes. Abandoned nodes that are no longer connected to a pipe are removed from the GIS node layer.

Node locations are established in the GIS database based on GPS-located appurtenances such as valves, reducers, tees, and crosses. Valves are typically not included in the model unless they are control valves or pressure zone separation valves. When the construction inspectors GPS-locate constructed assets as they are installed in the field, the surveyed asset data are then imported into the GIS database on a weekly basis. This process results in a more current GIS database and model, since as-built data can be received months to years later. Utility locators rely on the GIS database, so the database needs to be as current as possible.

Once the GPS-located data are in the GIS database, corresponding nodes and pipes are also digitized on a weekly basis. The pipes are always drawn in "from the system out." This procedure of editing from the system out allows the model to be updated at any point in time without introducing disconnected pieces. The typical protocol is shown in Figure 9-3.

For the modeling software used in this case study, the modeler has the option to prompt, overwrite, merge, ignore, update based on asset ID, update based on existing objects, or delete missing objects when importing the GIS data. If "delete missing objects" is used, pipes and nodes that have been removed from the GIS database are deleted from the model automatically. This is effective if there is only one database. In the case of four databases used to build an ACSA model, if "delete missing objects" is used when only one database is being imported, the other three databases are removed from the model.

Given this issue, another method for updating had to be performed. The GIS editors did not want to date their edits so that only new edits would be imported into the model. Therefore, the entire GIS database is imported every quarter using an overwrite feature. Any old, potentially out-of-date data that were previously imported would still be included in the model, along with the current updated system. For the import, ownership

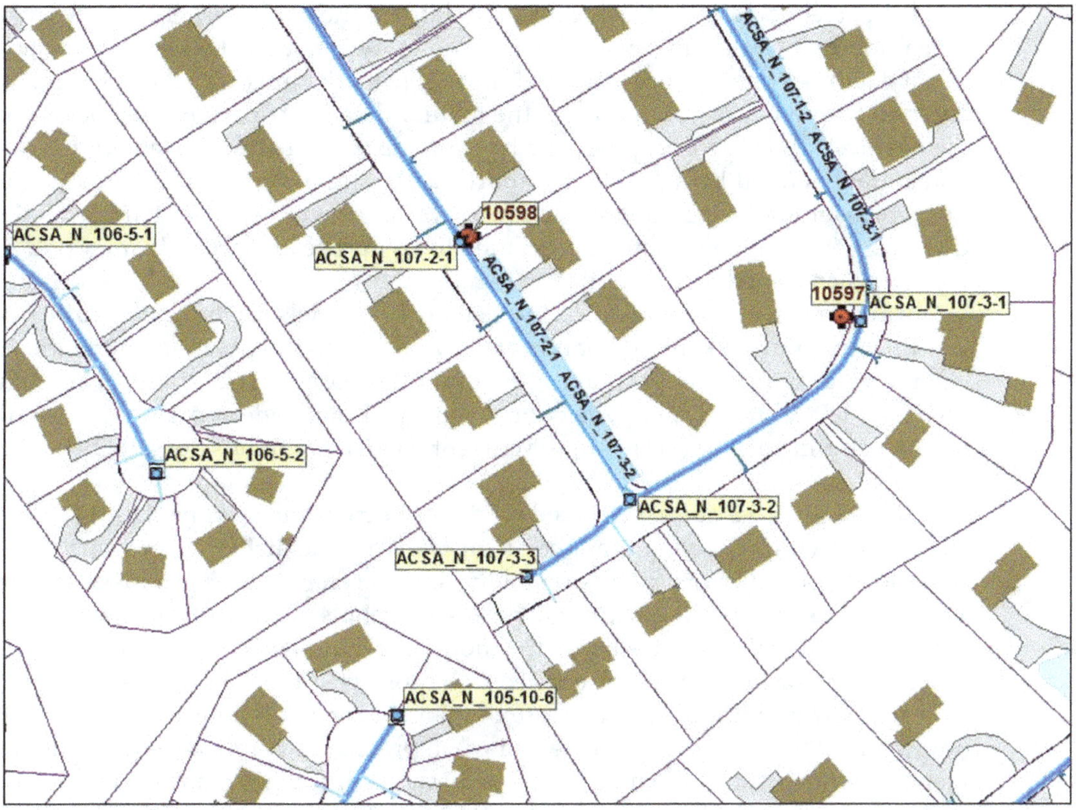

Figure 9-2 Geographic information system editing sessions, nodes, and pipe labels; mislabeling causes disconnects in the model

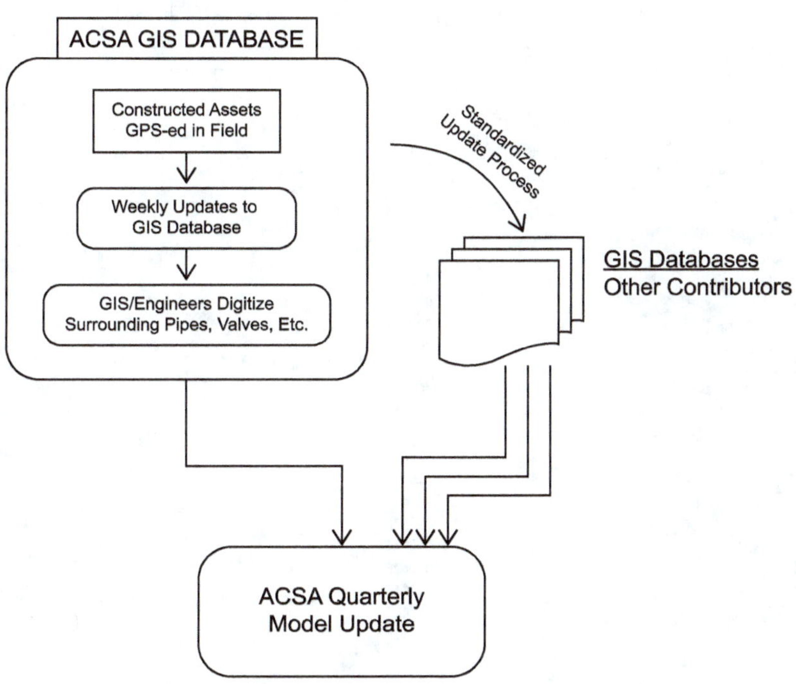

Figure 9-3 Typical model update protocol

Abbreviations: ACSA, Albemarle County Service Authority; GIS, geographic information system; GPS, global positional system.

and the current date are set as the default values in the two user fields for every node and pipe imported. "Delete missing data" functionality is not used when importing. The different databases of nodes are imported first, then all hydrants, and then all pipe databases. After all the new data are imported, the tabular data are sorted by user fields containing ownership and date. Node, pipe, and hydrant tables are sorted systematically, and abandoned data marked by a previous import date are removed from the model. Pump stations, pressure-reducing valves, and tanks are considered static and are located on other GIS layers. They are not routinely included in an update and are only imported and overwritten when something major has changed.

Barring GIS editing mistakes, the 18,000 pipe models can be successfully updated in less than an hour. Existing demands are reassigned to the new pipe network. In Figure 9-4, the pipes with old import dates (shown in the "User Text 2" column) are highlighted (in the far left column) for bulk deletion. Each pipe is not reviewed, since this would be too time consuming. If a pipe is not in the current database and thus does not have the current date, it is removed.

The new model build is checked for disconnection and missing data. With as many as eight people editing the four databases, mistakes are inevitable. If necessary, the model is corrected for small errors such as removing small orphaned pipe groups or lone hydrants. The GIS editors are notified via email with screenshots of any problems with their editing included. If the mistakes are large, the model is rebuilt again with a corrected GIS database. The minor problems are expected to be corrected before the next quarterly update, so that a new import does not recreate the same error again. In general, the GIS database (rather than the model data) is corrected when problems are found in the model. If the correction is only made in the model, the next import of the GIS database will introduce the error again. The only features added to the model and not to the GIS database are pipes and nodes for a what-if scenario.

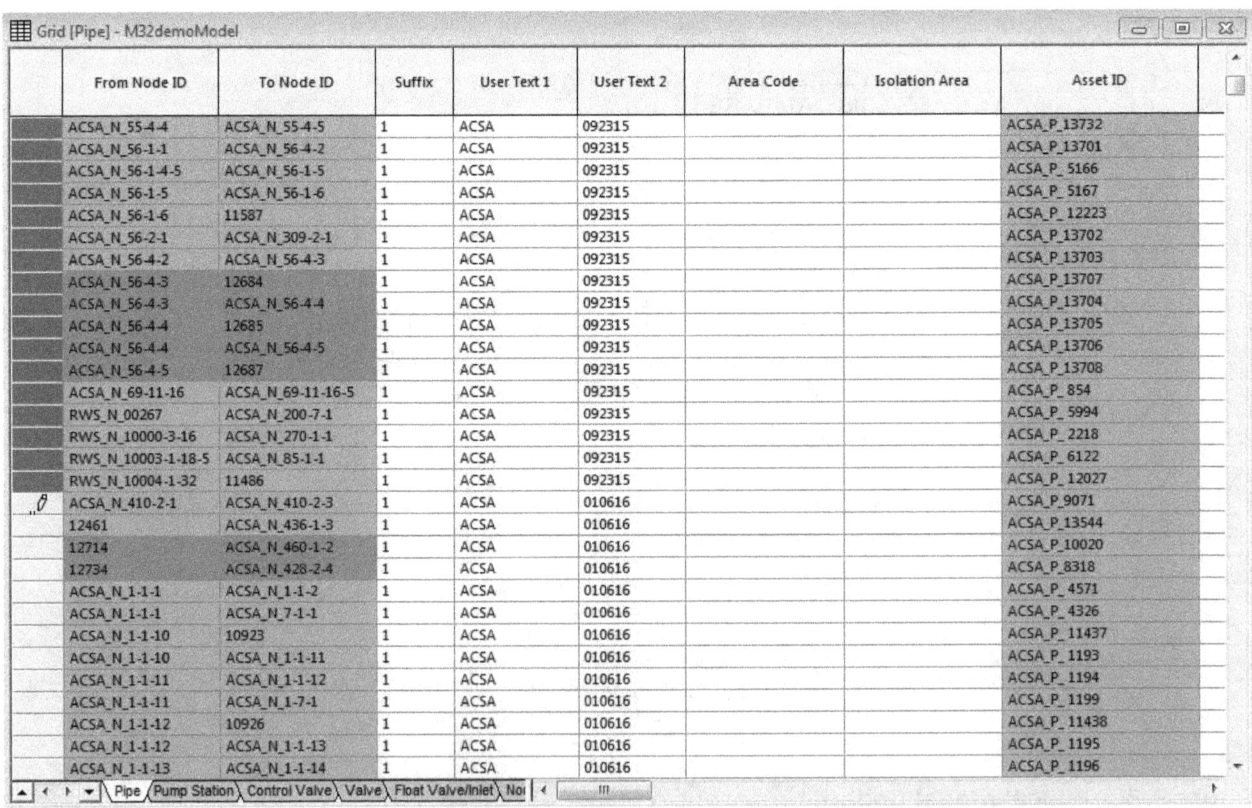

Grid [Pipe] - M32demoModel

From Node ID	To Node ID	Suffix	User Text 1	User Text 2	Area Code	Isolation Area	Asset ID
ACSA_N_55-4-4	ACSA_N_55-4-5	1	ACSA	092315			ACSA_P_13732
ACSA_N_56-1-1	ACSA_N_56-4-2	1	ACSA	092315			ACSA_P_13701
ACSA_N_56-1-4-5	ACSA_N_56-1-5	1	ACSA	092315			ACSA_P_ 5166
ACSA_N_56-1-5	ACSA_N_56-1-6	1	ACSA	092315			ACSA_P_ 5167
ACSA_N_56-1-6	11587	1	ACSA	092315			ACSA_P_ 12223
ACSA_N_56-2-1	ACSA_N_309-2-1	1	ACSA	092315			ACSA_P_13702
ACSA_N_56-4-2	ACSA_N_56-4-3	1	ACSA	092315			ACSA_P_13703
ACSA_N_56-4-3	12684	1	ACSA	092315			ACSA_P_13707
ACSA_N_56-4-3	ACSA_N_56-4-4	1	ACSA	092315			ACSA_P_13704
ACSA_N_56-4-4	12685	1	ACSA	092315			ACSA_P_13705
ACSA_N_56-4-4	ACSA_N_56-4-5	1	ACSA	092315			ACSA_P_13706
ACSA_N_56-4-5	12687	1	ACSA	092315			ACSA_P_13708
ACSA_N_69-11-16	ACSA_N_69-11-16-5	1	ACSA	092315			ACSA_P_ 854
RWS_N_00267	ACSA_N_200-7-1	1	ACSA	092315			ACSA_P_ 5994
RWS_N_10000-3-16	ACSA_N_270-1-1	1	ACSA	092315			ACSA_P_ 2218
RWS_N_10003-1-18-5	ACSA_N_85-1-1	1	ACSA	092315			ACSA_P_ 6122
RWS_N_10004-1-32	11486	1	ACSA	092315			ACSA_P_ 12027
ACSA_N_410-2-1	ACSA_N_410-2-3	1	ACSA	010616			ACSA_P_9071
12461	ACSA_N_436-1-3	1	ACSA	010616			ACSA_P_13544
12714	ACSA_N_460-1-2	1	ACSA	010616			ACSA_P_10020
12734	ACSA_N_428-2-4	1	ACSA	010616			ACSA_P_8318
ACSA_N_1-1-1	ACSA_N_1-1-2	1	ACSA	010616			ACSA_P_ 4571
ACSA_N_1-1-1	ACSA_N_7-1-1	1	ACSA	010616			ACSA_P_ 4326
ACSA_N_1-1-10	10923	1	ACSA	010616			ACSA_P_ 11437
ACSA_N_1-1-10	ACSA_N_1-1-11	1	ACSA	010616			ACSA_P_ 1193
ACSA_N_1-1-11	ACSA_N_1-1-12	1	ACSA	010616			ACSA_P_ 1194
ACSA_N_1-1-11	ACSA_N_1-7-1	1	ACSA	010616			ACSA_P_ 1199
ACSA_N_1-1-12	10926	1	ACSA	010616			ACSA_P_ 11438
ACSA_N_1-1-12	ACSA_N_1-1-13	1	ACSA	010616			ACSA_P_ 1195
ACSA_N_1-1-13	ACSA_N_1-1-14	1	ACSA	010616			ACSA_P_ 1196

Pipe / Pump Station / Control Valve / Valve / Float Valve/Inlet / Noi

Figure 9-4 Model update process

There is a true one-to-one relationship between the GIS and the model nodes and pipes. Quarterly updates are effective in keeping the comprehensive model current. Monthly imports were too frequent, and imports twice a year were too infrequent. Quarterly imports work well, and usually only a handful of errors need to be corrected.

Average demand data for winter and summer are imported into the model via customer point locations. Losses are calculated by averaging production and taking the difference between production and total usage for the same half-year time period. Losses are applied to nodes where usage is applied. This procedure results in two models: one cool-weather model and one hot-weather model. There are two network rebuilds for every one demand rebuild.

9.9.2. Case Study: City of Abbotsford, British Columbia, Canada

The Abbotsford Mission Water Sewer Commission (AMWSC) supplies bulk water to the city of Abbotsford (population approximately 138,000) and District of Mission (population approximately 36,000), which in turn distribute water to their respective water customers. In 2010, Abbotsford became the first Canadian municipality to implement a fixed base advanced metering infrastructure (AMI) system integrated with their model. The Abbotsford is located in the Fraser Valley, approximately 40 miles (65 km) east of Vancouver, British Columbia, Canada. The District of Mission is located immediately north of Abbotsford, with the Fraser River forming the dividing line between the two municipalities. AMWSC water comes from the three raw water sources listed in Table 9-1.

As the model started to become more complex (more than 28,000 pipes), Abbotsford recognized the importance of and need for a set of model standards and conventions. These model management and documentation protocols include applying best practices, following conventions, tracking model changes (capital and development related), and updating model documentation. These efforts have proven to be excellent ways to ensure quality and accurate results, increase the user's level of confidence in the model, and increase the user's understanding of the model. In the long run, the Abbotsford believes that a well-managed model will reduce the time and effort required for model maintenance while achieving more meaningful results.

Abbotsford manages an annual capital asset renewal program that includes the replacement of approximately 3.8 mi (6,200 m) of water mains, with pipes ranging in size from 4 in. (100 mm) to 24 in. (600 mm). Private development applications add an average of approximately 2.2 mi (3,500 m) of water mains each year. Abbotsford decided to integrate and coordinate the model maintenance and updates with their assets under construction (AUC) drawing review and approval process. A summary of the total length of water main installed between 2013 and 2015 for both capital projects and development applications is provided in Table 9-2.

Table 9-1 AMWSC water sources

Water Source	Maximum Capacity	
	Millions Liters per Day	Million Gallons per Day
Norrish Creek	89	23.5
Cannell Lake	9	2.4
Nineteen groundwater wells	45	11.9
Total	143	37.8

Abbreviation: AMWSC, Abbotsford Mission Water Sewer Commission.

Table 9-2 Water main installation summary

	Capital Projects		Development Projects	
Year	Meters	Feet	Meters	Feet
2013	6,470	21,227	2,485	8,153
2014	5,915	19,406	2,715	8,907
2015	6,125	20,095	5,185	17,011
Average	6,170	20,243	3,462	11,357

The overall model update process is comprised of five steps. The first three steps describe how data are moved from CAD and GIS to the modeling application. Steps 4 and 5 describe how the model is updated based on changes made to the GIS database when record or as-built drawings are received. Each step is discussed here.

Step 1: Perform the Engineering Drawing Review Process

Capital projects and development application projects follow the same process from drawing review through AUC model input file creation. Abbotsford reviews all design drawings related to capital- and development-related projects to ensure project feasibility and compliance with city design and construction standards. All comments and required changes are provided to the design engineer in PDF format for inclusion in the final CAD drawing submission. An example is shown in Figure 9-6. Only new and to-be-abandoned water main and hydrant layers are used for the model.

Step 2: Create and Process the AUC Model Input Files

A GIS shapefile that contains simple line work is created based on the "proposed_pipe" layer in the project "Issued for Construction" drawings submitted at the preconstruction meeting. This shapefile is used as a background layer when new pipes and hydrants are added to the model following the city modeling standards, conventions, and guidelines. The "Issued for Construction" drawing shapefile also identifies infrastructure to be removed as part of the project.

The model contains several user-defined fields including "Project_ID" and "CapDate" (Capitalization Date). The same fields are present in the GIS database and are used in the query-building process to identify model elements to be removed prior to importing the GIS database to the model.

Step 3: Create and Process GIS Model Input Files

A semiautomated process is used with modeling software tools to create the model elements from the "Issued for Construction" drawings background layer shapefile. The city modeling standards, conventions, and guidelines are followed when pipe roughness coefficients (C-factors), nodes, and elevations are added and when checking for data gaps and network connectivity issues. Some additional entities are added to the model in order to satisfy software requirements for system operation and continuity. The existing demands are added to the new nodes using data from the AMI system.

The naming convention used for pipes is "PPE-2014-13-xxx." The "PPE" is the pipe prefix, the "2014-13" is the capital project identifier, and the "xxx" is an incremental pipe number. Fire hydrant naming follows the same convention, with the prefix being "HYD." The model should represent a one-to-one match with the GIS database, complete with matching unique identification numbers for each element.

At this point, however, the model will contain pipes and hydrants that are not contained in the GIS database. These elements are used for what-if scenarios in the model, which are used to analyze the system, to assess the impact of future development applications and capital projects, and as an operational tool. When a fire hydrant is serviced and commissioned, the city operations department provides the GIS department with hydrant servicing details including the unique city hydrant number to be painted on the hydrant. Once the information is added to the GIS database, a model update shapefile is created and provided for import into the model, and the "Issued for Construction" hydrant element is deleted. The modeler then runs a fire flow analysis to determine the appropriate National Fire Protection Association 291 hydrant color codes and markings.

Step 4: Process the As-Built or Record Drawings.
When a capital or development project is completed, the as-built drawings are submitted for review and acceptance. Typically, capital project as-built drawings can be submitted, reviewed, and accepted and the update to the GIS database completed within three to six months after the infrastructure is in service. This has not always been the case for development projects. Historically as-built drawing submissions for development projects have not been submitted for review and acceptance until the final phase of the project has been completed, which can occur anywhere from six months to five years after the infrastructure has been commissioned and put into service.

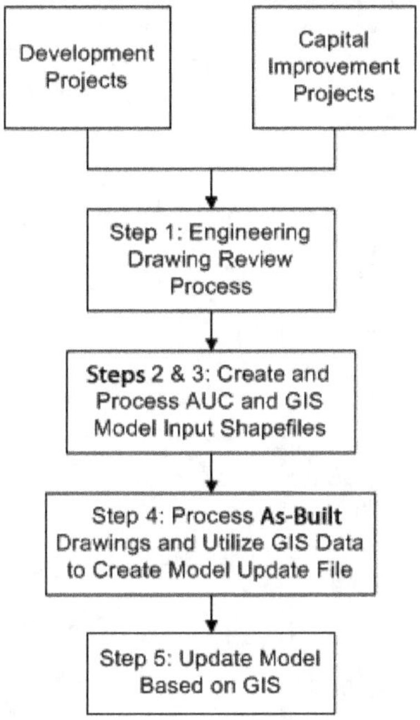

Figure 9-5 Work flow process flowchart

Abbreviations: AUC, assets under construction; GIS, geographic information system.

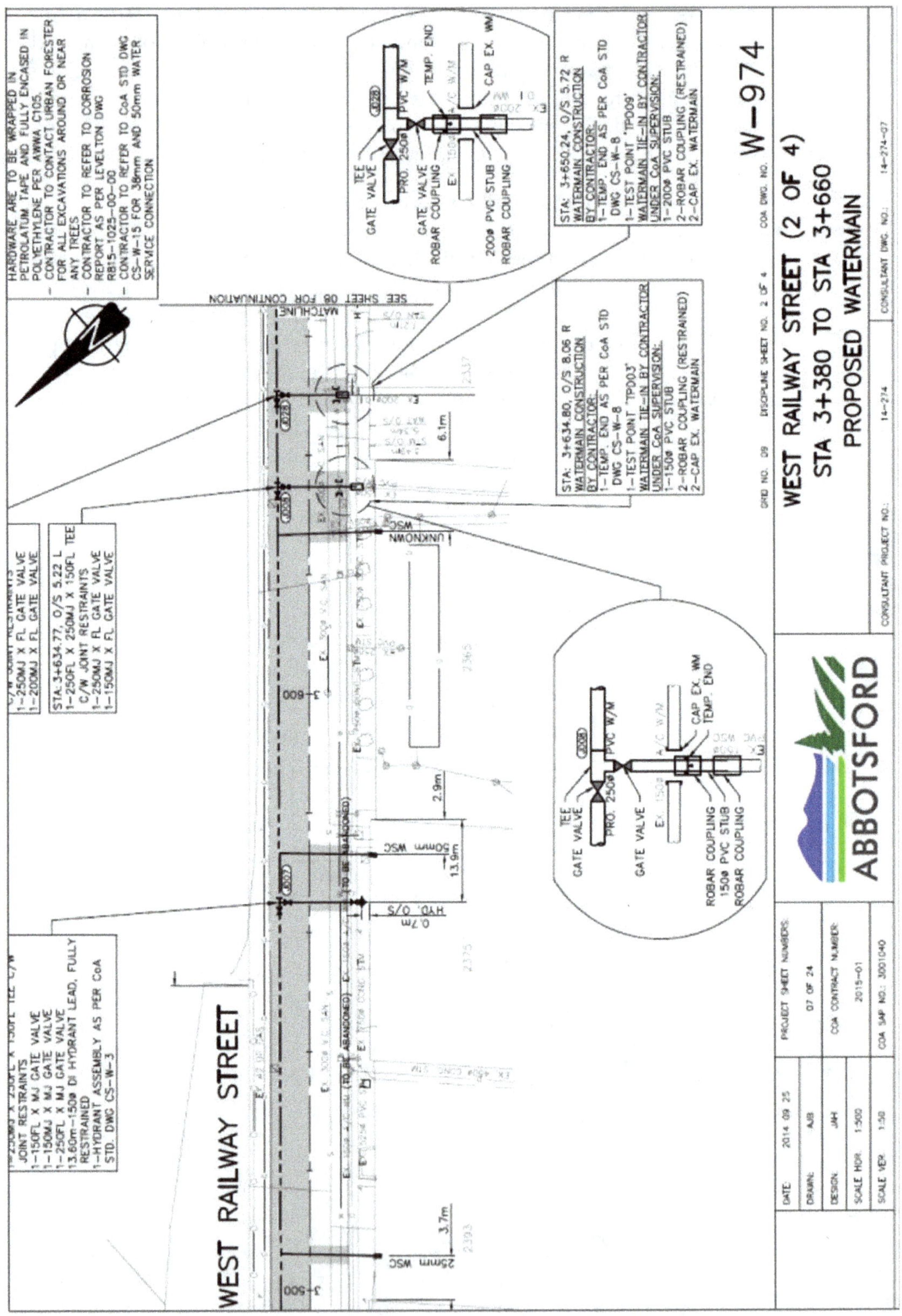

Figure 9-6 Example of the drawing review process

Step 5: Update the Model Based on GIS

Once the GIS has been updated to reflect the as-built records, a GIS model input file is created in a similar format as that identified in steps 2 and 3. The main difference is the GIS input file represents all pipes and hydrants added to the system and it contains additional database fields. This is also a semiautomated process that uses modeling software tools to identify, delete, and replace the AUC elements that match the new GIS elements. An example model representation is shown in Figure 9-7.

Maintaining a one-to-one relationship between the model and the GIS database is not always attainable or desirable, depending on the intended use of the model. For Abbotsford, the intent of the model is to reflect the real-world system as accurately as possible. This means tracking and managing capital projects, development projects, and changes in system demands and patterns as the system expands. Abbotsford ensures that when the model is updated to reflect GIS database changes, it is done on a quarterly basis.

The model will always be out of sync with the GIS database and contain AUC pipes and hydrants that have been constructed and commissioned. These AUC elements will remain in the model until the as-built drawings have been accepted and the GIS database has been updated. This ensures that any system changes and planned or unplanned events are modeled as accurately as possible. A model that more accurately represents the real-world system provides confidence in the model predictions needed to support the decision-making process. The ultimate goal is to ensure that the system will continue to be capable of delivering an acceptable level of water quantity and quality to the customers.

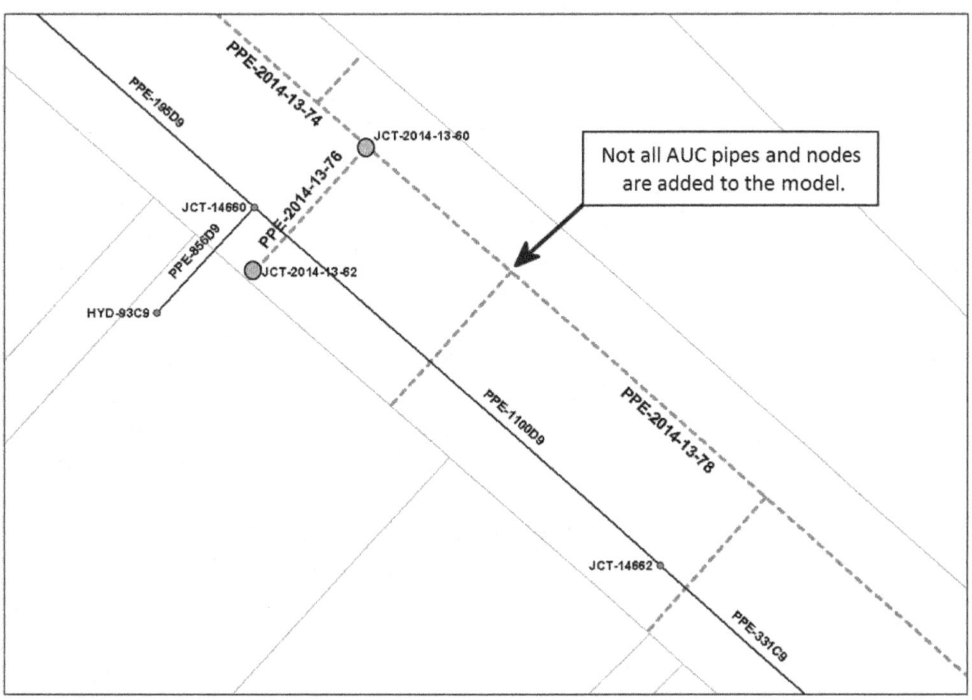

Figure 9-7 Model representation of proposed water main

Abbreviation: AUC, assets under construction.

This page intentionally blank.

Chapter 10

Transient Analysis

10.1. SYNOPSIS

Transients can introduce large pressure forces and rapid fluid accelerations into a piping system. These disturbances may result in pump and device failures, system fatigue or pipe ruptures, and backflow, with intrusion of untreated and possibly hazardous water. Some transient events can lead to water column separation, which can result in catastrophic pipeline failures. Thus, transient events can cause health risks and can lead to increased leakage, interrupted service, decreased reliability, and breaches in piping system integrity. Transient flow simulation has become an important tool for ensuring safety and the safe operation of drinking water supply and distribution systems.

► KEY USES OF HYDRAULIC TRANSIENT ANALYSES

- Developing understanding of the potential for extreme pressures
- Selecting materials that can withstand surge pressures
- Evaluating measures to reduce the magnitude of extreme pressures
- Training operators

This chapter introduces the concept and fundamentals of hydraulic transients, including the causes of transients, general rules to follow when determining whether or not the system may be exposed to unacceptable conditions under a transient event, governing transient equations, numerical solution methods, guidelines for control and suppression of transients, transient modeling considerations, and transient data requirements. Illustrative examples are also discussed, and conclusions are stated.

The chapter is geared toward engineers involved in the planning, design, and operation of water supply and distribution systems and engineers who need an insight into the most common causes of hydraulic transients and suitable methods that can be applied to alleviate their consequences. Such capabilities will greatly enhance a water utility's ability to evaluate cost-effective and reliable water supply protection and management strategies for preserving system integrity (physical, hydraulic, and water quality), preventing potential problems, and safeguarding public health.

10.2. INTRODUCTION

Most people have been in an older house with pipes that rattle when someone turns off a faucet. When the faucet handle turns, closing the valve almost instantaneously, the pipes rattle against the walls. The rattling is caused by the phenomenon known as *water hammer,* which is also referred to as a *surge* or a *hydraulic transient.* Water hammer refers to rapid and often large pressure and flow fluctuations that result from transient flow conditions in pipes that transport fluids. Transient flow analysis of the piping system is often more important than analysis of the steady-state operating conditions that engineers normally use as the basis for system design. Transient pressures are most significant when the velocity/momentum is changed rapidly, such as during rapid valve closures or pump stoppages. Such flow disturbances, whether caused by design or accident, may create traveling large pressure waves of excessive magnitude. These transient pressures are superimposed on the steady-state conditions present in the line at the time the transient occurs. The total pressure that is acting within a pipe is obtained by summing the steady-state and transient pressures in the line. The severity of transient pressures must be accurately determined so that the pipes can be properly designed to withstand these additional shock loads. In fact, pipes are often characterized by pressure ratings (or pressure classes) that define their mechanical strength and have a significant influence on their cost.

► POSSIBLE IMPACTS OF HYDRAULIC TRANSIENT EVENTS

- Equipment failure or pipe rupture from high pressure
- Pipeline collapse due to low or vacuum pressures
- Cavitation damage from vacuum pressures
- Release of sediments, biofilms, and tuberculation from pipe walls
- Intrusion of contaminants from groundwater surrounding a pipe
- Air binding of pipeline, increasing hydraulic resistance

Transient events may be associated with equipment failure, pipe rupture, separation at bends, and introduction of contaminated water into the distribution system via unprotected cross-connections or intrusion. Uncontrolled pump shutdown can lead to low pressures, resulting in undesirable occurrence of cavitation and water column separation, which can result in catastrophic pipeline failures due to severe pressure rises that follow the collapse of the vapor cavities. Vacuum conditions can create high stresses and strains that are much greater than those that occur during normal operating regimes. They can cause the collapse of thin-walled pipes or reinforced concrete sections, particularly if these sections were not designed to withstand such strains (e.g., pipes with a low pressure rating) and cause intrusion of nonpotable water into pipes.

Cavitation occurs when the local pressure is lowered to the value of vapor pressure at ambient temperature. At this pressure, gas within the liquid is released, and the liquid starts to vaporize. When the pressure recovers, liquid enters the cavity caused by the gases and collides with whatever confines the cavity (i.e., another mass of liquid or a fixed boundary), resulting in an additional pressure surge. In this case, both vacuum and strong pressure surges are present, a combination that may result in substantial damage. The primary difficulty here is that accurate estimates are difficult to achieve, particularly because the parameters that describe the process are not yet determined during design. Moreover, the vapor cavity collapse cannot be effectively controlled. In less drastic cases, strong pressure surges may cause cracks in internal lining or damage connections between pipe sections. In more serious cases, surges can destroy or cause deformation to equipment such as pipeline valves, air valves, or other surge-protection devices. Sometimes the damage is not realized at the time it occurs but results in intensified corrosion that, combined with repeated transients, may cause the pipeline to collapse in the future. Transient events in pipelines also damage seals, which often leads to increased leakage and significant water loss.

Transient events can have significant water quality and health implications. These events can generate high intensities of fluid shear and may cause resuspension of settled particles as well as biofilm detachment. Moreover, low pressure caused by transients may promote the collapse of water mains; leakage into the pipes at loose joints, cracks, and seals under subatmospheric conditions; back-siphonage at cross-connections; and potential intrusion of untreated, possibly contaminated groundwater into the distribution system. Pathogens or chemicals in close proximity to the pipe can become potential contamination sources, where continuing consumption or leakage can pull contaminated water into the depressurized main.

Recent studies have confirmed that soil and water samples collected immediately adjacent to water mains can contain varying levels of microorganisms that are indicators of fecal pollution (fecal coliforms, *Escherichia coli, Clostridium perfringens,* coliphages) and in some cases enteric viruses (Besner et al. 2008; Karim et al. 2003; Kirmeyer et al. 2001). This is especially significant in systems with leaking pipes below the water table. Problems with low or negative pressure transients have been reported (Walski and Lutes 1994; LeChevallier et al. 2003). Gullick et al. (2004) studied transient pressure occurrences in actual distribution systems and observed 15 surge events that resulted in a negative pressure. Hooper et al. (2006) and Besner et al. (2007) also reported such events in full-scale systems.

The most often identified cause for the events reported in the literature was the sudden shutdown of pumps, either unintentional (power failure) or intentional (pump tests). Using a pilot-scale test rig, Friedman et al. (2004) confirmed that negative pressure transients can occur in the distribution system and that the intruded water can travel downstream from the site of entry. Locations with the highest potential for intrusion were identified as sites that experience leaks and breaks, areas of high water table, and flooded air–vacuum valve vaults. Preliminary results by Besner et al. (2007) showed that significant concentrations of indicator microorganisms could be detected in the water found in flooded air–vacuum valve vaults. In the event of a large intrusion of pathogens, the chlorine residual normally sustained in drinking water distribution systems may be insufficient to disinfect contaminated water, which could lead to serious health effects. A case study in Kenya (Ndambuki 2006) showed that in the event of a 0.1 percent raw sewage contamination, the available residual chlorine within the distribution network would not render the water safe.

Transient events that can allow intrusion to occur are caused by sudden changes in water velocity due to loss of power, sudden valve or hydrant closure or opening, a main break, or an uncontrolled change in on–off pump status (Boyd et al. 2004). Transient-induced intrusions can be minimized by knowing the causes of pressure surges, defining the system's response to surges, and estimating the system's susceptibility to contamination when surges occur (Friedman et al. 2004). Therefore, water utilities should never overlook the effect of pressure surges in their distribution systems. Even some common transient protection strategies such as relief valves or air chambers, if not properly designed and maintained, may permit pathogens or other contaminants to find a back door route into the potable water distribution system. Any optimized design that fails to properly account for pressure surge effects is likely to be, at best, suboptimal and, at worst, completely inadequate.

Pressure transients in water distribution systems are inevitable and will normally be most severe at pump stations and control valves, in areas of high elevation, in locations with low static pressures, and in remote locations that are far from overhead storage (Fleming et al. 2006; Friedman et al. 2004). All systems will, at some time, start up, switch off, undergo unexpected flow changes, and likely experience the effects of human errors, equipment breakdowns, earthquakes, or other risky disturbances. Although transient conditions can result in many abnormal situations and breaches in system integrity, the engineer is most concerned with those that might endanger the safety of a treatment plant and its personnel, that have the potential to cause equipment or device damage, or that result in operational difficulties or pose a risk to public health.

Transient pressures are difficult to predict and are system dependent, including specific system layout, configuration, design, and operation. Engineers must carefully consider all potential dangers when designing their pipe systems and estimate and eliminate the weak spots. They should then perform a detailed transient analysis to make informed decisions on how best to strengthen their systems and ensure safe, reliable operations (McInnis and Karney 1995; Karney and McInnis 1992).

Generally, computer programs are needed to perform transient analysis, unless the system being evaluated is very simple. When conducting transient analysis by numerical methods, the speed of execution, accuracy, stability, and reliability are all important considerations. However, extreme precision is not required or justified for numerical solutions specific to transient analysis, because water system data are not generally known for high accuracy.

10.3. CAUSES OF TRANSIENTS

Transient events are disturbances in water flow that are created during a change in operation, typically from one steady-state or equilibrium condition to another (Figure 10-1). The principal components of the disturbances are pressure and flow changes at a point that cause propagation of pressure waves throughout the distribution system. The pressure waves travel with the velocity of sound within the pipe and liquid (acoustic or sonic speed), which depends on the water's compressibility and the pipe walls' elasticity. As these waves propagate, they create transient pressure conditions. Over time, damping actions and friction reduce the magnitude of the waves until the system stabilizes at a new steady state. Only through extremely slow flow regulation will the transition from one steady state to another be smooth without large fluctuations in pressure.

► PRIMARY CAUSES OF HYDRAULIC TRANSIENTS

- Sudden shutdown of pumps
- Pump startup
- Pipe break
- Rapid valve closure
- Introduction of air into the system

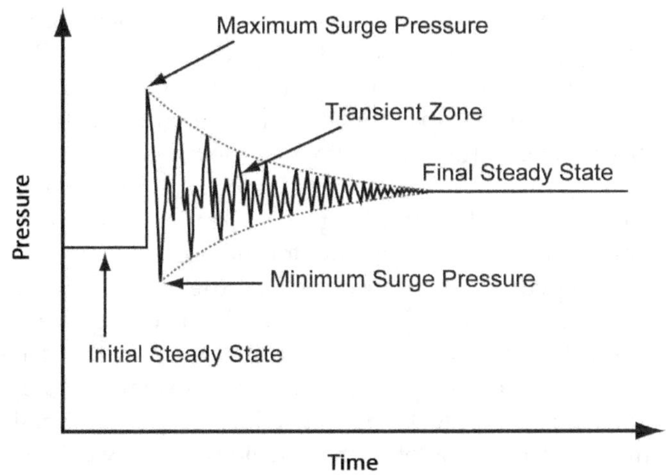

Figure 10-1 Example steady-state transition after a period of rapid transients

10.3.1. Transient Definitions

Surge: the significant pressure changes in pipeline systems commonly referred to as *transient events* that result from operational changes such as pump starting and stopping, valve closing, and pipe breaks.

Pressure wave velocity (also known as celerity): the speed at which a pressure wave moves through the pipes. It depends on the compressibility of materials, which implies that one part of a system cannot affect another part of a system faster than the time it takes for the pressure waves to travel through the material from one part of the system to another. The pressure wave velocity affects the pressure and velocity at a given point, which can change when other pressure waves arrive at that point.

Cavity formation/column separation (or cavitation): water vapor formation in water pipes when the static pressure drops to the value of the vapor pressure. At this stage, the vapor cavity can grow larger or smaller. However, when the cavity collapses, the liquid–vapor interface moves at the pressure wave velocity. When the interface hits a solid boundary such as the pipe wall, valve, disk, or pump impeller, pressure increases dramatically. Cavitation is a major cause of large pressure increases and damage in a water system.

Resonance: occurs when a periodic hydraulic or mechanical event coincides with the system's natural period of oscillation. Periodic events such as flow change or moving parts in hydraulic machines can be considered separately, as hydraulic transient modeling software generally does not account for them.

10.3.2. Basics—Rapid Changes in Velocity

In general, any disturbance of the flow of water generated during a change in mean flow conditions will initiate a sequence of transient pressures (waves) in the pipe system. Disturbances usually originate from changes or actions that affect fluid devices or boundary conditions.

Both high and low surge pressures may cause problems. Figures 10-2 through 10-5 illustrate four typical hydraulic transient problems. The problem of shutting down a pump is illustrated in Figure 10-2. When the pump is suddenly shut down, the momentum of the moving water carries it away from the pump, causing the pressure at the discharge side of the pump to rapidly decrease and form a wave of reduced pressure that begins to propagate down the pipeline toward the downstream reservoir. When the negative pressure wave reaches the high point in the pipe (which already has a relatively low pressure due to the higher elevation), the pressure can drop below atmospheric to reach vapor pressure. On subsequent cycles of the transient when the pressure recovers, the cavity can collapse, generating a large pressure surge spike. On the suction side of the pump, the solid sloping line represents the initial hydraulic grade and the dashed straight line depicts the final hydraulic grade; startup transients are not shown.

When the pipeline velocity reverses and the water column returns toward the pump, it is suddenly stopped by the check valve on the discharge side of the pump, causing very high pressures. If no check valve exists, the pump can spin backward, possibly reaching speeds that can damage the equipment. During normal pump operation, the magnitude of pressure transients can typically be controlled or reduced by using slow-closing and slow-opening pump control valves or with variable-frequency drives (VFDs) and soft start–stop controllers on the pump motors.

The problem of pump startup transient is illustrated in Figure 10-3. When a pump is started, the pressure at the discharge side of the pump rises, sending a positive pressure wave (which increases pressure) down the pipeline toward the downstream reservoir. The resulting peak pressure can cause the pipe to rupture or fail if the pipe's pressure rating is less than the maximum surge pressure. When the initial positive pressure wave reaches the downstream reservoir, it is reflected as a negative pressure wave that propagates back to the pump and may induce cavitation. On the suction side of the pump, the solid straight line represents the initial hydraulic grade and the dashed sloping line depicts the final hydraulic grade; shutdown transients are not shown.

Pipe filling is especially vulnerable to pressure transients when the pipeline is empty and large amounts of air are expelled during startup. The air is expelled at a high velocity through the valve orifice. When the air is fully exhausted and the liquid reaches the valve face, the valve closes. In some cases, the valve may slam closed, leading to secondary pressure surges. Pump startup transients in pressurized pipelines are not commonly a problem because the pumps can be started against a closed pump control valve, and then the valve can be slowly opened or VFDs can be used on the pump motors.

Opening a valve and closing a valve too quickly can also result in severe hydraulic transients; this is illustrated in Figures 10-4 and 10-5, respectively. When a valve is rapidly opened (Figure 10-4), a negative pressure wave is initiated at the valve as the potential energy in the water is converted to kinetic energy and the wave propagates upstream toward the reservoir, decreasing the pressure in the pipe. Similar to the pump shutdown scenario, the initial negative surge can drop to vapor pressure, causing cavitation in the pipe. In the second example (Figure 10-5), rapidly closing the downstream valve generates a positive pressure wave at the valve that propagates toward the upstream reservoir, increasing the pressure in the pipe.

In municipal systems, opening and closing hydrants too quickly can sometimes cause unacceptable transients. In Puerto Rico, the testing of a 60-in. (152-cm) butterfly valve caused a 72-in. (183-cm) pipeline to rupture due to a rapid valve closure (Figure 10-6a). Walski (2009) showed that increasing hydrant closing time can make a dramatic difference in pressure surges in the distribution system, greatly decreasing the impacts of hydrant closure.

Sudden and complete failure of a single pipe may also cause other unacceptable pressures elsewhere in the water system. A pipeline rupture sends low pressure waves propagating in both directions from the break. This can produce negative pressures and cavitation at higher elevations. Sometimes a pipe rupture on a thin-walled pipeline will cause a pipe to collapse elsewhere in the pipeline. The low pressure waves will get reflected from nearby tanks and reservoirs as positive pressure waves that, in turn, can also damage pipes in other locations. Other real-life examples of catastrophic failures due to surge pressures are shown in Figures 10-6b and 10-6c.

Pipe systems must be designed to handle both normal and abnormal operating conditions. If an analysis indicates that severe transients could occur, the main techniques generally used to mitigate transient conditions are (Boulos et al. 2006; Wood et al. 2005a; Walski et al. 2003) as follows:

- Installation of stronger (higher pressure class) pipes,
- Rerouting of pipes,
- Improving valve and pump control/operation procedures,
- Limiting pipeline velocity,
- Reducing wave speed (e.g., different pipe material),
- Increasing pump inertia (e.g., fitting a flywheel between the pump and motor), and
- Designing and installing surge protection devices.

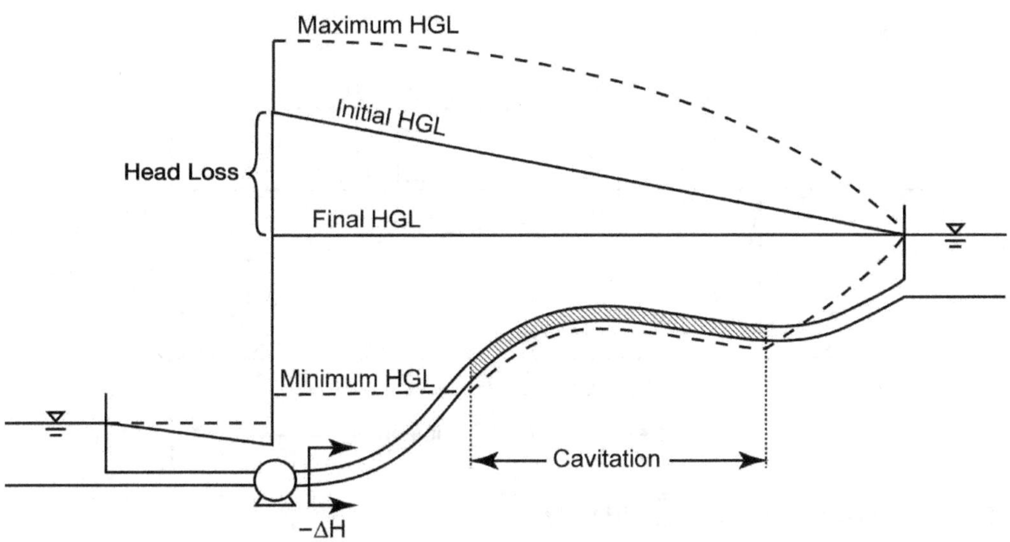

Figure 10-2 Transient caused by pump shutdown

Abbreviation: HGL, hydraulic grade line.

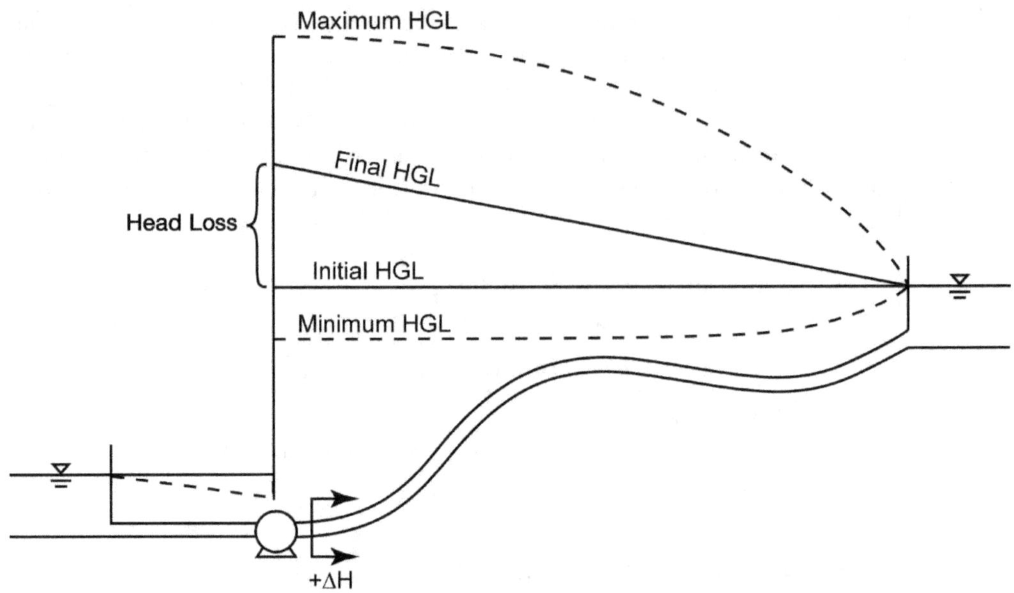

Figure 10-3 Transient caused by pump startup

Abbreviation: HGL, hydraulic grade line.

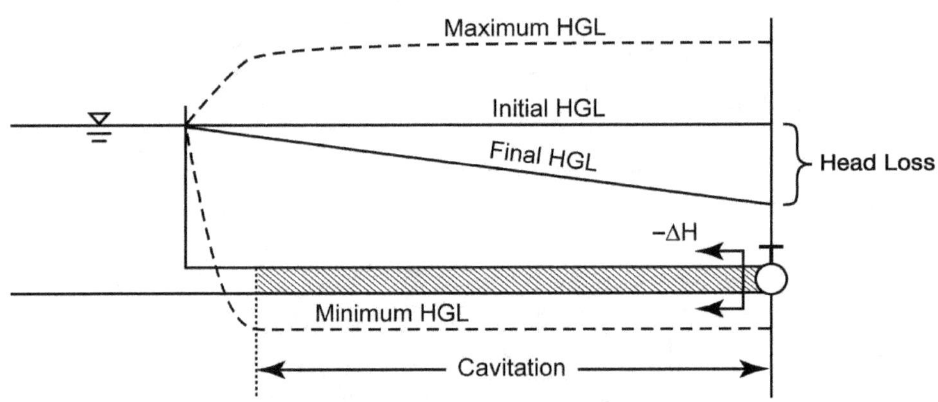

Figure 10-4 Transient caused by rapid valve opening

Abbreviation: HGL, hydraulic grade line.

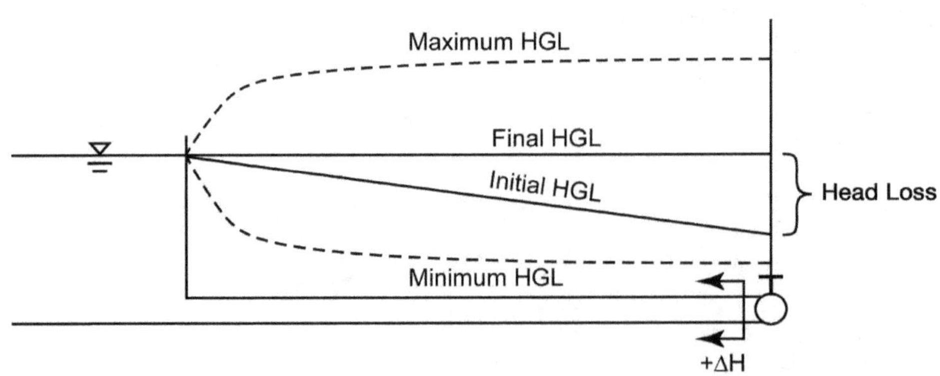

Figure 10-5 Transient caused by rapid valve closing

Abbreviation: HGL, hydraulic grade line.

Source: Super Aqueduct of Puerto Rico

Figure 10-6a Rupture caused by valve closure

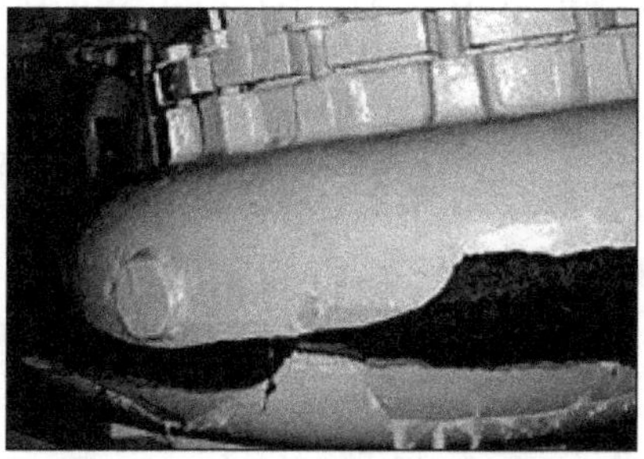

Figure 10-6b Damaged pump bowl

Figure 10-6c Broken air admission valve

10.3.3. Contributing Factors

The potential for and severity of surge problems depend on contributing factors such as pipeline length, shape of the pipeline profile, location of changes in the pipeline slope (also called knees), static head, and initial or steady-state velocities. Figure 10-7 compares varying pipeline profiles, showing the best profile with the least potential for surge problems compared to pipeline profile shapes that have increasingly higher potential for surge problems.

During pump startup, some axial pumps require large starting torques and have high shutoff head compared to normal operational head. If the startup time is short for these types of pumps, pressure surges can develop in the system.

For long pipeline systems, the inertia of fluid inside the pipe may be much larger than the inertia of the pump. In this case, loss of power can cause pressure waves to travel along the pipeline and move to areas of high elevation. The pipeline may then develop vapor pressure, depending on the elevation of pipeline outlet relative to the elevation of pump station and its reservoir.

Air entrainment into a water system can cause the air pockets to travel to pumps and turbines and cause a sudden change in head. Air entrainment is often caused by filling operations, as the result of vortices and air introduction at inlets; and operation of air valves during surge. Air in pipeline systems can form pockets at high elevation points that can then separate the water into two columns, one on each side of the air pocket. The air pocket reduces the cross-sectional area of flow and increases velocity and head losses. Interactions of two or more air pockets can also lead to local surges.

Pipe failure can lead to a rapid drop in head, and the resulting pressure waves that travel along the pipe can lead to vapor pressures at some locations. Under these conditions, cavities form and then collapse, resulting in surges.

10.3.4. Rules of Thumb for Identifying Vulnerable Systems

The following indicators can serve as rules of thumb and are intended to aid in identifying one or more possible conditions that tend to cause surge problems in a water distribution system:

- Pipelines without demands or water takeoffs with lengths greater than 1,000 ft (300 m),
- Static head greater than 40 ft (12 m),
- Steady-state velocities greater than 5 ft/s (1.5 m/s),
- Pipeline profiles with knees and many high and low points,
- Inline booster pump stations with long suction lines, and
- Air release valves that are float operated, allowing rapid discharge of air during power failure.

Systems with high velocities (greater than 5 ft/s [1.5 m/s]) can also be vulnerable even with much shorter pipelines. It should be noted that exceeding any general rule of thumb results in the potential for surge problems; prudent design should include a detailed surge analysis. Jung et al. (2007a) studied the need for comprehensive transient analysis of water distribution systems. They concluded that only systematic and informed surge analysis can be expected to resolve the complex transient characterizations and adequately protect water distribution systems from the vagaries and challenges of rapid transient events. An example of a potentially vulnerable system is shown in Figure 10-8 with measured pressure variation shown in Figure 10-9.

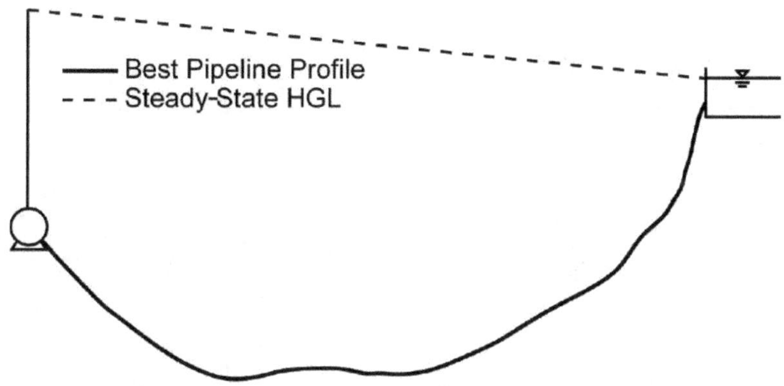

(a) Best pipeline profile (surge pressures will be lower for this type of pipeline profile)

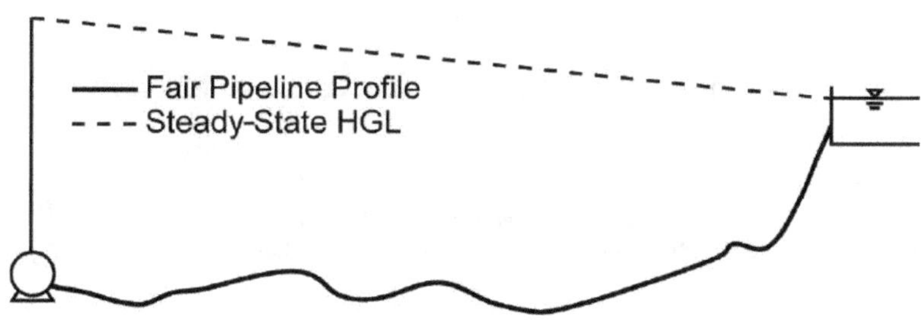

(b) Fair pipeline profile (some potential for surge problems due to varying terrain)

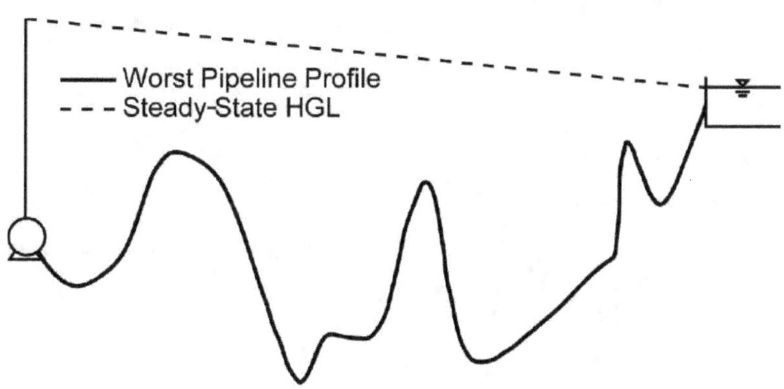

(c) Worst pipeline profile (knees and high points cause more severe surge problems)

Figure 10-7 Varying pipeline profiles

Abbreviation: HGL, hydraulic grade line.

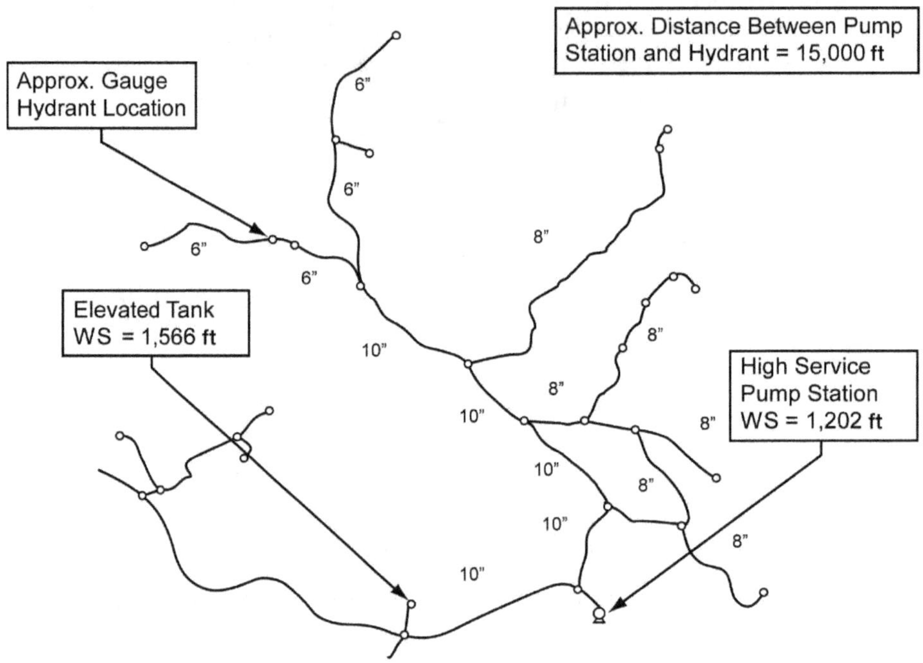

Figure 10-8 Example network schematic of a potentially vulnerable system

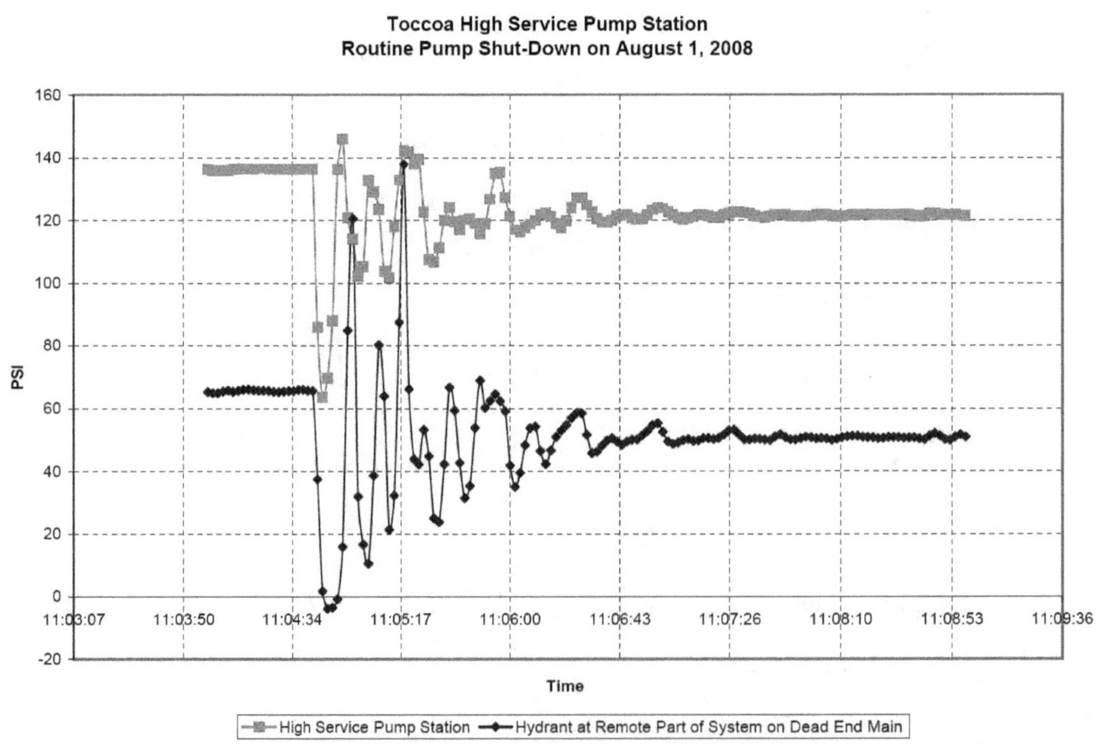

Figure 10-9 Pressure surge fluctuations (field measurements) following routine pump shutdown

10.3.5. Dead-End Pipelines

Dead-end pipelines are generally needed to allow supply for future developments. Although initial velocities are relatively low, these dead ends can sometimes be a source of severe transients as hydrants are opened and closed for flushing. Dead ends, which may also be caused by closure of check valves, lock pressure waves into the system in a cumulative fashion, and wave reflections will double both positive and negative pressures.

10.4. BASIC PRESSURE WAVE RELATIONS

10.4.1. Wave Action in Pipes

The relationship between pressure change (ΔP) and flow change (ΔQ), which is associated with the passage of a pressure wave, defines the transient response of the pipe system and forms the basis for the development of the required mathematical expressions (Boulos et al. 2006; Wood et al. 2005a; Walski et al. 2003; Wylie and Streeter 1993; Chaudhry 1979). Figure 10-10 shows flow and pressure conditions that exist a short time Δt apart, as a pressure wave of magnitude ΔP propagates a distance Δx in a liquid-filled line. During the short time Δt, the pressure on the left side of the wave front is $P + \Delta P$, while the right side of the wave front is P. This unbalanced pressure causes the fluid to accelerate. The propagation speed of the pressure wave is equal to the sonic velocity (c) in the system if the mean velocity of the liquid in the line is neglected. Because the mean velocity of the liquid is usually several orders of magnitude smaller than the sonic velocity, this is acceptable. The change in pressure can be expressed as:

$$\Delta P = \frac{\rho c \Delta Q}{A} \quad (10\text{-}1)$$

or in terms of pressure head:

$$\Delta H = \frac{c \Delta Q}{gA} = \pm \frac{c \Delta V}{g} \quad (10\text{-}2)$$

where:

ΔP = change in pressure, psi (kPa)
ΔH = change in head, ft (m)
ρ = density of water, lb/ft^3 (kg/m^3)
c = wave velocity, ft/s (m/s)
ΔQ = change in flow rate, ft^3/s (m^3/s)
A = cross-sectional area of pipe, ft^2 (m^2)
g = acceleration of gravity, 32.17 ft/s^2 (9.81 m/s^2)

The resulting head rise equation is referred to as the *Joukowsky relation*, sometimes the *fundamental equation of water hammer*. The equation is derived with the assumption that head losses due to friction are negligible and no interaction takes place between pressure waves and boundary conditions at the end points of the pipe. The negative sign in this equation is applicable for a disturbance that propagates upstream and the positive sign for one that moves downstream.

The Joukowsky relation is a simple way to estimate if there is a potential for damaging transient pressures. However, it tends to overestimate pressure extremes. If excessively high or low pressures are indicated by this relation, further evaluation is needed.

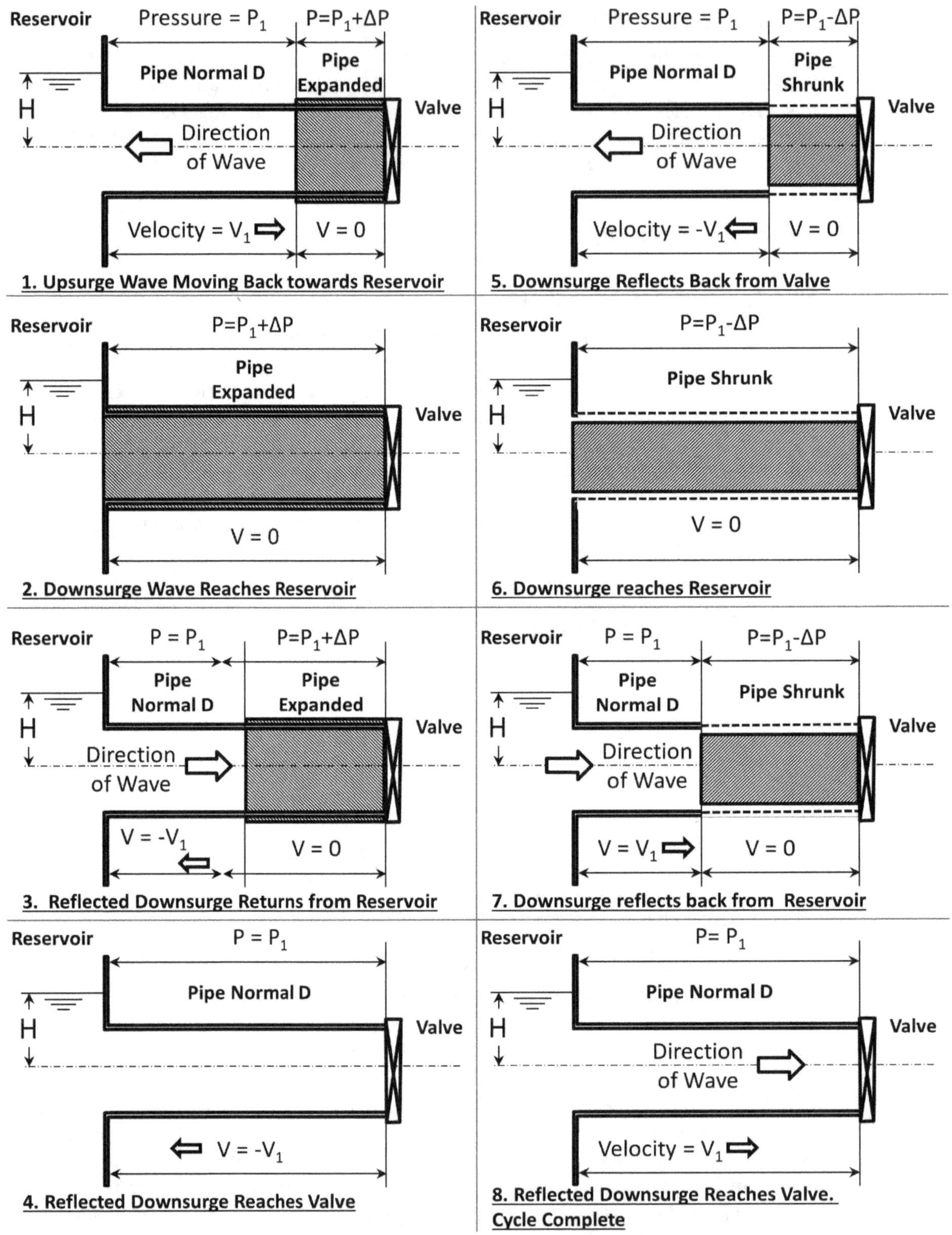

Figure 10-10 Pressure wave propagation in a pipe

Because values of wave speed in many pipelines are in the range of 3,000 to 4,000 ft/s (900 to 1,200 m/s), typical values of c/g in Eq 10-2 are large, often 100 or more. Thus, this relationship predicts large values of head rise and highlights the importance of transient analysis. For example, if an initial velocity of 3 ft/s (1 m/s) is suddenly arrested at the downstream end of pipeline and c/g equals 100 m/s, a head rise of 300 ft (100 m) will result. However, there are many factors that attenuate these pressures. The Joukowsky relation should only be used as an indicator of the need for further evaluation.

10.4.2. Wave Speed

The basic equation for the speed of a pressure wave in a fluid is given by:

$$C = \sqrt{\frac{E_f g}{\rho}} \tag{10-3}$$

where c is the wave speed in ft/s (m/s) and E_f is the bulk modulus in psi (kPa). The wave speed of a liquid flowing within a pipeline is also influenced by the elasticity of the line wall (Korteweg 1878) and how the pipe is restrained against longitudinal pipe movement. For a pipe system with some degree of axial restraint, a good approximation for the wave propagation speed is obtained using the following equation (Wylie and Streeter 1993):

$$C = \sqrt{\frac{E_f / \rho}{\left(1 + \frac{E_f D}{E_p t} \Psi\right)}} \tag{10-4}$$

where E_f and E_p are the elastic modulus of the fluid and pipe material, respectively; D is the pipe diameter; t is the pipe thickness; and ψ is the coefficient of restraint for longitudinal pipe movement. Typically, the following three types of pipeline support are considered for restraint:

Case a: The pipeline is restrained at the upstream end only.

$$\psi = 5/4 - \mu \tag{10-5}$$

Case b: The pipeline is restrained throughout.

$$\psi = 1 - \mu^2 \tag{10-6}$$

Case c: The pipeline is unrestrained (has expansion joints throughout).

$$\psi = 1 - \mu/2 \tag{10-7}$$

where μ is the Poisson's ratio for the pipe material. Table 10-1 lists physical properties of common pipe materials.

Entrained air within the flowing liquid can lower the effective wave speed of the fluid–air mixture, but the estimation of this impact is difficult. The effect of entrained air can be neglected during initial evaluations.

Table 10-1 Physical properties of common pipe materials

Material	Young's Modulus E_p (GPa)	Young's Modulus—E_p (psi) × 10^6 (Typical Values)	Poisson's Ratio μ
Asbestos cement	23–24	3.5	0.2–0.3
Cast iron	80–170	13–15	0.25–0.27
Concrete	14–30	2.9–4.4	0.1–0.15
Reinforced concrete	30–60	4.4–8.7	0.25
Ductile iron	172	24–25	0.28–0.3
Fiberglass	50	7.3	0.35
High-density polyethylene	0.8–1.1	0.1–0.2	0.4–0.5
Polyvinyl chloride	2.4–3.5	0.4–0.5	0.45–0.46
Steel	200–207	30	0.30
Stainless steel	193	28	0.3–0.31

Abbreviation: GPa, gigapascal.

10.4.3. Wave Action at Pipe Junctions

In a piping system, junction nodes have a significant impact on the direction and movement of pressure waves in the system. The effects of a pipe junction on pressure waves can be evaluated using conservation of mass and energy at the junction. Energy losses at the junction usually cause only minor effects and are neglected.

A wave of magnitude ΔH impinging on one of the junction legs, j_{in}, is transmitted equally to each adjoining leg (Figure 10-11). The magnitude of the waves is $T_{jin}\Delta H$, where the transmission coefficient, T_{jin}, is given by:

$$T_{j_{in}} = \frac{2\left(\frac{g_{jin} A_{jin}}{c_{jin}}\right)}{\sum \frac{g_j A_j}{c_j}} \quad (10\text{-}8)$$

where the summation j refers to all pipes connecting at the junctions (incoming and outgoing). A reflection back in pipe j_{in} occurs and is of magnitude $R_{jin}\Delta H$ where:

$$R_{jin} = T_{jin} - 1 \quad (10\text{-}9)$$

For the simultaneous impingement of waves arriving in more than one leg, the effects are superimposed. The magnitude of the superimposed waves depends on the signs of the waves.

Equation 10-8 provides the basis for evaluating the effect of wave action at two special junction cases: dead-end junctions and open ends or connections to reservoirs. A dead end is represented as a two-pipe junction with A_2 equal to zero. With A_2 equal to zero, T_{jin} equals 2 and R_{jin} is 1, which indicates that the wave is reflected positively from the dead end. This condition implies that the effects of pressure waves on dead ends can be of significant importance in transient consideration. If the pressure wave reaching the dead end is positive, the wave is reflected with twice the pressure head of the incident wave. Figure 10-12 illustrates the action of a positive wave at a dead-end junction. If the pressure wave reaching the dead end is negative, the wave reflection will cause a further decrease in pressure that can lead to the formation and collapse of vapor cavity. For a reservoir connection, A_2 is infinite so T_{jin} is zero and R_{jin} equals –1, which indicates that a negative reflection occurs at a reservoir. Figure 10-13 illustrates the action of a positive wave at a reservoir.

10.4.4. Control Elements

A general analysis of pressure wave action at a control element (e.g., pump, valve, orifice) in a pipe system is described below (Boulos et al. 2006; Wood et al. 2005a). This analysis provides relations to account for a variety of situations.

Figure 10-14 shows a general situation at a control element where pressure waves ΔH_1 and ΔH_2 are impinging. At the same time, the characteristics of the control element may be changing. It is assumed that the relationship between flow through the control element, Q, and the pressure head change across the control element, ΔH, always satisfies a head–flow equation for the control element having the general form:

$$\Delta H = A(t) + B(t)Q + C(t)Q \qquad (10\text{-}10)$$

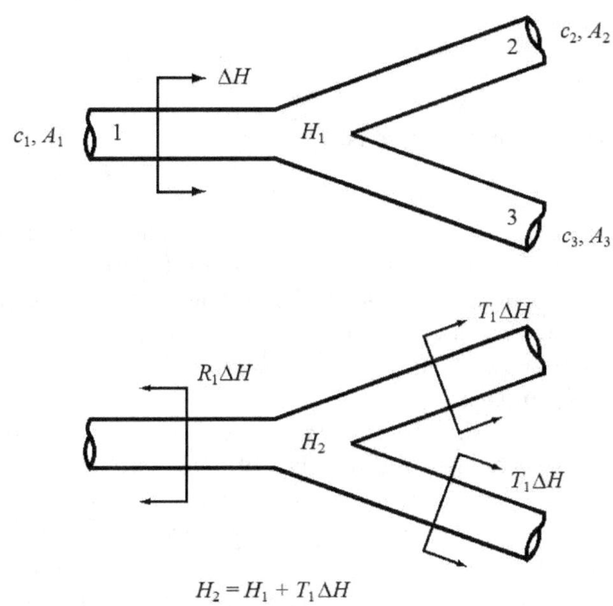

Figure 10-11 Effect of a pipe junction on a pressure wave

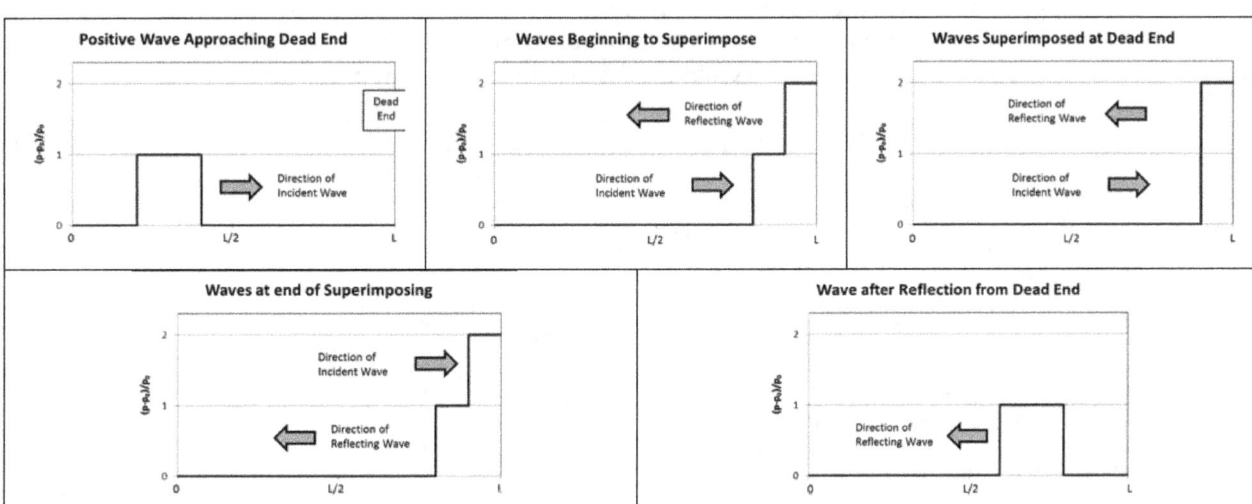

Figure 10-12 Effect of a dead end on a pressure wave

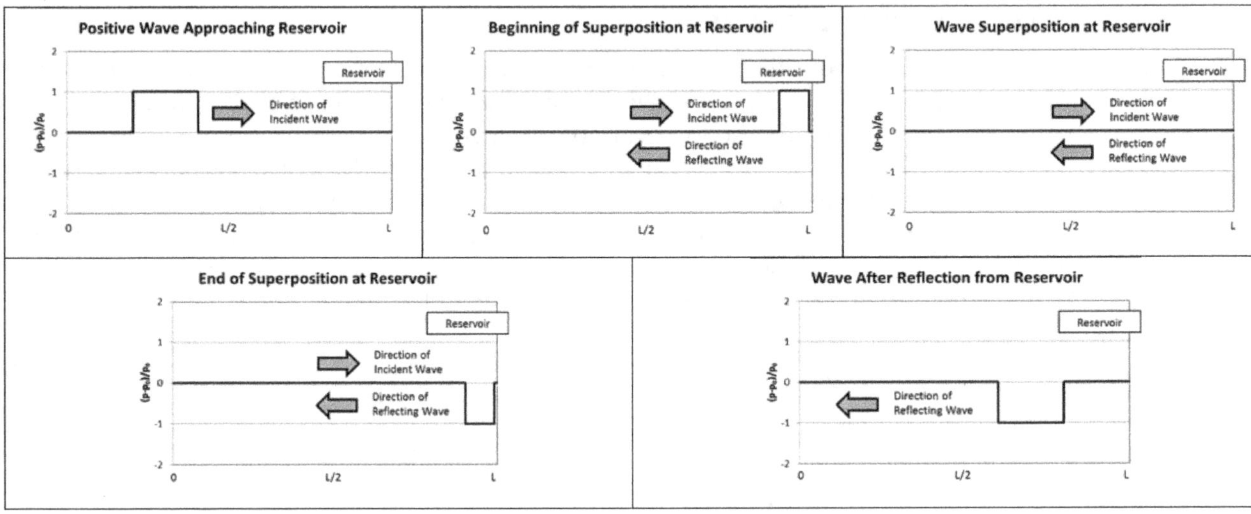

Figure 10-13 Effect of a reservoir on a pressure wave

The terms *A*, *B*, and *C* represent the coefficients for a general representation of the control element head–flow equation. These coefficients may be time dependent but will be known (or can be determined) at all times. The absolute value of *Q* is used so that the resistance term is dependent on the flow direction. This representation applies to both passive resistance elements such as valves, orifices, fittings, and friction elements and active elements such as pumps.

For passive resistance elements, however, the terms *A* and *B* are both zero. Thus, only the coefficient *C*, representing the effect of irreversible loss, is not zero, and this coefficient represents the ratio of the head loss to the square of the flow through the control element. For hydraulic considerations, this type of square law relationship is appropriate. The sign of the pressure head change is dependent on the direction of flow through the control element; this necessitates the use of the absolute value of the flow rate in Eq 10-12.

In Figure 10-14, subscripts 1 and 2 denote conditions on the left and right side of the control element before the impinging waves arrive, while subscripts 3 and 4 designate these conditions at the control element after the wave action. Here, Q_b and Q_a are the flows before and after the wave action, respectively.

The basic transient flow relationship for pressure–flow changes is applied to incoming and outgoing waves to yield the following for the outgoing waves:

$$\Delta H_3 = \Delta H_1 + \beta_1 (Q_b - Q_a) \tag{10-11}$$

$$\Delta H_4 = \Delta H_2 + \beta_2 (Q_a - Q_b) \tag{10-12}$$

Where:

$$\beta_1 = \frac{c_1}{gA_1} \text{ and } \beta_2 = \frac{c_2}{gA_2} \tag{10-13}$$

Pressure heads after the action are given by:

$$H_3 = H_1 + \Delta H_1 + \Delta H_3 \tag{10-14}$$

and

$$H_4 = H_2 + \Delta H_2 + \Delta H_4 \tag{10-15}$$

The characteristic equation relating the pressure head change across and the flow through the control element after the action is:

$$H_4 - H_3 = A(t) + B(t)Q_a + C(t)\, Q_a|Q_a| \tag{10-16}$$

The coefficients of the characteristic equation, $A(t)$, $B(t)$, and $C(t)$, represent the values at the time of the wave action and may vary with time.

Substituting Eqs 10-14 and 10-15 into Eq 10-16 and rearranging results in a quadratic relationship for Q_a or:

$$\begin{aligned} &C(t)\, Q_a|Q_a| + (B(t) - \beta_1 - \beta_2)\, Q_a \\ &+ A(t) + H_1 + 2\Delta H_1 - H_2 - 2\Delta H_2 + (\beta_1 + \beta_2)\, Q_b = 0 \end{aligned} \tag{10-17}$$

Equation 10-17 can be solved directly for Q_a using the quadratic formula or iteratively using the Newton–Raphson method. Equations 10-11 and 10-12 are then solved to give the magnitude of the pressure waves produced by the action, and Eqs 10-14 and 10-15 yield the pressure head after the action takes place.

This general analysis represents a wide variety of control elements that can be subject to a range of conditions.

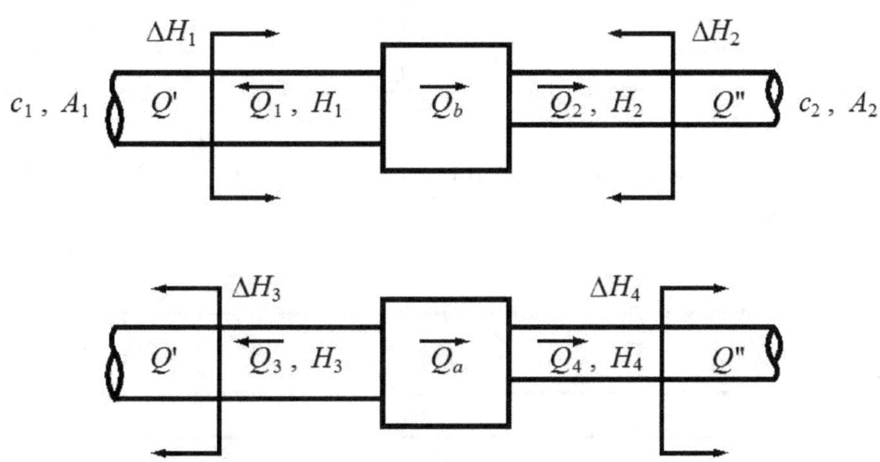

Figure 10-14 Condition at a control element before and after action

10.4.4.1. Control Element Characteristics. The coefficients of the control element characteristic equation (Eq 10-8) are determined using head–flow operating data for the control element. Some control elements such as pumps will use all three coefficients to represent the head–flow variation. In some cases, the characteristic equation will be based on data that represent the head–flow relationship for a relatively small range of operations. For these applications, the coefficients used for the control element analysis will be based on data valid for the operation in the vicinity of the operating point and will be recalculated as the operating point changes. This is true for the analysis of variable-speed pumps and for pumps using data representing a wide range of operating conditions, including abnormal situations such as flow reversal.

Many control elements such as valves can be modeled using only the *C* coefficient. These are referred to as *resistive control elements,* where the head–flow relation is adequately described by a single resistive term. For this application, the coefficient $C(t)$ is defined as the control element resistance. The term *resistance* is defined as the head drop divided by the square of the flow ($\Delta H/Q^2$). Here, the head drop is in feet (meters) and the flow is in ft^3/s (m^3/s).

The control element resistance is directly related to other resistive parameters such as minor loss (K_M), valve flow coefficient (C_v), sprinkler constant (K_s), and others that characterize the head–flow characteristic of a resistive control element.

10.4.4.2. Wave Propagation With Friction. Because all pipeline systems contain friction, the pressure wave is attenuated as it travels down a line. Line loss can be simulated by concentrating the losses in length *L* at an orifice, as shown in Figure 10-15. This orifice will then partially transmit and reflect pressure waves and account for the effect of wall shear. The friction orifice will therefore attenuate a pressure wave in a manner similar to the total attenuation that will occur as the wave travels the length *L* in the pipe.

In this representation, the loss at the orifice is:

$$\Delta H = H_2 - H_1 = \left[-\frac{fL}{2gDA^2}\right]Q^2 = CQ^2 \tag{10-18}$$

where f is the friction factor, g is the acceleration of gravity, D is the pipe diameter, and A is the pipe area.

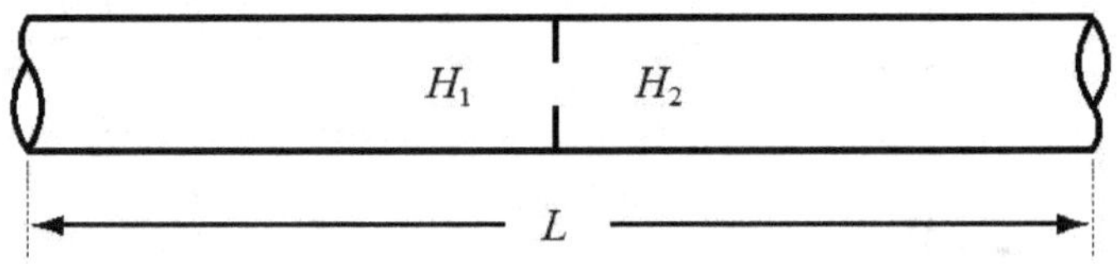

Figure 10-15 Wave propagation in a pipe section considering friction

In this case, the coefficients of the characteristic equation (Eq 10-12) for the line friction orifice are:

$$A(t) = B(t) = 0 \tag{10-19}$$

and

$$C(t) = -\frac{fL}{2gDA^2} \tag{10-20}$$

A simplifying assumption is that the friction factor can be determined using the flow rate through the orifice prior to the wave action. Although it is true that some approximation errors will be introduced if excessively long reaches are used, these errors are generally very small and can be minimized or eliminated using shorter pipe reaches. Ramalingam et al. (2009) developed sound guidelines in the form of an error study for selecting the optimal number of friction orifices to ensure accurate results.

The accuracy of the one-dimensional models of transition flow is affected by the friction coefficients used in the model. Traditionally, transient models estimated this coefficient to be unchanging (steady) over the course of the simulation and are based on the steady-state velocity in the system. The assumption of steady viscous losses may be satisfactory for slow transients where the wall shear stress has a quasi-steady behavior. However, large discrepancies in attenuation, shape, and timing of pressure traces have been noted when computational results for rapid transients were compared with measurements. The magnitude of these discrepancies depends on flow conditions (severity of transients; laminar, transitional, or turbulent flow) and liquid properties (viscosity). The discrepancies can be related to significant increases in turbulent shear that are not accounted for. Thus, quasi-steady or unsteady approaches to determination of friction coefficients may be more accurate methods for estimating the attenuation of transient energy. An illustration of the impact of friction method is provided in Figure 10-16.

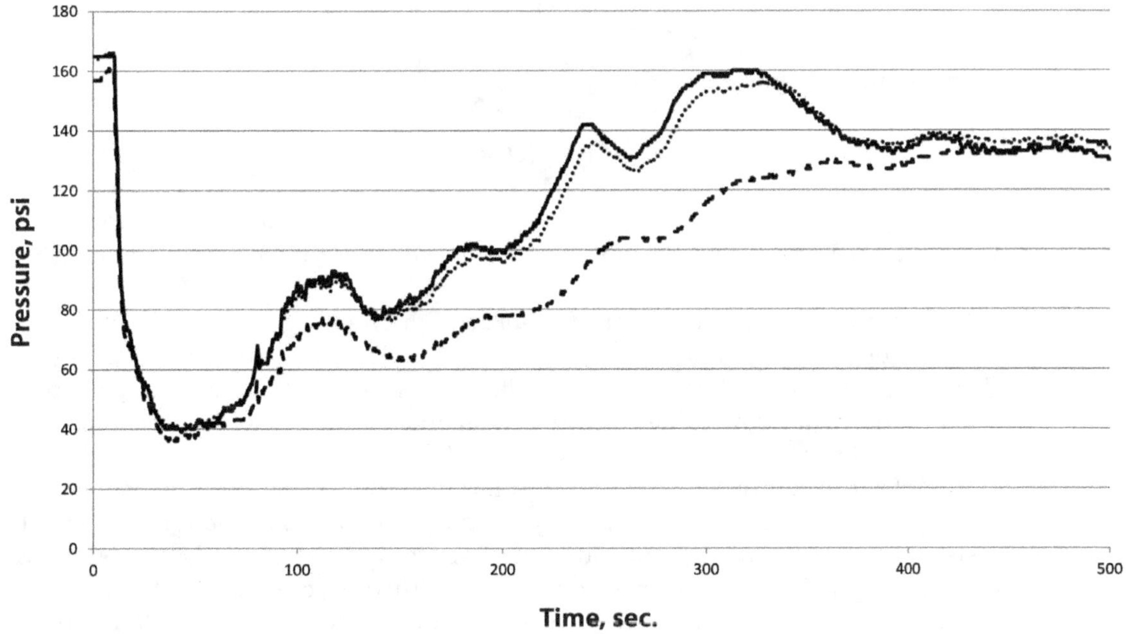

Source: Bentley Systems 2015.

Figure 10-16 Impact of friction factor method on transient modeling results

In the steady friction approach, a constant value of the Darcy–Weisbach friction factor is assumed. In the quasi-steady approximation, friction losses are estimated using formulas derived for steady-state flow conditions, but the friction loss coefficient is based on the instantaneous transient velocity (as opposed to the pretransient velocity). This approach also has a limitation in that the shear stress at the wall is not in phase with the mean velocity during transient conditions, and the velocity profile can be completely different from a uniform flow profile. Therefore, friction losses computed using steady-state relationships are inaccurate in transient laminar and turbulent flow. To address this issue, it is possible to algebraically add unsteady-flow terms to the quasi-steady resistance term of one-dimensional models for both laminar and turbulent flows. There are numerous groups of unsteady friction modeling approaches that differ on how the shear stress is estimated (Brunone et al. 1991; Zielke 1968; Vardy and Brown 1996; Vitkovsky et al. 2000). While these methods do provide a better approximation of pressure wave attenuation due to unsteady friction, there is a cost in terms of computation times. Close attention should be paid to how the software used for transient modeling handles friction during the transient event.

10.5. GOVERNING EQUATIONS

The fundamental equations that describe hydraulic transients in liquid pipeline systems are developed from the basic conservation relationships of physics or fluid mechanics. They can be fully described by Newton's second law (equation of motion) and conservation of mass (kinematic relation). These equations can incorporate typical hydraulic devices and their interactions with the wave conditions in the pipes.

By applying these basic laws to an elementary control volume, a set of nonlinear hyperbolic partial differential equations can be derived. If x is the distance along the pipe centerline, t is the time, and partial derivatives are represented as subscripts, the governing equations for transient flow can be written as:

Continuity:

$$H_t + \frac{c^2}{gA} Q_x = 0 \tag{10-21}$$

Momentum (dynamic):

$$H_x + \frac{1}{gA} Q_x - f(Q) = 0 \tag{10-22}$$

where $f(Q)$ is a pipe resistance (nonlinear) term that is a function of flow rate.

The preceding equations are first-order, hyperbolic, partial, differential equations in two independent variables (space and time) and two dependent variables (head and flow). The solution of these equations with appropriate initial and boundary conditions will give head and flow values in both spatial and temporal coordinates for any transient analysis problem.

Unfortunately, no exact analytical solution exists for these equations except for simple applications that neglect or greatly simplify the boundary conditions and the pipe resistance term. When pipe junctions, pumps, surge tanks, air vessels, and other hydraulic components are included, the basic equations are further complicated. As a result, numerical methods are used to integrate or solve the transient flow equations (Wylie and Streeter 1993; Tullis 1989; Chaudhry 1979).

10.6. NUMERICAL SOLUTIONS OF TRANSIENTS

Several approaches have been taken to numerically model pressure transients in water distribution systems (Boulos et al. 2006; Wood et al. 2005a, 2005b; Walski et al. 2003). The two most widely used and accepted methods are the Lagrangian wave characteristic method (WCM) and the Eulerian fixed-grid method of characteristics (MOC). Each method assumes that a steady-state hydraulic solution is available that gives initial flow and pressure distributions throughout the system. The main difference between the two numerical methods is in the way pressure waves are tracked between the pipe boundaries (e.g., reservoirs, tanks, dead ends, partially opened valves, pumps, junctions, surge control devices, vapor cavities). The MOC tracks a disturbance in the time–space grid using a numerical method based on characteristics, while the WCM tracks the disturbance based on wave propagation mechanics. Both methods have been well documented in the literature (Ramalingam et al. 2009; Jung et al. 2007a; Boulos et al. 2005, 2006; Wood et al. 1966, 2005a, 2005b; Walski et al. 2003; Wylie and Streeter 1993) and have been implemented in various computer programs for pipe system transient analysis. The methods will, for the most part, produce the same results at the network nodes when using the same data and model to the same accuracy. A brief description of each method follows.

10.6.1. Method of Characteristics

In MOC, the governing partial differential equations are converted to ordinary differential equations and then to a difference form for solution by a numerical method. Solution space comprises two equations referred to as the *characteristic equations* along with two compatibility equations for any point in a space–time grid. The method divides the entire pipeline into a fixed number of segments, writes the characteristic and compatibility equations for every grid location, and then solves these equations for head and flow at all grid locations. The line friction of the entire pipeline is distributed in each of these segments. The various boundary conditions are handled by combining the appropriate characteristic equation with the equations defining the boundary.

10.6.2. Wave Characteristic Method

The WCM is based on the concept that transient pipe flow results from generation and propagation of pressure waves that occur as a result of a disturbance in the pipe system. The method essentially tracks the movement of pressure waves as they propagate throughout the system and computes new conditions at either fixed time intervals or only at times when a change actually occurs. It requires the calculation of the effects of these pressure waves impinging on the network junctions and control elements along with the appropriate boundary conditions. The entire line friction is modeled as an equivalent orifice situated at the midpoint of a pipeline or multiple orifices distributed uniformly throughout the pipeline (i.e., distributed friction profile). Pressure and velocity or flow rate time histories are computed for any point in the network by summing with time the contributions of incremental waves.

10.7. METHODS FOR CONTROLLING TRANSIENTS

The methods for controlling pressure transients in water distribution systems generally depend on whether the initiating event results in an *upsurge* (e.g., a high-pressure event caused by a shutdown of a downstream pump or valve) or a *downsurge* (e.g., a low-pressure event caused by the shutdown of an upstream pump or valve). Upsurge events can lead to pipe ruptures, and downsurge events can lead to the undesirable occurrence of water column separation (cavitation) that can result in severe pressure surges following the collapse of a vapor cavity or intrusion of contaminated water through a leak or other opening.

A number of surge protection devices are commonly used to help control high- and low-pressure transients in water distribution pipe systems. Small systems are just as vulnerable as large systems. No two systems are identical; hence, the ultimate choice of surge protection devices and operating strategies usually differ. Of course, it is always best, whenever possible, to avoid rapid flow changes. A transient analysis should be carried out to predict the effect of each individually selected device. As a result of the complex nature of transient behavior, a device intended to suppress or fix a transient condition could result in a worsening of the condition if the device is not properly selected or located in the system. Designers must evaluate the relative merits and shortcomings of all protection devices that they select. A combination of devices may prove to be the most desirable and economical. A brief overview of various commonly used surge protection devices and their functions is provided here. Additional details are available in Boulos et al. (2005, 2006), Wood (2005a), Walski et al. (2003), and Thorley (1991).

10.7.1. Devices and Systems

10.7.1.1. Simple Surge Tank: Open. Open surge tanks or standpipes can be an excellent solution to both upsurge and downsurge problems. These tanks can be installed only at locations where either the normal static pressure heads are small or where tall tanks are acceptable. They serve two main purposes: to prevent high pressures during pump startup conditions or valve shutdown conditions by accepting water, and to reduce downsurge or prevent cavitation during conditions of pump shutdown or valve closure by providing water to a low-pressure region.

10.7.1.2. Surge Vessel: Air Chamber, Closed Surge Tank, Bladder Tank, Hybrid Tank. Surge vessels (or air chambers), which are pressure vessels partially filled with air, have the advantage that they can be installed anywhere along a line regardless of normal pressure head. It should be noted that under very low static conditions or downhill pumping, special care is required to keep the surge tank from completely draining. These vessels are normally positioned at pump stations (downstream of the pump delivery valve) to provide protection against a loss of power to the pump. They serve the same function as an open surge tank but respond faster and allow a wider range of pressure fluctuation. Their effect depends primarily on their location, vessel size, entrance resistance, and initial gas volume and pressure. Closed surge vessels are normally equipped with an air compressor to control the initial gas volume and to supply makeup gas, which is absorbed by the water. Some closed surge tanks are equipped with a precharged pressurized bladder (bladder surge tanks) that eliminates the need for an air compressor. The precharge pressure for bladder tanks must be accurately and strategically determined. If the precharge pressure is too low, the surge tank will have very little to no effect on system transient pressures, whereas if the precharge pressure is too high, the surge tank will completely drain and may not adequately dampen transient pressures. Hybrid tanks are equipped with an air vent that admits air when the pressure goes below atmospheric pressure.

10.7.1.3. Feed Tank: One-Way Surge Tank. Feed tanks prevent initial low pressures and potential water column separation by admitting water into the pipe subsequent to a downsurge. They can be either open or closed, have a check valve to allow flow only into the pipe system, and can be installed anywhere on the line.

10.7.1.4. Pressure-Relief Valve. A pressure-relief valve (PRV) ejects water out of a side orifice to prevent excessive high-pressure surges. In pump stations, this valve may sometimes be routed back to the suction side of the pumps. PRVs are activated when the line pressure at a specified location (not necessarily at the valve) reaches a preset value. The valve opens and closes at prescribed rates over which the designer often has some degree of control. It can eject water into the atmosphere or a pressurized region or into an open or closed surge tank.

10.7.1.5. Surge Anticipation Valve. A surge anticipation valve is much like a PRV but it gets triggered to open on a downsurge in pressure (sensed at a specified location) in anticipation of an upsurge to follow. This valve, when activated, follows and completes a cycle of opening and closing based on valve opening and closing rates. For systems where water column separation is not expected to occur, the surge anticipation valve can solve the problem of upsurge at the pump due to reverse flow or wave reflection. However, this valve must always be used with caution because it can make low-pressure conditions in a line worse than they would be without this type of valve.

10.7.1.6. Air-Release/Vacuum Valve. Air-release/vacuum-breaking valves are installed at high points in a pipeline to prevent low pressure (cavitation) by admitting air into the pipe when the line pressure drops below atmospheric conditions. The air is expelled (ideally at a lower rate) when the line pressure exceeds atmospheric pressure. Two-stage air valves release air through a smaller orifice to prevent the "air slam" that occurs when all the air is released and the water column rejoins. A three-stage air valve can be designed to release air through a second (smaller) orifice to further reduce the air slam.

10.7.1.7. Check Valve. A check valve allows flow in one direction and closes when flow reversal is imminent. For transient control, check valves are usually installed with other devices such as a pump bypass line, as described in section 10.7.1.8. Pumps are often equipped with a check valve to prevent flow reversal. Because check valves do not close instantaneously, it is possible that substantial backflow could occur before closure; this can produce additional and sometimes large surges in the system. Check valve modeling includes a time delay between check valve activation and complete closure of the check valve. The check valve is often treated as a valve that closes in a linear fashion, is activated by flow reversal, and closes completely over the delay period. Check valves can also be used to isolate high-pressure waves from reaching a section of pipeline. A great advantage of a check valve is that it can prevent pipes from draining; keeping pipes full of fluid tends to reduce startup transients.

10.7.1.8. Pump Bypass Line. In low-head pumping systems that have a positive suction head, a bypass line around the pumps can be installed to allow water to be drawn into the discharge line following the downsurge that results from pump shutoff or power failure. Bypass lines are generally short line segments equipped with a check valve (nonreturn valve); they prevent backflow (from the pump discharge to the suction side) and are installed parallel to the pump in the normal flow direction. Bypass lines are activated when the pump suction head exceeds the discharge head. They help prevent high-pressure buildup on the pump suction side and low-pressure downsurge on the pump discharge side. Bypass lines typically are not activated in high-head pumping systems, as the suction head may never overcome the discharge head following a pump shutoff or power outage.

10.7.1.9. Flywheel. Increasing the pump rotational inertia by attaching a flywheel (Figure 10-17), which is a large-diameter steel plate, to the pump motor is sometimes a useful surge control device, especially for sewage pumping systems, because the flywheel is not in contact with the foul water. When power fails, the rotational energy provided by the flywheel reduces pump speed gradually, allowing the pumping system to slowly come to rest, thus avoiding unacceptable transient pressures. There are three problems related to flywheel installations:

- Flywheels must be large and heavy enough to provide additional rotating inertia to be effective. Pump and motor bearings and supports must be designed to accommodate the extra weight, and additional space is required.
- Pump startup requires extra power to overcome the inertia of flywheels.
- Pump startup must be gradual to keep the motors from burning out.

Nonetheless, flywheels are useful for surge control and can be found in many pumping installations.

10.7.1.10. Booster Pump Control Valve. Booster pump control valves are slow-opening and slow-closing valves. They are installed on discharge pumps used to gradually increase and decrease the flow during pump startup and shutdown, respectively. The opening and closing duration can range on the order of seconds to minutes, depending on the valve design. One drawback of pump control valves is that they provide no downsurge protection in the event of a power outage. It should be noted that the effectiveness of pump control valves may vary depending on system conditions. Therefore, care should be invested in the selection and setup of pump control valves in order to achieve the desired surge protection benefit.

10.7.1.11. Inline Pump Control Valve. For large pumping systems with large-diameter headers and high flows, inline pump control valves are a viable option. These valves are operated similarly to surge anticipator valves, which open when line pressure drops below a specified set point, remain open for a preset period of time, and then close in a manner that prevents the high pressures that result from rapid valve closure. When properly installed, these valves prevent high pressures at pump stations. However, the valve

Figure 10-17 Flywheels to be installed in a large pump station

movements must be set with care; otherwise, more severe problems can occur. Four quadrant pump curves have to be analyzed in conjunction with the other transient parameters to develop a solution. Of course, these valves do not prevent low pressures and column separation from occurring in the downstream pipelines.

10.7.1.12. Variable-Speed Drives. Variable-speed drives (VSDs) can be used on pumps to slowly ramp pumping speeds up or down. As the speed ramps up or down, the flow through the pump gradually increases or decreases, thereby avoiding the creation of large upsurges or downsurges. However, it should be noted that in certain systems, depending on the pump's size and design and the system curve, a VSD may not be able to stably ramp up or down at a slow enough rate to avoid creation of sizeable transient pressures. In addition, VSDs provide no surge protection in the event of a power failure. Added benefits associated with VSDs include potential energy savings and increased life of pump bearings.

10.7.1.13. Soft Starters. Soft starters offer another method for controlling a pump's startup and shutdown speed. Soft starters are designed to apply voltage to the motor at an increasing or decreasing rate, with the intent of gradually increasing or decreasing flow through the pump. Different soft-starter technologies are available. Care should be taken when selecting and setting up soft starters in order to achieve the desired results. Soft starters will not provide surge protection in the event of a power failure.

10.7.2. Choice of Surge Protection Strategy

A number of techniques can be used to control and suppress transients in municipal water distribution systems. Some involve system design and operation, while others are related to the proper selection of surge protection devices. For example, PRVs, surge-anticipation valves, surge vessels, surge tanks, pump bypass lines, or any combination of these can be used to control high pressures. Low pressures can be controlled by increasing pump inertia or by adding surge vessels, surge tanks, air-release/air-vacuum valves, pump bypass lines, or any combination of these. The primary objectives are to reduce the rate at which changes to flow occur and to attenuate the extremes of pressure that occur in response to a transient event.

Surge protection devices are normally installed at or near the point where the disturbance is initiated, such as at the pump discharge or by the closing valve (with the exception of air-relief/vacuum-breaking valves and feed tanks). Figure 10-18 illustrates typical locations for the various surge protection devices in a water distribution system. A comprehensive transient flowchart for considering the transient protection of the system is shown in Figure 10-19. Boulos et al. (2005, 2006) discuss this flowchart in detail. When developing a protection strategy, it must be recognized that no two systems are hydraulically the same; hence, no general rules or universally applicable guidelines are available to eliminate unacceptable pressures in any given water distribution system. Any surge protection devices and/or operating strategies must be chosen accordingly (Boulos et al. 2005, 2006; Wood et al. 2005a; Walski et al. 2003; Thorley 1991).

The final choice is based on the initial cause and location of the transient disturbance(s), the system itself, the consequences if remedial action is not taken, and the cost of the protection measures. A combination of devices may prove to be the most effective and most economical. A final check of the proposed solution's adequacy and efficacy should be conducted and validated using a detailed transient analysis.

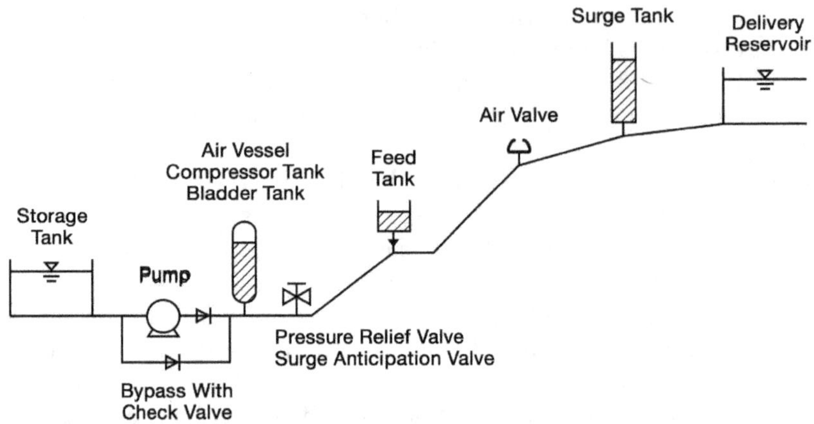

Figure 10-18 Typical locations in a water distribution system for various surge-protection devices

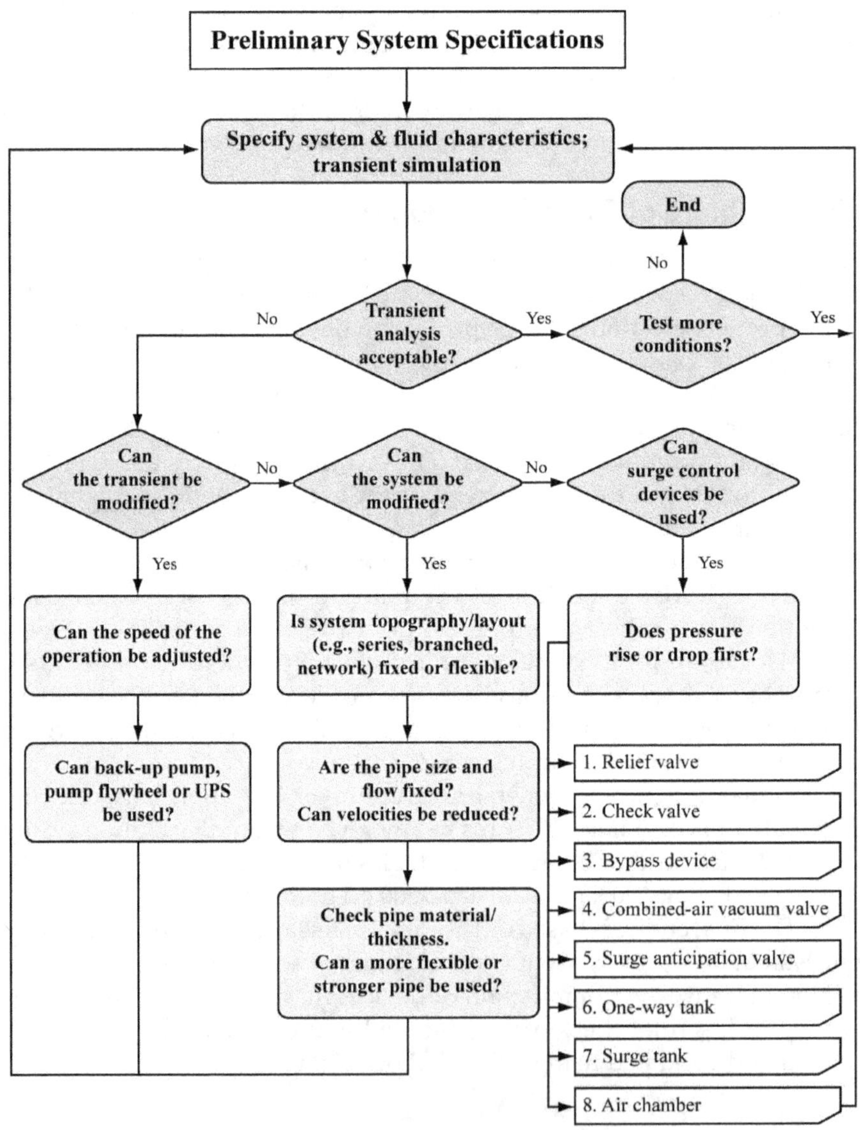

Figure 10-19 Flowchart for surge control in a water distribution system

Abbreviations: UPS, uninterruptible power supply.

10.7.2.1. Pump Station and Downstream Pipeline Protection. Because pump stations are aboveground, surge failures and remedial solutions can be readily observed. However, failures of buried downstream pipelines are sometimes undetected and become subject to unseen and overlooked problems. It is important to recognize that some surge control solutions protect only the pump station and do not necessarily protect the downstream pipelines from column separation and negative pressures. Both the pump station and the downstream pipelines should be protected.

10.7.2.2. Float-Operated Air Valves Versus Surge-Resistant Air Valves. A number of vacuum-relief valves are often needed for filling and draining pipelines. Also, additional air valves may be required to protect the pipelines from extreme low pressures due to surge transients. The total number of air valves depends on pipeline profile and initial pipeline velocities. Of course, the number of air valves should be limited and carefully considered, because adding large volumes of air to pipelines can create other operational problems such as air binding. If pipeline profiles are too steep and the initial velocities are too high, numerous air valves would be required. In that case, other types of surge control devices should be considered.

Because water is required to close the float in float-operated air valves, some pipe systems with high static head require surge-resistant air valves to prevent damage to the float. Damage to the air valve and high pressures occur when a water column slams the float closed, pushing the low-density air out around the float.

A safer air valve solution would be the so-called surge-resistant air valve that traps the air inside the air valve body, using it as a cushion, and then slowly releases the air to prevent high pressures and damage to the valves.

10.7.2.3. Normal Pump Operation Protection Versus Power Failure Protection. Normal pump operation for a surge-vulnerable system can include slow-closing and slow-opening pump control valves or VFDs on the pump motors. These are used to eliminate undesirable surges during daily operation.

Power failure on surge-vulnerable systems requires surge control devices. Backup power and generators do not come on line quickly enough to prevent transients from occurring. It should be noted that the use of surge chambers requires quick-closing check valves immediately downstream of the pumps. These valves prevent the surge chamber from spilling water back through the pumps during pump shutdown. Slow-closing pump control valves would not be used in conjunction with a surge chamber. If a surge chamber is required for surge control, it would also be used for normal pump shutdown and startup, and pump control valves would not be required.

In addition to pumps, there are other control strategies such as educating of operators to slowly open and close valves, adjusting the speed of mechanically operated valves, and not operating tanks such that the altitude valve closes and cannot absorb the transient.

10.8. TRANSIENT MODELING CONSIDERATIONS

Transient analysis is essentially based on a number of simplified equations that govern the movement of pressure waves throughout a municipal water system. This type of network analysis requires a significant number of calculations and is an extremely demanding computational exercise. This is usually carried out using highly specialized transient modeling software. Although the system should be described in full, some simplifications and assumptions are needed for its model representation. This is necessary to reduce the network model complexity and computational run times and because some of the basic transient flow data required will not be available (Thorley 1991).

Having considered some of the detail of transient analysis and protection, it is helpful to include some tips and guidelines that would assist in preparing computer simulation files. In essence, the good news is that transient modeling uses much of the same data that steady-state modeling does. A steady-state analysis of the initial conditions for the transient analysis is required. A number of additional considerations for developing a transient analysis model include the following:

- The location (including elevation) of hydraulic devices (e.g., pumps, control valves, check valves, regulating valves) is required for the model.
- A transient model should carry out calculations at all local high and low points because the pressure extremes often occur at these locations.
- Cavitation must be modeled for transient analysis. If cavitation occurs at any location in the distribution system, it can greatly affect the transient analysis results (Jung et al. 2009b).
- Skeletonization can simplify networks for transient analyses, but the guidelines for skeletonization are significantly different than those for steady-state analysis. Dead-end pipes, for example, will have a very significant effect on transient analysis and will have no effect on steady-state analysis. Jung et al. (2007b) performed a detailed study of the issues associated with water distribution model skeletonization for surge analysis. They concluded that skeletonization can introduce some significant error when estimating pressure extremes and can overlook water column separation and subsequent collapse at vulnerable locations in the distribution system. However, where applicable for transient analyses within localized areas of a distribution system, the model can be "clipped" where the distribution mains intersect larger transmission mains, which can often dampen the transient event.
- Series and parallel pumps with similar characteristics may be modeled as a single equivalent pump if none of the individual pumps to be combined are subjected to a pump trip, startup, or speed change.
- Extreme pressure (positive or negative) surges can drastically alter the local pressure and affect the demand magnitude that can be extracted at any given node in the system. Pressure-sensitive demands versus fixed (constant or pressure independent) nodal outflows should be used to assess the impact of pressure changes and produce more accurate transient results (Jung et al. 2009a).
- It is good practice to allow a transient model to operate initially at steady state for a short period before the transient is initiated. This provides additional assurance that the transient model is operating correctly.

For transient analysis, it is necessary to use a computational time interval such that the pressure wave travel times for all pipe segments are a multiple of this time increment and this integer multiple is calculated for each line segment. The time-step, *dt*, can be estimated using the following equation:

$$dt = 2L/c \qquad (10\text{-}23)$$

where L is the length of the shortest segment in the sytem and c is the pressure wave velocity.

Some adjustment is required to obtain a time interval that is not unreasonably small. This involves analyzing a model of the piping system with pipe lengths (or wave speeds) that are different from the actual values. Also, a sensitivity check should be made to ensure that the time interval chosen is acceptable. The time interval required for accurate transient analysis may be quite small (0.01 to 0.001 s). For example, if the wave speed is 4,000 ft/s (1,220 m/s), a time increment of 0.01 s represents a travel distance (and therefore, length accuracy) of 40 ft (12.2 m) for the model. A time increment of 0.001 s represents a travel distance (and length accuracy) of 4 ft (1.2 m) for the model. The length accuracy of the model (maximum difference between actual and model pipe lengths) must be sufficient to generate an accurate solution. However, increasing the accuracy requires a longer computational time. It should be noted that the shortest pipe in the network model plays a dominant role in determining the computational time interval. Consideration should be given to merging very short pipes into longer pipe segments or simply removing short pipes from the network model, but only when their resulting effect on the system transient behavior is negligible.

10.9. TRANSIENT MODEL INPUT DATA REQUIREMENTS

The data requirements for setting up a transient model include all data necessary to perform a steady-state hydraulic analysis. A steady-state analysis for the starting conditions is required. Initial conditions of flow in all pipe segments and static pressure head (or pressure) at all junctions and components must be calculated before initiating the transient analysis. Note that the required initial pressure head (or pressure) conditions are static heads (P/γ) and not total heads or hydraulic grade lines (elevation + P/γ). The static pressure is the pressure that a pressure gauge attached to the line at that position would read. Knowing static pressure is necessary to address cavitation during transient conditions.

In addition to the usual pipe data (length, diameter, and roughness), the wave speed must be calculated for each pipe. This requires the wall thickness of each pipe, material, Poisson's ratio and elastic (Young's) modulus, and the bulk modulus and density of the fluid. In addition, information on the degree of restraint of the pipes is required. Because many of these parameters may not be readily available, estimates for typical pipe materials are often used.

Characteristic data (or an equation) that relate the pressure change and flow to the operating point must be used for all hydraulic components and for the full range of operating conditions during the transient analysis. Initially, all system components (e.g., pumps, valves) should be in a balanced state (the initial pressure change across the component and flow through the component should be compatible with the characteristic relationship for that component). The characteristic data for a component must account for changes in the set point of the component (e.g., valve opening, pump speed) that occur during the transient analysis. Normally, the steady-state characteristics are used. Figure 10-20 shows typical valve closure characteristics for a linear stem movement for several types of valves (Boulos et al. 2006; Wood et al. 2005a). It is assumed that the valves follow the characteristics on these curves during transient conditions.

Pump steady-state, head-flow speed curves are often used to model pump operation. However, if during the transient the pump operates in abnormal zones (e.g., turbining), it is necessary to use a more detailed four-quadrant pump operating characteristics (i.e., ± speed and ± flow). Of the many methods developed for this purpose, the Suter curve is the most widely used (Thorley 1991; Suter 1966). A typical four-quadrant pump characteristic curve is shown in Figure 10-21. In this figure, the *x*-axis represents the four quadrants of pump operation, and the *y*-axis represents the corresponding head and torque characteristics. These data are normally not provided by pump manufacturers. However, many of these four-quadrant curves are available for a range of pump-specific speeds. Typically, one of these curves is selected by matching the pumps' specific speeds to the available curves. The additional pump data required for transient analysis include efficiency and total inertia (pump, motor, and shaft). If this information is not available, the moments of inertia for the pump and motor can be estimated using the following relationships, which are based on the work of Thorley (1991).

In English units, the moment of inertia for larger and older pumps is given by:

$$I = 0.6674\left[\frac{P}{\left(N/_{1000}\right)^3}\right]^{0.9556} \tag{10-24e}$$

The moment of inertia for small, lightweight pumps is given by:

$$I = 0.6244\left[\frac{P}{\left(N/_{1000}\right)^3}\right]^{0.844} \tag{10-25e}$$

The moment of inertia for motors is given by:

$$I = 0.0648\left[\frac{P}{\left(N/_{1000}\right)^3}\right]^{1.48} \tag{10-26e}$$

where I is the moment of inertia (lb ft^2), P is the pump power (horsepower), and N is the pump speed (revolutions per minute).

In metric units, the moment of inertia for larger and older pumps is given by:

$$I = 0.03768\left[\frac{P}{\left(N/_{1000}\right)^3}\right]^{0.9556} \tag{10-24m}$$

The moment of inertia for small, lightweight pumps is given by:

$$I = 0.03407\left[\frac{P}{\left(N/_{1000}\right)^3}\right]^{0.844} \tag{10-25m}$$

The moment of inertia for motors is given by:

$$I = 0.0043\left[\frac{P}{\left(N/_{1000}\right)^3}\right]^{1.48} \tag{10-26m}$$

where I is the moment of inertia (kilograms per square meter), P is the pump power (kilowatts), and N is the pump speed (revolutions per minute).

A characteristic equation that relates the pressure change and flow to the operating point of all hydraulic surge control devices along with their locations must also be available. Initially all such devices (e.g., surge tanks, air valves) should be in a balanced state (the initial pressure change across the component and flow through the component should be compatible with the initial characteristic relationship for that component). The characteristic equation for a surge control device must account for changes in the set point of the device (e.g., relief valve, nonslam air valve) that occur during the transient analysis.

Finally, the exact nature of the disturbance and its intended time scale must be specified. The disturbance can be a change in the open area ratio for a valve or in the speed ratio for a pump, or any other change in operation (e.g., check valve closing, pipe rupture) that results in a change in flow.

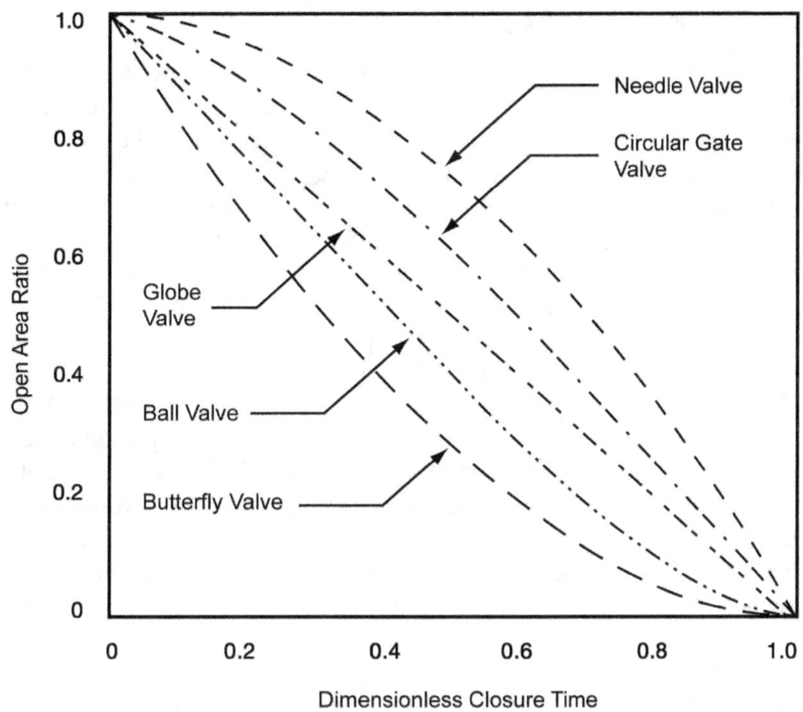

Figure 10-20 Representative valve closure characteristics

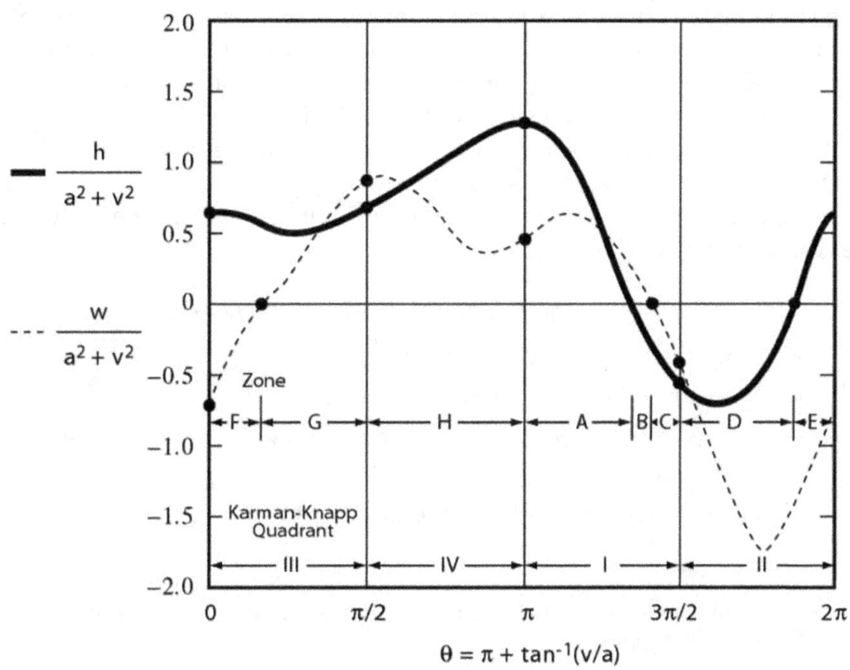

Figure 10-21 Typical pump four-quadrant characteristics (the Suter curve)

10.10. TRANSIENT MODEL CALIBRATION

As with any other type of model, hydraulic transient models are most reliable, accurate, and useful when they are calibrated. Because a well-calibrated transient model offers a near representation of the actual system, the engineer will be able to use it to understand the transient dynamics of the existing system, test the effects of making alterations to the system, determine optimal valve or pump operations, and select the sizes and locations of various types of surge protection devices. Although these tasks are all feasible with an uncalibrated model, a calibrated model allows the engineer to identify problems and make recommendations with a much higher degree of certainty and confidence.

The most fundamental and perhaps the most critical part of model calibration is calibration of the steady-state model. Because the transient model is built on the steady-state model, it is essential that the steady-state model be well calibrated and that it closely represents the system (see chapters 4 and 5 for steady-state model calibration). The engineer should ensure that the steady-state model network and results closely represent field conditions because, as noted previously, simplifications of the system network may introduce significant error in transient model results. An extended-period model is not necessary for transient analysis model, as an extended-period model will be calibrated over a period of many hours or days, whereas the transient model only needs to be calibrated over a period of seconds.

Once the transient model is set up with the required input data (see section 10.9), field testing should be conducted by installing high-frequency pressure loggers in the system. Because of the fast nature of pressure transient waves, the engineer should ensure that the logger has a sample rate high enough to capture the high and low points of the pressure transient waves. A rate of 20 Hz is typically sufficient to obtain a good approximation of the system's transient dynamics. However, for sharp surges (e.g., surges created by vapor cavity collapses), a faster sample rate may be necessary to capture the full characteristics of the pressure waves and accurate maximum and minimum pressure values. Figure 10-22 illustrates a pressure logger with a sample rate that is too slow to accurately represent the pressure waves in the system. This can lead to an erroneous understanding of the transient dynamics of the system and poor surge protection recommendations. It is also important to install transient pressure loggers in strategic locations that will reveal an adequate picture of the transient dynamics in the system. The following locations may be considered:

- On the discharge and/or suction lines of the operating pump(s) of interest,
- Near operating valve(s) of interest,
- On dead-end pipes, especially if long,
- At local high points (with low pressure) and low points (with high pressure) in the system, and
- At any other potentially vulnerable point in the system.

The uncalibrated surge model may be used as a rough guide when identifying potential weak points in the system that may experience unacceptably high or low pressures (Friedman et al. 2004).

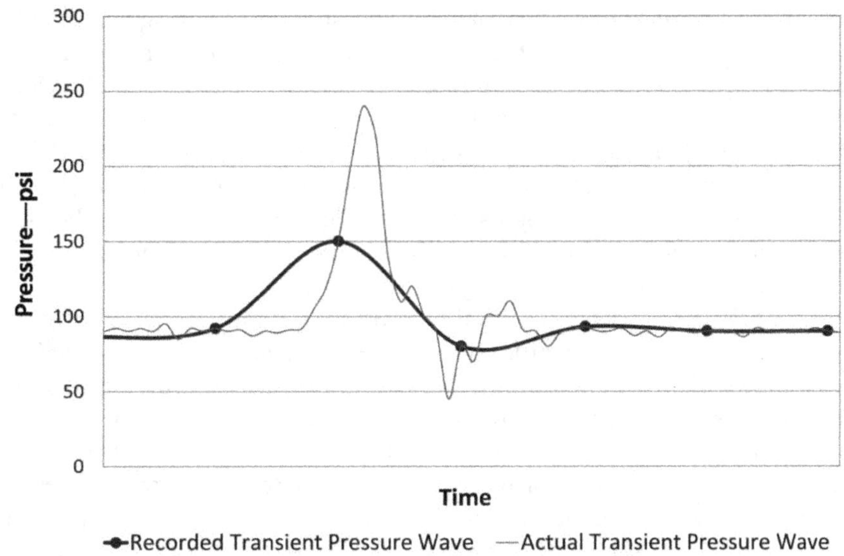

Figure 10-22 Example of pressure logger recording at insufficient rate

Once the field data for the desired control events (e.g., pump startup and shutoff or valve closure) are collected, the engineer should compare the field data with the uncalibrated model results in the key locations of interest and at vulnerable points and make adjustments to the model, as necessary. In general, if the period or timing of the wave does not match the field data, the wave speeds should be adjusted; if the model peaks have the wrong shape, the control event (e.g., pump speed, pump inertia, or valve closure rate) should be adjusted; and if there are secondary peaks or missing peaks, the engineer may need to ensure that critical loops or diversions in the field are accurately represented in the model (Walski et al. 2003). In models that have been skeletonized for the purpose of conducting a transient analysis, the engineer should make sure the velocities and pressures in the skeletonized representation of the system match those in the nonskeletonized version. If they do not match, some pipes removed during skeletonization may have to be reinserted. The engineer should make adjustments to the model so that the model results represent the field data as closely as possible in order to avoid both underdesigning and overdesigning the surge protection system. An underdesigned surge-protection system will not protect the system to the desired or necessary extent, and an overdesigned system will incur added cost and, in some cases, may provide even less surge protection than an underdesigned system (Jung et al. 2007b).

10.11. SUMMARY

When flow conditions in the water pipeline system are changed, such as by closing a pump or a valve or by starting a pump, a series of pressure waves are generated. These disturbances propagate at the velocity of sound within the medium until they are dissipated down to the level of the new steady state by the action of some form of damping or friction. In the case of flow in a water distribution network, these transients produce velocity and pressure changes. When sudden changes take place, the results can be dramatic because pressure waves of considerable magnitude can occur that are capable of destroying equipment and pipelines. Only if flow regulation occurs very slowly is it possible to go smoothly from one steady state to another without large fluctuations in pressure head or flow velocity.

Clearly, flow control actions can be extremely important, and these actions have implications not only for design of the hydraulic system but also for other aspects of system operation and protection. Problems such as selecting the pipe layout and profile, locating control elements within the system, and formulating operating rules as well as the ongoing challenges of system management are all influenced by the details of the control system. A rational and economic operation requires accurate data, carefully calibrated (static or extended-period simulation) models, ongoing predictions of future demands, the response of the system to transient loadings, and correct selection of both individual components and remedial strategies. These design decisions cannot be considered an afterthought to be appended to a nearly complete design. Transient analysis is a fundamental and challenging part of rational system design.

Transient analysis is essential to good design and operation of piping systems. Transient modeling provides the most effective and viable means of predicting potentially negative impacts of hydraulic transients under a number of worst-case scenarios, identifying weak spots, and evaluating how they may possibly be avoided and controlled. The basis of surge modeling is the numerical solution of conservation of mass and linear momentum equations. A number of widely used computer codes based on Eulerian (MOC) and Lagrangian (WCM) numerical solution schemes are currently available and have been successfully validated against field data and exact analytical solutions. However, surge analysis computer models can only be effective and reliable when used in conjunction with properly constructed and well-calibrated hydraulic network models. Poorly defined and calibrated hydraulic network models may result in poor prediction of pressure surges and locations of vapor cavity formation and, thus, defeat the purpose of the surge modeling process.

Looped water distribution systems that include short lengths of pipes may be less vulnerable to problems associated with hydraulic transient than a single long pipe system. This is because wave reflections (e.g., at tanks, reservoirs, junctions) tend to limit further changes in pressure and counteract the initial transient effects. For networks with long pipelines and irrespective of whichever numerical basis is used, a good transient model will have nodes along the pipes that define the important high and low points to ensure accurate calculations are made at those critical locations. An important consideration is dead ends (which may be caused by closure of control or check valves) that lock pressure waves into the system in a cumulative fashion (wave reflections double both positive and negative surge pressures). As a result, the effects of dead ends need to be carefully evaluated in transient analysis.

Proper selection of components for surge control and suppression in water distribution systems requires a detailed surge analysis in order to be effective and reliable. In addition, good maintenance, pressure management, and routine monitoring programs are essential components of transient protection. With these capabilities, water utility engineers can greatly enhance their ability to better understand and estimate the effects of hydraulic transients and to conceive and evaluate efficient and reliable water supply management strategies, safeguard their systems and public health with maximum effectiveness, and forge closer ties with their customers. It is understanding complexity through simplicity.

10.12. GLOSSARY OF NOTATIONS

A cross-sectional area $[L^2]$

A, B, C coefficients of control element characteristic equation

c sonic wave speed $[LT^{-1}]$

D pipe diameter $[L]$

E_c elastic modulus of the pipe $[ML^{-1}T^{-2}]$

E_f	elastic modulus of the fluid [$ML^{-1}T^{-2}$]
f	pipe friction factor
$f(Q)$	pipe resistance (nonlinear) term that is a function of flow rate [L]
g	acceleration of gravity [LT^{-2}]
H	head [L]
K_r	coefficient of restraint for longitudinal pipe movement
L	pipe length [L]
P	pressure [$ML^{-1}T^{-2}$]
Q	volumetric flow rate [L^3T^{-1}]
R	surge wave reflection coefficient at pipe junction
T	time [T]
T	surge wave transmission coefficient at pipe junction
t_l	pipe thickness [L]
V	flow velocity [LT^{-1}]
Δt	time interval [T]
Δx	spatial grid size [L]
μ_P	Poisson's ratio for the pipe material
ϱ	liquid density [ML^{-3}]

10.13. REFERENCES

Besner, M.C., Lavoie, J., Payment, P., & Prevost, M., 2007. Assessment of Transient Negative Pressures at the Site of the Payment's Epidemiological Studies. In *Proceedings of the AWWA Water Quality Technology Conference.* Denver, CO: AWWA.

Besner, M.C., Lavoie, J., Morissette, C., & Prevost, M., 2008. Effect of Water Main Repairs on Water Quality. *Journal of the American Water Works Association,* 100(7):95–109.

Boulos, P.F., Lansey, K.E., & Karney, B.W., 2006. *Comprehensive Water Distribution Systems Analysis Handbook for Engineers and Planners.* Broomfield, CO: MWH Soft.

Boulos, P.F., Karney, B.W., Wood, D.J., & Lingireddy, S., 2005. Hydraulic Transient Guidelines for Protecting Water Distribution Systems. *Journal of the American Water Works Association,* 97(5):111–124.

Boyd, G.R., Wang, H., Britton, M.D., Howie, D.C., Wood, D.J., Funk, J.E., & Friedman, M.J., 2004. Intrusion Within a Simulated Water Distribution System Due to Hydraulic Transients. 1: Description of Test Rig and Chemical Tracer Method. *Journal of Environmental Engineers,* ASCE, 130(7):774–783.

Brunone, B., Golia, U.M., and Greco, M., 1991. Some Remarks on the Momentum Equations for Fast Transients. *International Meeting on Hydraulic Transients with Column Separation,* 9th Round Table, IAHR, Valencia, Spain, 201–209.

Chaudhry, M.H., 1979. *Applied Hydraulic Transients.* New York: Van Nostrand Reinhold Co.

Fleming, K.K., Gullick, R.W., Dugandzic, J.P., & LeChevallier, M.W. 2006. *Susceptibility of Distribution Systems to Negative Pressure Transients*. Denver, CO: AwwaRF.

Friedman, M., Radder, L., Harrison, S., Howie, D.C., Britton, M.D., Boyd, G.R., Wang, H., Gullick, R.W., Wood, D.J., & Funk, J.E., 2004. *Verification and Control of Pressure Transients and Intrusion in Distribution Systems.* Denver, CO: AwwaRF.

Gullick, R.W., LeChevallier, M.W., Svindland, R.C., & Friedman, M.J., 2004. Occurrence of Transient Low and Negative Pressures in Distribution Systems. *Journal of the American Water Works Association*, 96(11):52–66.

Hooper, S.M., Moe, C.L., Uber, J.G., & Nilsson, K.A., 2006. Assessment of Microbiological Water Quality After Low Pressure Events in a Distribution System. In *Proceedings 8th Annual International Water Distribution Systems Analysis Symposium*. Cincinnati, Ohio.

Jung, B.S., Karney, B.W., Boulos, P.F., & Wood, D.J., 2007a. The Need for Comprehensive Transient Analysis of Water Distribution Systems. *Journal of the American Water Works Association*, 99(1):112–123.

Jung, B.S., Boulos, P.F., & Wood, D.J., 2007b. Pitfalls of Water Distribution Model Skeletonization for Surge Analysis. *Journal of the American Water Works Association*, 99(12):87–98.

Jung, B.S., Boulos, P.F., & Wood, D.J., 2009a. Effect of Pressure Sensitive Demand on Surge Analysis. *Journal of the American Water Works Association*, 101(4):100–111.

Jung, B.S., Boulos, P.F., Wood, D.J., & Bros, C.M., 2009b. A Lagrangian Wave Characteristic Method for Simulating Transient Water Column Separation. *Journal of the American Water Works Association*, 101(6):64–73.

Karim, M.R., Abbaszadegan, M., & LeChevallier, M.W., 2003. Potential for Pathogen Intrusion During Pressure Transient. *Journal of the American Water Works Association*, 95(5):134–146.

Karney, B.W., & McInnis, D., 1992. Efficient Calculation of Transient Flow in Simple Pipe Networks. *Journal of Hydraulic Engineering*, 118:7:1014–1030.

Kirmeyer, G.J., Friedman, M., Martel, K., Howie, D., LeChevallier, M., Abbaszadegan, M., Karim, M., Funk, J., & Harbour, J., 2001. *Pathogen Intrusion Into the Distribution System*. Denver. CO: AWWA and AwwaRF.

Korteweg, D.J., 1878. Über die Fortpflanzungsgeschwindigkeit des Schalles in Elastischen Röhren. *Annalen der Physik* 241:12:525–542.

LeChevallier, M.W., Gullick, R.W., Karim, M.R., Friedman, M., & Funk, J.E., 2003. The Potential for Health Risks From Intrusion of Contaminants Into Distribution Systems From Pressure Transients, *Journal Water Health*. 1:1:3–14.

McInnis, D.A., & Karney, B.W., 1995. Transients in Distribution Networks: Field Tests and Demand Models. *Journal of Hydraulic Engineering*, ASCE, 121:3:218–231.

Ndambuki, J.M., 2006. Water Quality Variation Within a Distribution System: A Case Study of Eldoret Municipality, Kenya. In *Proceedings of the Environmentally Sound Technology in Water Resources Management*. Gaborone, Botswana.

Ramalingam, D., Lingireddy, S., & Wood, D.J. 2009. Using the WCM for Transient Modeling of Water Distribution Networks. *Journal of the American Water Works Association*, 101:2:75–89.

Suter, P., 1966. Representation of Pump Characteristics for Calculation of Water Hammer. *Sulzer Tech. Rev.*, 66:45–48.

Thorley, A.R.D., 1991. *Fluid Transients in Pipeline Systems*. Herts, UK: D&L George.

Tullis, J.P., 1989. *Hydraulics of Pipelines*. New York: John Wiley & Sons.

Vardy, A.E. and Brown, J.M., 1996. On Turbulent, Unsteady, Smooth-Pipe Friction. *7th International Conference on Pressure Surges and Fluid Transients in Pipelines and Open Channels*. BHR Group, April, Harrogate, England, 289–311.

Vitkovsky, J., Lambert, M., Simpson, A., & Bergant, A., 2000. Advances in Unsteady Friction Modeling in Transient Pipe Flow. 8th International Conference on Pressure Surges, BHR, The Hague, The Netherlands, 587–597.

Walski, T.M. ,2009. Not So Fast! Close Hydrants Slowly. *Opflow*, 35:2:14–16.

Walski, T.M., & Lutes, T., 1994. Low Pressure Problems Caused by Hydraulic Transients. *Journal of the American Water Works Association*, 86:12:24–32.

Walski, T.M., Chase, D.V., Savic, D.A., Grayman, W.M., Beckwith, S., & Koelle, S. 2003. *Advanced Water Distribution Modeling and Management*. Waterbury, CT: Haestad Press.

Wood, D.J., Dorsch, R.G., & Lightner, C., 1966. Wave Plan Analysis of Unsteady Flow in Closed Conduits. *Journal of Hydraulics Division*, ASCE, 92:83–110.

Wood, D.J., Lingireddy, S., & Boulos, P.F., 2005a. *Pressure Wave Analysis of Transient Flow in Pipe Distribution Systems*. Broomfield, CO: MWH Soft.

Wood, D.J., Lingireddy, S., Boulos, P.F., Karney, B.W., & McPherson, D.L. 2005b. Numerical Methods for Modeling Transient Flow in Distribution Systems. *Journal of the American Water Works Association*, 97:7:104–115.

Wylie, E.B., & Streeter, V.L., 1993. *Fluid Transient in Systems*. Englewood Cliffs, NJ: Prentice Hall.

Zielke W., 1968, Frequency-Dependent Friction in Transient Pipe Flow. *Journal of ASME*, 90:109-115.

This page intentionally blank.

Chapter 11

Advanced Modeling Applications

11.1. INTRODUCTION

A completed and calibrated model delivers value across the utility, well beyond traditional high-level master planning and facility sizing. The industry's increased focus on asset management and operational efficiency have highlighted model applications for advanced use cases, including system criticality, asset prioritization, pump energy and pressure zone management, system flushing, nonrevenue water (NRV), real-time modeling, and water system optimization.

Utilities and consultants are advancing model applications for their predictive analytics and risk-based decision support. Advanced model applications are creating higher-value capital replacement projects for utility engineers, while highlighting opportunities to reduce energy costs and NRV for operations and maintenance (O&M) departments.

11.2. RELIABILITY AND CRITICALITY

Most water distribution systems work well when all components are operating. However, their reliability is tested when components fail due to pipe breaks, power outages, tank cleaning, and other events that cause a system outage. Measuring the impact on the water system under a simulated break or shutdown is typically referred to as a *reliability* or *criticality analysis*. Modelers represent this event by "turning off" or "closing" a model element that represents a system asset, which could include a pipe, pump, or valve. Changing the status of these model elements simulates the removal of a pipeline, section of the system, or facility from system operations to help determine the predicted hydraulic impact.

A model can be used to analyze the impacts of outages and shutdowns by simulating pump shutoffs, pipes closing, increasing demands, or tank isolations within the model one at a time either manually or automatically. The resulting impacts on pressures and flows can then be predicted. The model can support the design process by adding pipe loops, backup pumps and power, more valves, and larger tanks that can be tested for their effectiveness. The model can support operations by evaluating alternative shutdown processes, water bypassing, and/or use of standby power and interconnections with neighboring utilities and pressure zones.

Use of steady-state model runs can help measure the instantaneous hydraulic impacts of the shutdown. Extended-period simulations can predict how well storage tanks can supply flows and the longer-term impacts of shutdowns.

The difficulty in isolating a segment is highly dependent on the number of operable valves that must be closed to achieve a successful shutdown. Often, even in looped systems, the lack of adequate operable valves may result in a significant number of customers being out of water due to a single outage. Overlaying the model with geographic information system (GIS) layers that include customers and/or parcels on a background map can help identify customers who may be out of service or have their service impaired.

By modeling and measuring the impacts of shutdowns, models can indicate where additional redundancy is needed to improve the reliability of supply. Valve criticality tools also exist to determine their impact on the system. Metrics can be used to measure reliability, such as the shortfall of water supplied versus water demanded or the reduction in pressure or available fire flow rates. Shortfall can be expressed in units of flow for a steady-state run and in volume or as a percent of demand satisfied for an extended-period simulation (EPS).

The model can also be used to plan shutdowns by analyzing the impacts of taking system components out of service. For example, the available fire flow may be significantly lower in an area while a tank is out of service for cleaning and painting. Simulating shutdowns before they occur can help mitigate impacts and lead to an understanding of which customers would need to be notified during an event.

11.3. SYSTEM REHABILITATION AND PRIORITIZATION

Addressing the prioritization of rehabilitation and replacement of water distribution assets has become a priority for many water utilities. Risk-based approaches that consider a combination of the probability and consequences of failure of individual assets (e.g., pipe, valves, hydrants) have become common. The model becomes a valuable tool for applying risk-based criteria and for calculating appropriate risk-scoring values.

Common pipeline criteria for likelihood of failure provided by the model may include pipes with high and fluctuating pressures, pipes with high velocities and flows, high head loss (evidenced by low pipe roughness coefficients), and other hydraulic pipeline criteria. Example criteria for consequences of failure of water assets that the model can help with include the following:

- Flow: Pipes that carry more flow tend to be more critical.
- Demand: Outages that affect a greater amount of demand or impact supply to wholesale customers tend to be more critical.
- Redundancy: Branched systems with zero or little redundancy are more seriously affected than looped systems.

A calibrated model can often represent the best version of this data, as the calibration process tends to identify data quality issues that surface when comparing model results to field results. While models can be an important tool for asset management systems, the details of asset management systems are beyond the scope of this manual.

11.4. PUMP ENERGY MANAGEMENT

Energy used for pumping is a major operating cost for many water utilities. Models can be used both in pump selection and pump operations to reduce energy use, energy cost, and greenhouse gas emissions.

A typical operational use of the model is to simulate the pump's delivered head and flow given system demand and system losses. As part of the model solution, the pump's head curve is used to calculate the pump operating point, which is solved simultaneously with the other hydraulic equations.

If the pump efficiency curve is also provided, the model can calculate the efficiency of the pump at any given operating point, as well as the power generated for that operating point. For any pump (at a given speed with a given impeller), there is a best efficiency point. In terms of operation, the model can show how close the installed pump operates to its best efficiency point.

Once the model calculates the pump's operating points and energy use, these values can be compared with the measured properties from the pumps to determine if the pumps are being operated efficiently and if they are still operating on their curves. Pumps may not operate on their curves because of normal wear or as a result of a mechanical problem. Pumps that operate on their curves but do not run efficiently are not (or are no longer) matched well with system hydraulics.

The modeler can test measures to improve energy use, such as replacing pumps and/or motors, modifying or replacing impellers, avoiding inefficient pump combinations, avoiding high demand charges, taking advantage of time-of-day pricing, using storage more efficiently, and installing variable-speed drives. These measures can be tested in the model before being implemented. Usually, inexpensive operational changes should be tried before more capital intensive projects are investigated. In general, improvements to pumping operation can show significant cost savings and relatively quick payback periods.

Energy calculations can be accurately performed on most models as long as key facilities are included. The sizing of transmission mains, system constrictions that create significant head loss, high system demand, and more can impact energy costs. A model can be used to test the hydraulics of different-sized piping designs under varying operating conditions and the energy impacts of changing diameters.

Numerous references on energy use and pumping are available, including AWWA (2016), Walski (2011), WEF (2009), and WRF (2011). This topic is discussed further in chapter 6.

11.5. PRESSURE ZONE MANAGEMENT

Pressure zones are used in water distribution systems to support delivery of acceptable pressures across varied topography with significant changes in elevation. Low pressures are corrected by pumping, while high pressures are controlled with pressure-reducing valves (PRVs). Models are useful for calculating system pressures and identifying areas that may be best served from new or revised pressure zones. Models can identify high and low pressures in each zone and where these pressures occur.

The model can color-code pipelines and other model elements within a pressure zone with the same color to indicate the same zone. An overall system pressure zone map can be made using color-coded pressure zones. This map is created based on calculated pressures or hydraulic grade lines.

System pressures can be adjusted in the model by modifying pressure regulators, opening or closing zone isolation valves to redefine new pressure zone boundaries, updating pump curves, and performing other hydraulically impactful model changes. These types of updates would recalculate hydraulic grade lines and resulting system pressures until system-wide pressures fall within targeted customer service level criteria.

Adjusting system pressure zones can serve as an effective strategy for managing system pressures and hydraulic gradient. This, in turn, could reduce pumping costs by allowing pumps to operate within a range of optimal efficiency on their pump curve.

11.6. FLUSHING

Water utilities perform flushing for a variety of reasons, which can be grouped into two general categories: to remove settled or attached solids and to move water of higher quality into a portion of the distribution system. When done correctly, flushing can accomplish both to some extent. If the primary goal of flushing is solids removal, the flushing should be conducted to maximize shear stress (or system velocity as an indicator of shear stress). If the goal is to move higher-quality water into an area, shear stress is less important, and planning is focused on locating and moving water of different quality.

If the intent of flushing is to remove stale water with low disinfectant residual or some suspected contamination, conventional flushing, evaluated with EPS water quality models, is the recommended tool for flushing. With correct initial conditions, the model can show how higher-quality water can displace poor-quality water with the correct operation of hydrants or blowoffs. Movement of water is not always intuitive, and EPS can sometimes provide surprising insights.

Unidirectional flushing (UDF) is a procedure for cleaning water distribution system pipelines by flowing relatively cleaner water in one direction through the mains, typically starting at a source (e.g., pump station or PRV) and then moving systematically outward through the system. This is accomplished by closing specified valves and by flushing strategically identified hydrants. As opposed to conventional flushing, where hydrants are flushed systematically from a clean source but without closing valves to produce one-directional flow, UDF provides higher flushing velocities and typically uses less water, as shown in Figure 11-1.

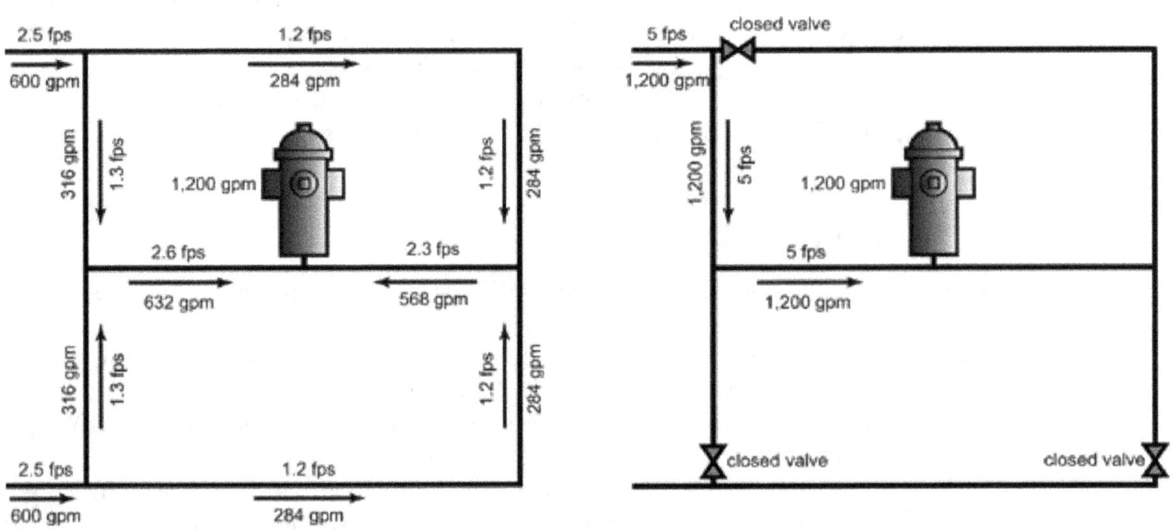

Conventional flushing in a looped system results in water flowing toward the hydrant from all directions, generating lower flow velocity and less scouring of the pipes.

Unidirectional flushing results in water flowing toward the hydrant in only one direction, resulting in higher flow velocity along with less water use.

Figure 11-1 Conventional versus unidirectional flushing

UDF increases the velocity of water in the mains, maximizing shear stress near the pipe wall. This produces a scouring action that effectively removes sediment deposits and biofilm. If conventional flushing cannot achieve a minimum scouring velocity, typically UDF is required.

A model can be used to evaluate the series of pipes and valve operations required to control flow direction and maximize velocity. These are termed *design sequences*. A model is useful in creating UDF design sequences to evaluate velocities, validate that minimum system pressures are met during flushing, and optimize the number of valves operated. In order to model UDF, hydrant and valve locations for the flushing area must be represented in the model or referenced by the model. Commercial packages are available to help manage the development of UDF design sequences as well as efficiently develop field forms and maps that present the order and location of hydrants and valves to be operated and allow field operators to record field measurements. The time needed to develop a UDF program is considerable. However, efficiencies in implementation and required flushing times and water volumes are known to decrease over time compared to a non-UDF program.

The overall benefit of flushing is improvement in water quality, because flushing restores the disinfectant residual, reduces bacterial regrowth, dislodges biofilms, removes sediments and deposits, and restores system capacity (i.e., flows and pressures), potentially prolonging the water system's life expectancy. A corresponding benefit is that flushing helps ensure water system assets (e.g., hydrants, valves, and pipes) are routinely assessed and field tested. Moreover, data collected during these efforts can support further model calibrations or validations.

11.7. NONREVENUE WATER AND DISTRICT METER AREAS

NRW consists of many categories where real losses and apparent losses occur. A portion of NRW is water that is produced by the system and is "lost" before it reaches a customer. This water leaves the system without being metered or is inaccurately metered and, therefore, does not generate the desired water revenue it was intended to, and is lost revenue to the utility. Modeling attempts to account for avoidable NRW initially by allocating the customer meter demand to model junctions as accurately as possible. This is discussed in the demand allocation section in chapter 2. The measurement between modeled demands and resulting flows, compared with flowmeters strategically placed in the water distribution system, can help identify locations of water loss, or NRW.

Flowmeters are often used to help identify the extent of NRW in district metered areas (DMAs). These areas are used to partition the water system into discrete areas that are more manageable and measurable for model calibration and comparison. DMAs with high water demand per capita, especially at night, are areas where higher quantities of NRW are likely to be found. The model can be used to help establish DMAs, site flowmeter locations, and calculate predicted flows at flowmeters when evaluating NRW within a given area.

DMAs are useful in identifying NRW by metering all of the flow into relatively small areas. DMAs with high NRW can be targets for intensive leak-detection activities. To accomplish this, isolation valves are often closed to force all of the water entering a DMA to pass through a single meter. While useful for analysis, DMAs typically reduce the metered area's hydraulic performance, especially for fire flows. Models can be used to test these impacts. DMAs with higher NRW can then be targeted for improvements.

Before creating a DMA, it is advisable to simulate pressures and fire flows in the model. If the model does not meet operational criteria with the DMA, alternative DMA configurations or multiple feeds may be required. Models in which demands are loaded from meter data can indicate what the demand should be in a DMA. If the actual inflow differs significantly, water loss or a breakdown in the boundary isolation can be suspected.

DMAs can be fed through PRVs to reduce leakage flows. The settings on these PRVs can be adjusted in the model to determine their effects on pressures and fire flows. In some systems, pressure may be lowered during night hours (when pressures normally tend to increase), which can decrease leakage.

High pressures have been found to induce NRW due to the added stress at pipeline joints. In general, average pressures should be in the range of 60 to 90 psi (413.69 to 620.53 kPa). Pressure management can be used to reduce loss due to leakage at pipelines, joints, and customer meters. The amount of water lost to leakage can be estimated using a modified orifice equation of the form:

$$Q = aP^n \tag{11-1}$$

where:

Q = leak flow
a = coefficient that depends on leak size
P = pressure
n = exponent

This type of water loss can be modeled using a pressure-dependent demand in the model. The exponent depends on the nature of the leak. Most leaks such as circular holes have an exponent of 0.5. However, cracks in viscoelastic pipes may have higher values, but not greater than 1.5 (Cassa et al. 2010).

11.8. REAL-TIME MODELING

The traditional application of the model within the utility has been focused on planning and engineering activities. The use of models to assist in day-to-day system operations is not yet widespread, primarily because operators are skeptical of using a model that does not incorporate real-time or near real-time data. The source of this limitation comes from a historic inability to easily incorporate dynamic boundary conditions and demands into a model to accurately represent real-time field conditions. Advances in utility information technology, supervisory control and data acquisition (SCADA) systems, and industry technology have helped overcome many of those technical hurdles.

The most common applications of operational real-time models are the following:

- Ongoing daily energy management;
- Managing pump station, tank, and other complicated maintenance tasks;
- Creating response plans for planned outages and maintenance;
- Responding to main breaks and emergency events; and
- Troubleshooting water quality issues.

The benefits of implementing real-time models vary as listed above, based on the applications desired by the utility, and can range from planning to more efficient operations and forecasting. For example, operators can analyze the effect of a predicted low storage tank level and pinpoint all customers who will be negatively affected by low pressures. They can then conceive and formulate alternative operating scenarios that can be quickly and accurately analyzed and compared in order to determine their respective levels of improvement. The most desirable scenario is then selected and implemented. Operators can assess the effects of main breaks; pump, valve, and reservoir shutdowns; other scheduled maintenance or repair; and any planned or unplanned incidents. They can then respond swiftly with appropriate countermeasures.

A real-time model begins with a calibrated EPS model. A model becomes real time when connected to the operator's SCADA information system and other real-time telemetry data in an automated fashion such that boundary conditions (e.g., tank water levels) and operational statuses (e.g., pump speeds or on–off status, valve settings) can be applied to the model with little or no manual intervention. Real-time models can be programmed to "run automated," which refers to simulations that are run on a regular basis without human intervention in order to update the model based on real-time boundary conditions and perhaps system demands. This automation can occur on a continuous basis (e.g., every 15 minutes, 1 hour, 1 day) or on a more periodic basis, such as monthly, bimonthly, or yearly. The most common association with the term *real-time modeling* is in the frequency of every 15, 30, or 60 minutes to continually predict system flows, pressures, and water quality results.

An overall benefit of implementing a real-time model is that model calibration and/or validation can be streamlined and be more frequent than traditional non–real-time models. Since boundary and operational conditions are connected to the real-time model and continuously compared to real-time SCADA data, data anomalies will present themselves more frequently. This presents the opportunity to identify field differences, and/or adjust model parameters, and/or expose incorrect SCADA readings. The outcome is a model that is continually being used for predictive modeling and is continually field tested, in turn providing a high level of confidence for operators and modelers to make reliable, informed decisions.

11.9. OPTIMIZATION FOR PLANNING AND SYSTEM OPERATIONS

Models serve as the basis for more advanced network analyses such as optimization studies to better plan system capital improvements and system operations. Optimization is simply the process of finding the best solution to a problem that has many possible solutions. There are numerous possible solutions to water distribution system planning and operations problems. The challenge for utility modelers and decision makers is to develop a preferred solution or plan that meets the utility's specified performance criteria or service levels for a range of anticipated future conditions while also minimizing overall costs and limiting water rate increases.

The traditional approach to water distribution planning and operational problems is to use a model to perform a deficiency analysis and identify where improvements or operational changes are needed. These steps are necessary for the modeler to understand the problem. Typically, the modeler creates a trial solution and then tests and modifies it using engineering judgment and system experience until the performance criteria are met. This can be extremely challenging, and often not all criteria are fully met nor is the solution likely to be the most cost effective. If carefully applied, grounded, and validated by practical engineering judgment, optimization as a decision support tool can be helpful to modelers and decision makers in developing more comprehensive solutions with less effort.

Today, water utilities and consultants are using numerous mathematical optimization techniques to develop improved solutions to complex water distribution problems. It is important to note that these optimization techniques rely on calibrated models as well as the engineer's proper understanding of the water distribution system to yield accurate and useful solutions. For most applications, potential improvement alternatives such as new facilities or pipes, performance criteria, and/or penalties to test the solutions against must be considered and included as part of these optimization routines. The optimized decision support provided by these advanced techniques must always be tempered with engineering judgment and experience.

11.9.1. Mathematical Optimization Formulation

An explanation of how modelers take the step from a calibrated model to formulating a useful optimization model may be helpful. All mathematical optimization problems are defined by three components:

- *Objective function* can be a single-objective statement, such as to minimize project capital improvement cost or system pumping cost or project life-cycle cost. The objective function for a multiobjective optimization might be to minimize system pumping costs and also water age, which leads to an optimized trade-off curve (or Pareto optimal front) of cost versus water age. An example trade-off curve for a multiobjective optimization problem is shown in Figure 11-2. Each point on the curve represents a distinct solution that is created and evaluated using the model.
- *Decision variables* are the specified variables to be manipulated by the optimization, such as sizes for new or replacement pipes, tanks, pumps, valves, and supply sources, as well as operational set points or trigger levels for pumps and regulating valves. Capital and operating or O&M costs are prepared for the complete range of allowable sizes and settings to allow for an evaluation of the objective function.
- *Constraints* are used to represent the utility's specified design and performance criteria that would be met by any feasible solution. If maximum or minimum pressures, or maximum velocities, or minimum tank drawdowns during an extended period simulation are not met, that particular trial solution is not strictly feasible and can be rejected or modified.

Using these three components, the optimization problem might be stated as: "Find the set of decision variables such that all constraints are met while achieving the lowest possible cost." The absolute lowest possible cost is unlikely to be achieved, but the optimization can guide the modeler to find one or more near-optimal solutions that may exhibit different features. One strength of the optimization is that both system performance and cost are evaluated simultaneously for numerous trial solutions using the model.

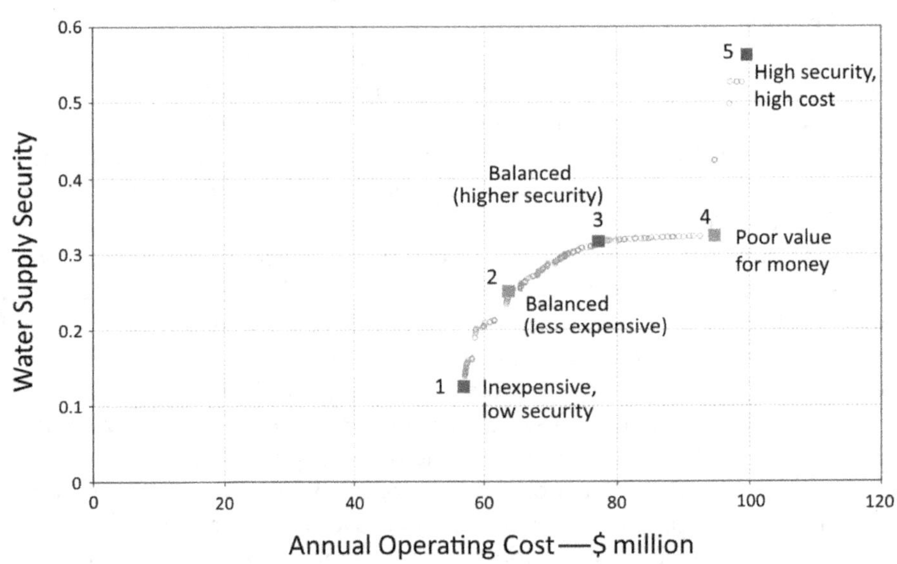

Figure 11-2 Example trade-off curve generated by a multiobjective optimization run

11.9.2. Water Distribution System Optimization Applications

The optimization problem formulation just described can be adapted to any number of applications for planning and operations of water networks. Examples of water distribution system optimization applications include the following:

- Water master plans and capital improvement plans for limited areas or the entire system;
- Integrated planning for water supply, wastewater, and reclaimed-water systems;
- Locating and sizing pipes or facilities as part of rehabilitation planning;
- System improvements to meet fire flow requirements;
- Evaluation of pressure zone boundaries and settings for better distribution and pressures;
- Pressure reduction to minimize leakage volumes;
- Operational planning to reduce O&M costs and improve service levels and/or water quality;
- Blending of source waters to meet regulatory requirements; and
- Emergency planning for forced or planned outages of facilities and improved redundancy (AWWA [2016], Walski [2011], WEF [2009], and WRF [2011]).

11.10. REFERENCES

American Water Works Association (AWWA), 2016. *Energy Management for Water Utilities.* Denver, CO: AWWA.

Cassa, A.M., Van Zyl, J.E., & Laubscher, R.F., 2010. A Numerical Investigation Into the Effect of Pressure on Holes and Cracks in Water Supply Pipes. *Urban Water Journal,* 792:109–120.

Walski, T., 2011. "Practical Tips for Reducing Energy Use." *Eleventh International Conference on Computing and Control for the Water Industry.* Exeter, UK.

Water Environment Federation, 2009. *Energy Conservation in Water and Wastewater Facilities.* Alexandria, VA: Water Environment Federation.

Water Research Foundation, 2011. *Energy Efficiency Best Practices for North American Drinking Water Utilities.* Denver, CO: Water Research Foundation.

11.11. ADDITIONAL RESOURCES

Dandy, G.C., Simpson, A.R., & Murphy, L.J., 1996. An Improved Genetic Algorithm for Pipe Network Optimization. *Water Resources Research,* 32:2:449–458.

Jacobsen, L., Diallo, N., & Pribila, C., 2015. Genetic Algorithm Optimization of Master-Planned Facilities. *AWWA Water Infrastructure Conference,* Bethesda, Md.

Liner, B., & Maier, H., 2015. *H2Optimization: An Introduction to Optimization and Operations Research for Infrastructure Professionals*. Granite Bay, CA: Aldera–Water Anthology Press.

Schaake, J., & Lai, D., 1969. *Linear Programming and Dynamic Programming Application to Water Distribution Network Design*. Report 116. Cambridge: Massachusetts Institute of Technology Hydrodynamics Laboratory.

Walski, T., 1993. Water Distribution Valve Topology for Reliability Analysis. *Reliability Engineering & System Safety*, 42:2:21–27.

Index

Note: *f.* indicates figure; *t.* indicates table

D

E

N

O

P

Q

R

This page intentionally blank.

AWWA Manuals

- M1, *Principles of Water Rates, Fees, and Charges*, #30001
- M2, *Instrumentation and Control*, #30002
- M3, *Safety Management for Water Utilities*, #30003
- M4, *Water Fluoridation Principles and Practices*, #30004
- M5, *Water Utility Management*, #30005
- M6, *Water Meters—Selection, Installation, Testing, and Maintenance*, #30006
- M7, *Problem Organisms in Water: Identification and Treatment*, #30007
- M9, *Concrete Pressure Pipe*, #30009
- M11, *Steel Pipe—A Guide for Design and Installation*, #30011
- M12, *Simplified Procedures for Water Examination*, #30012
- M14, *Backflow Prevention and Cross-Connection Control: Recommended Practices*, #30014
- M17, *Fire Hydrants: Installation, Field Testing, and Maintenance*, #30017
- M19, *Emergency Planning for Water Utilities*, #30019
- M20, *Water Chlorination/Chloramination Practices and Principles*, #30020
- M21, *Groundwater*, #30021
- M22, *Sizing Water Service Lines and Meters*, #30022
- M23, *PVC Pipe—Design and Installation*, #30023
- M24, *Planning for the Distribution of Reclaimed Water*, #30024
- M25, *Flexible-Membrane Covers and Linings for Potable-Water Reservoirs*, #30025
- M27, *External Corrosion Control for Infrastructure Sustainability*, #30027
- M28, *Rehabilitation of Water Mains*, #30028
- M29, *Water Utility Capital Financing*, #30029
- M30, *Precoat Filtration*, #30030
- M31, *Distribution System Requirements for Fire Protection*, #30031
- M32, *Computer Modeling of Water Distribution Systems*, #30032
- M33, *Flowmeters in Water Supply*, #30033
- M36, *Water Audits and Loss Control Programs*, #30036
- M37, *Operational Control of Coagulation and Filtration Processes*, #30037
- M38, *Electrodialysis and Electrodialysis Reversal*, #30038
- M41, *Ductile-Iron Pipe and Fittings*, #30041
- M42, *Steel Water-Storage Tanks*, #30042
- M44, *Distribution Valves: Selection, Installation, Field Testing, and Maintenance*, #30044
- M45, *Fiberglass Pipe Design*, #30045
- M46, *Reverse Osmosis and Nanofiltration*, #30046
- M47, *Capital Project Delivery*, #30047
- M48, *Waterborne Pathogens*, #30048
- M49, *Butterfly Valves: Torque, Head Loss, and Cavitation Analysis*, #30049
- M50, *Water Resources Planning*, #30050
- M51, *Air Valves: Air-Release, Air/Vacuum and Combination*, #30051
- M52, *Water Conservation Programs—A Planning Manual*, #30052
- M53, *Microfiltration and Ultrafiltration Membranes for Drinking Water*, #30053
- M54, *Developing Rates for Small Systems*, #30054
- M55, *PE Pipe—Design and Installation*, #30055
- M56, *Nitrification Prevention and Control in Drinking Water*, #30056
- M57, *Algae: Source to Treatment*, #30057
- M58, *Internal Corrosion Control in Water Distribution Systems*, #30058
- M60, *Drought Preparedness and Response*, #30060
- M61, *Desalination of Seawater*, #30061
- M62, *Membrane Applications for Water Reuse* #30062
- M63, *Aquifer Storage and Recovery*, #30063
- M65, *On-Site Generation of Hypochlorite*, #30065
- M66, *Cylinder and Vane Actuators and Controls—Design and Installation*, #30066
- M68, *Water Quality in Distribution Systems*, #30068

This page intentionally blank.

www.ingramcontent.com/pod-product-compliance
Lightning Source LLC
LaVergne TN
LVHW080029210926
848481LV00011B/3142